Superplasticity
of Alloys, Intermetallides
and Ceramics

Oscar A. Kaibyshev

Superplasticity of Alloys, Intermetallides and Ceramics

With 129 Figures

Springer-Verlag
Berlin Heidelberg New York
London Paris Tokyo
Hong Kong Barcelona
Budapest

Professor Oscar A. Kaibyshev

Russian Academy of Sciences, Institute of Metal Superplasticity Problems
Khalturin Street 39, 450001 Ufa, Russia

ISBN-13:978-3-642-84675-5 e-ISBN-13:978-3-642-84673-1
DOI: 10.1007/978-3-642-84673-1

Library of Congress Cataloging-in-Publication Data. Kaibyshev, O. A. (Oscar Akramovich) Superplasticity of alloys, intermetallides and ceramics / Oscar A. Kaibyshev. p. cm. Translated from the Russian. Includes bibliographical references and index.ISBN-13:978-3-642-84675-5(alk. paper)1. Superplasticity. I. Title. TA418.14.K35 1992 92-2880

Typesetting: Springer TEX in-house system
54/3140 - 5 4 3 2 1 0 - Printed on acid-free paper

Preface

This book deals with the problems of superplastic deformation of materials. First of all, it is of interest due to its practical emphasis and the extensive treatment of superplasticity in alloys, intermetallics and ceramics. A significant conclusion which can be made by the reader is that this phenomenon bears the universal character and can be observed in virtually any crystalline material.

Much attention has been paid to the physical nature of the phenomenon. Various deformation processes operating in superplastic flow, i.e. grain boundary sliding, dislocation slip and diffusion creep have been experimentally studied.

A new point which is important for the theory of the superplastic effect is the establishment of interaction between intragranular slip and grain boundary sliding in superplastic deformation processes.

The present book substantiates the assumption that interaction between lattice dislocations and grain boundaries is significant in explaining the dominant contribution of sliding along grain boundaries during superplastic deformation. On the basis of the latest concepts in the physics of high-angle boundaries, it is shown that dislocations enter the grain boundaries and adopt a non-equilibrium structure under certain temperature-rate conditions typical of superplasticity. As a result, grain boundary sliding and migration are stimulated.

The main feature distinguishing this book from others is that superplastic deformation is considered, for the first time, not only as a phenomenon enabling large deformations to be produced without fracture, but as a new kind of materials processing permitting one to fix the fine-grained and homogeneous microstructures in the materials. This enables one to produce products not only of intricate shape but with good serviceability as well.

Dynamic recrystallization features allowing one to form micro- and submicrocrystalline structures in materials are also considered. It is shown that this processing enables one to refine the microstructure of virtually any kind of material and to transfer it into the superplastic state.

The main advantage of the present book is its in-depth consideration of the nature of the phenomenon and its exposition of the theoretical basis of the technology relevant to putting the results into practice. Commercial methods of refining the microstructure of materials are described. Superplastic deformation modes are given for the specific materials. The effect of superplastic processing on the properties of materials is evaluated. These concepts and methods can be useful in solving the problem of increasing the ductility of low-ductility materials.

In spite of the fact that this book presents data compiled from many studies, it is not a review of the literature but rather the result of the author's experimental and theoretical investigations. The author and his colleagues hope that this book will be of value both as an aid to learning and from a practical point of view.

Ufa, September 1992 *Oscar A. Kaibyshev*

Contents

List of Abbreviations and Symbols

CSL	– coincident site lattice
DC	– diffusion creep
GBD	– grain boundary dislocation
GBS	– grain boundary sliding
hcp	– hexagonal close-packed (lattice)
HTMT	– high-temperature thermomechanical treatment
IDS	– intragranular dislocation slip
PFZ	– precipitate-free zones
SP	– superplasticity (superplastic)
SPDHT	– superplastic deformation-cum-heat treatment
SPDT	– superplastic deformation treatment
TLD	– trapped lattice dislocations
$m = d \log \sigma / d \log \dot{\varepsilon}$	– coefficient of strain-rate sensitivity of the flow stress
σ	– flow stress
σ_u	– ultimate tensile strength
$\sigma_{0.2}$	– offset yield stress
ε	– strain
$\dot{\varepsilon}$	– strain rate
δ	– relative elongation up to fracture
d	– grain size

Introduction

Progress in materials science depends to a great extent on new data on the relationship between materials' structure and their properties. This conventional approach might seem rather unpromising, but the latest studies have confirmed that this old formula is still valid.

At present, much progress can be made in understanding the nature of ductility and its control. Ductility is the ability of materials to change their shape without failing under the effect of external stresses. This property is widely used in technology, in particular, in metals forming, and is the most important parameter in determining the application of metals in mechanical engineering. Materials bearing mechanical load and used in various structures must have a certain resource of ductility.

There are three main classes of materials which show marked differences in their ductility, viz., metallic materials, intermetallics, and ceramics. In metals and metal-based alloys, the relative elongation can vary from a few per cent to several dozens of a per cent, while in intermetallics it is hardly 1%, and in ceramics almost zero; and it usually fails without any signs of plastic deformation.

Investigations carried out in recent years radically changed the concept of ductility. It has been shown that the ductility of all classes of materials can be increased dozens and even hundreds of times by converting them to a superplastic (SP) state. The existing division of materials into ductile and nonductile refers only to the properties of materials under conventional test conditions and with ordinary structures. In an SP state, cast alloys, e.g., cast irons, can be even more ductile than deformable alloys, i.e., steels. Superplasticity has also been observed in intermetallics and ceramics.

Superplasticity, if understood as the ability for materials to have an abnormally large elongation, comprises a fairly wide range of phenomena, the most typical being structural superplasticity observed in microcrystalline materials and superplasticity observed during the development of phase transformations in the course of deformation, thermocycling, or under radiation. The first group is the most interesting, because, in this case, superplasticity is a universal state of materials observed at a certain microstructure, temperature and rate of deformation. This book deals only with structural superplasticity.

The main features of SP flow are (the abrupt growth of the relative elongation, which can attain hundreds or even thousands of a per cent, a remarkable decrease in the flow stress and practically no work hardening during deformation, and high strain rate sensitivity of flow stress. It is understood that these peculiarities of SP flow are quite favourable for the process of materials forming. However, to apply this phenomenon successfully, the following problems must be solved. How can an SP state in commercial materials be obtained? How will SP treatment affect the materials performance? Is it possible to apply SP treatment in combination with other techniques to alter the physical and mechanical properties of materials?

Regarding many materials, the answers are already available; in other cases, solutions of the problems can be outlined. This book presents a detailed discussion of the factors affecting superplasticity in materials. It is shown that there is a possibility of converting alloys most widely used in mechanical engineering, including conventionally low-ductile materials, such as titanium and nickel-based superalloys as well as intermetallics and ceramics to superplasticity. Special attention is given to the effect of the structure and area of grain boundaries on the properties of materials.

The mechanism of SP flow is thoroughly analyzed and a physical model of the phenomenon explaining the variation of both the mechanical properties and structure during deformation is proposed. This model can also be used to predict new effects. Based on current knowledge of the physical nature of high-angle boundaries, a principally new postulate is offered, viz., dislocations running into grain boundaries under the temperature-rate conditions of SP flow alter the structure of grain boundaries and become nonequilibrium. As a result, grain boundary sliding which is the main mechanism of SP flow is stimulated.

The microcrystalline structure is a factor indispensable for the realization of superplasticity. The conditions for creating such a microstructure in real materials posed another important problem discussed in this book. Systematic investigations in a wide range of materials indicated that by employing the peculiarities of static and dynamic recrystallization and phase transformations, easily applicable methods of the microstructure refinement can be proposed.

Materials with micro-, submicro-, and nanocrystalline structures are of considerable interest. The book presents data on the properties of such materials and basic opportunities for "grain boundary hardening", reduction of the temperature, or increase of the rate of SP deformation.

The properties of materials after their processing in the SP state are of practical interest. The book presents systematic data on the effect of such treatment on the properties of commercial alloys. A specific microstructure is formed in materials processed under SP conditions. When intricately shaped components are manufactured by conventional forming, the strain level is different in different areas of the blank, depending on the shape. Hence, the deformation is nonuniform, which results in the microstructure heterogeneity and, consequently, in the heterogeneity of the mechanical properties of materials that is retained and in some cases even aggravated during the subsequent heat treatment. Under SP flow, the strain dependence of the material microstructure is negligible, and the material undergoes intensive "mixing" due to the development of grain boundary sliding and diffusion processes providing for a microstructure with a high structural and chemical homogeneity. The absence of cold working and texture and the retention of the microcrystalline structure enhance the ductility, impact strength and stability of mechanical properties and decrease their anisotropy at room and lower temperatures. Thus, we can conclude that the SP flow conditions are favourable not only for the materials forming, but also for the controlled variation of their structure and physico-mechanical properties.

This book was compiled by a joint effort of several researchers and is based on results obtained by independent investigations performed in the Institute for Metals Superplasticity Problems of the USSR Academy of Sciences, Ufa. it follows the approach first described in [1.1]. However, all sections have been updated in accordance with new data reported in the world literature.

Chapters 1 and 2 were written in cooperation with R.Z. Valiev; Chapters 3, 6, and 7 with G.A. Salishchev; Chapter 4 with R.O. Kaibyshev; Chapter 5 in cooperation with M.Kh. Rabinovich; Chapter 9 with R.M. Imayev; and Chapter 10 in cooperation with N.G. Zaripov.

The author would like to acknowledge the help provided by N.V. Markushev, R.Ya. Lutfullin, R.G. Zaripova, and V.A. Valitov in the work on Chapters 5–8, respectively.

The author is also grateful to K.G. Farchutdinov and O.Sh. Gainutdinova for her assistance in preparing the manuscript for publication.

1. Structural Superplasticity

There are three conditions required for transforming a material into a superplastic (SP) state, viz., the presence of a stable microcrystalline structure ($d < 10$–$15\,\mu$m), a deformation temperature exceeding $0.4\,T_m$, and a specific range of strain rates (usually from 10^{-4} to $10^{-1}\,s^{-1}$). In virtually all alloys the development of the SP effect has common features because of which it can be treated as a peculiar type of plastic deformation that is characterized by the specific phenomenology. Recent studies show that typical features of superplasticity can be observed not only in metals, but in intermetallic compounds and ceramics as well.

The most important difference between conventional deformation and superplastic deformation (SPD) is that in the former case, the flow stress depends on the degree of deformation and not so significantly on the strain rate. On the other hand, under the conditions of SP flow, the flow stress varies slowly as a function of the degree of deformation but depends strongly on the strain rate. Thus, structural superplasticity can be defined as the ability of metals and alloys to undergo a high degree of plastic deformation without failing under the conditions of high strain rate sensitivity of the flow stress.

1.1 Phenomenology and Influence of the Initial Microstructure

Numerous books and reviews [1.1–9] detail the typical behaviour of materials in the SP state. In this section we report the most significant results of phenomenological studies of superplastic flow which are essential for understanding the deformation mechanisms and the properties of specific industrial alloys.

One of the most distinguishing features of a specimen undergoing SP deformation is its high deformation stability, i.e., its high resistance to neck formation and grain growth during tensile deformation. According to the concepts developed by *Hart* (1967), the reason for such a deformation stability can be understood on the basis of the dependence of the flow stress on the strain and the strain rate. Under isothermal conditions, this relationship assumes the form

$$\sigma = A\varepsilon^n\dot{\varepsilon}^m \tag{1.1}$$

where A is an empirical constant; and n and m are the parameters of the material that determine the dependence of strengthening on the degree of deformation and the strain rate.

The parameter of strain hardening of the materials whose flow stress is a slowly varying function of the strain rate (i.e., at m close to zero) can be written as

$$n = d \ln \sigma / d \ln \dot{\varepsilon} \ . \tag{1.2}$$

For the materials exhibiting insignificant (weak) strain hardening, (1.1) becomes

$$\sigma = A\dot{\varepsilon}^m \tag{1.3}$$

and the rate-sensitive flow stress parameter is given by

$$m = d \ln \sigma / d \ln \dot{\varepsilon} \ . \tag{1.4}$$

The stability of plastic deformation is evaluated using the aforementioned parameters and by taking into account the condition of neck formation in the specimen. In the region of localization of the deformation, the strain rate and the degree of deformation are larger than those observed in the rest of the specimen. This region (i.e. neck) of the specimen is correspondingly hardened as determined by the parameters n and m. Thus, a highly stable plastic deformation can be realized either by work hardening or by strengthening due to strain rate sensitivity, i.e., by ensuring high values of the coefficients n and m.

In order to reveal the differences between superplastic flow and the usual plastic deformation, let us examine the mechanical properties of the alloys tested under different conditions.

It is known that plastic deformation is usually accompanied by continuous hardening during the process of tensile deformation which is followed by localized deformation, neck formation, and fracture. On the other hand, an analysis of the true tensile stress-strain curves under the conditions of SP deformation shows that soon after the initial stage, plastic deformation occurs at low and constant stresses virtually without hardening up to several hundreds and thousands of percent; sometimes, softening is observed. In this case, localization of deformation in the form of necks either ceases or occurs slowly during the entire process of tensile loading.

Furthermore, the strain rate dependence of the mechanical properties of superplastic alloys also differs significantly from that observed during usual deformation (Fig. 1.1). SP deformation is characterized by a strong strain rate dependence of the flow stress which has a sigmoidal form in the ln – ln plot (Fig. 1.1a). Such a form of the $\sigma - \dot{\varepsilon}$ curve makes a division into three regions possible. At low strain rates, one observes a relatively mild dependence of σ on $\dot{\varepsilon}$ (Fig. 1.1a, region I) and low values of m and σ (Fig. 1.1b,c). With increasing $\dot{\varepsilon}$, the strain rate dependence of the flow stress becomes stronger (more distinct), the values of m and δ increase, and a transition to region II occurs where the effect of SP attains its maximum. The range of strain rates corresponding to region II is somewhat different for different alloys, but it is usually from 10^{-4} to $10^{-2}\,\mathrm{s}^{-1}$.

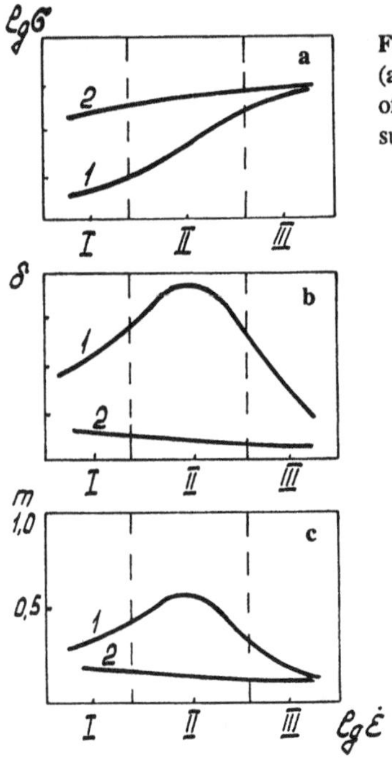

Fig. 1.1. Typical strain-rate dependences of the flow stress σ (a), the relative elongation δ (b), and the coefficient m (c) of alloys in the superplatic state (*1*) and in the usual (non-superplastic) condition (2). (*I–III*) characteristic regions

Subsequent increase of strain rate leads to a decrease of m and δ. In region III, at high strain rates, the relative elongation δ, the flow stress σ, and the parameter m approach the characteristic values of typical ductile materials. Such a shape of the $\ln \sigma - \ln \dot{\varepsilon}$ curve [usually referred to as a curve of superplasticity (SP)] showing three clearly distinct regions is typical for various materials existing in the superplastic state. We note that in [1.10–11], an increased strain rate sensitivity (rather than a decreased sensitivity) was observed at low strain rates where the parameter m approaches unity. Based on this, the authors proposed that the usually observed decrease of the parameter m in region I is nothing but an apparent effect due to the grain growth occurring during the process of SP deformation that is particularly rapid at low strain rates (Sect. 1.2). However, a careful study of the mechanical properties and the structure of the alloys does not support this hypothesis [1.12]. *Grivas* et al. [1.12] believe that in earlier studies there was an error due to the measurement of the parameter m at extremely low strains (levels of deformation) that are not characteristic of the stage of stable SP flow. Confirming the $\ln \sigma - \ln \dot{\varepsilon}$ relationship at low strain rates is an important aspect requiring additional studies.

The correlation between the strain rate dependences of δ and m is an important experimentally established fact. The results obtained when testing numerous alloys show that the transition to SP occurs at m exceeding 0.3 and that lower values of m retain the usual deformation during which σ and δ are slowly varying functions of the strain rate.

We note that Hart's mechanical model of visco-elastic flow offers only a qualitative explanation for the behaviour of superplastic materials, and the attempts made to establish the quantitative relationship between the coefficient m and ductility have given an ambiguous result [1.1–4]. Nevertheless, the qualitative relationship between δ and the coefficient m that forms the most important characteristic of superplastic materials has been established for almost all important metals and alloys in their superplastic state.

There have been numerous studies concerning the methods for determining the coefficient m [1.1–5, 13]. It is determined from the slope of $\ln\sigma - \ln\dot{\varepsilon}$ curve or according to the method of abruptly changing the strain rates during the process of deformation using different methods of calculations and extrapolation of the loading power versus deformation curves using the test data concerning stress relaxation. This problem has received a lot of attention because of the difficulties in determining the "true" value of m. Frequently, the values of m determined by different methods are qualitatively correlative with relative elongation, but, quantitatively, significnat differences are observed.

The differences observed in the absolute value of the coefficient m measured according to different methods are due to a number of factors which include the initial microstructural state of the material, the structural changes occurring during the process of tensile deformation, and the level of deformation (strain) at which m is determined. Our results [1.13] show that the shape of the stress-strain curves has great significance under the conditions of SP flow. The point is that the measured value of m is significantly affected by the magnitude and the sign of the coefficient of the strain hardening n which depends on the shape of the true deformation curves. The experiments show that, when determining the coefficient m, it is necessary to analyze the true curves of tensile deformation at different strain rates and carry out measurements only at the points where the coefficient n has the same sign for both the comparable strain rates. This situation requires a detailed study of the true curves of tensile deformation at different strain rates. However, if one opts to carry out the entire set of experiments, the very essence of determining the coefficient m as a parameter simplifying the evaluation of the superplastic behaviour of materials is lost from a practical standpoint.

There is another problem in measuring the "true" value of the coefficient m, namely, the nonuniformity of SP deformation. In many studies the formation of one or several necks and their "wandering" along superplastically deformed samples were observed. The localization of deformaiton at the level of individual grains is also typical of SP flow. The studies performed on samples with networks marked on their surfaces give evidence of a significant nonuniformity of deformation of a few dozens of microns when the strains of the neighbouring areas can differ by 50–60% from the average strain of the specimen. The local values of strain rate and coefficient of strain rate – stress sensitivity also vary in this case. The measured values of m are average; hence, they depend on the distribution of local deformations.

The facts show that at the present time no universal method can be selected for determining the absolute value of the coefficient of strain rate sensitivity of the flow stress, and the measured coefficient m must be treated not as a material constant, but as a structure-sensitive parameter whose specific value depends not only on the procedure followed for its determination, but also on the strain of the specimen and the strain rate. Thus, SP flow is characterized by the following features.

1) Large values of relative elongation of the specimens subjected to tensile loading and a significantly stable plastic deformation process owing to the increased strain rate sensitivity of the flow stress;
2) a relatively low flow stress and the absence of noticeable strain hardening during the process of deformation;
3) a considerable strain rate dependence of the flow stress, the elongation, and the parameter m.

The parameters of superplastic flow σ, δ, and m and the location of the optimum range of strain rates are determined by the structure of the alloy and the temperature of deformation. An analysis of the effect of various factors on SP flow is given below.

Temperature

SP deformation is characterized by a significant dependence of the mechanical properties on the deformation temperature. Increasing the temperature leads to a decreased flow stress and to a displacement of the optimum strain rate range of SP deformation (Fig. 1.1, region II) towards higher values of $\dot{\varepsilon}$. However, such an effect of temperature is observed only in the range corresponding to a stable microcrystalline structure. If a significant grain growth occurs with increasing temperature or the stability of the microstructure is lost as a result of the transformation of the material into a single phase condition, the SP effect decreases abruptly or disappears completely.

The temperature dependence of the rate of superplastic deformation can be described by Arrhenius' equation

$$\dot{\varepsilon} = c\exp(-Q/RT) , \tag{1.5}$$

where c is a constant; Q is the activation energy; and R is the universal gas constant.

A similar relationship can be used in describing the effect of temperature on the flow stress

$$\sigma = k\exp(Q/RT) , \tag{1.6}$$

where k is a constant.

The activation energy is an important parameter of SP flow that reflects the nature of the thermally activated processes controlling the deformation rate. The magnitude of Q is usually determined by tests at a constant stress or at a constant strain rate within a certain temperature range of deformation. The activation energy is experimentally determined from the slope of the $\ln\dot{\varepsilon} = 1/T$ or $\ln\sigma - 1/T$ plots, depending on the test conditions. However, the preexponential terms in (1.5, 6) usually depend on the temperature and, consequently, the so-called

Table 1.1. Values of the activation energy for SP flow at three ranges of strain rates and activation energy for volume Q_{vol} and grain boundary Q_{gb} diffusion in kJ/mole for some alloys. Data for the self diffusion of the indicated component of the alloy is given. From [1.1]

Alloy	Activation energy for SP flow in the regions			Q_{vol}	Q_{gb}
	I	II	III		
Zn-0.4%Al	59	42	84	(Zn)82	63
Zn-22%Al	74	79	–	(Zn)94	51
	96	70	–	(Al)143	72
	119	78	–		
Sn-38%Pb	80	50	80	(Sn)104	40
	84	53		(Pb)107	65
Sn-35%Bi	–	46	88	(Sn)104	40
Cd-5%Pb	59	42	–	(Cd)95	42

apparent activation energy which somewhat differs from the true magnitude of Q. In spite of this conditional nature, the measurements of the activation energy are of great importance in analyzing the nature of SP flow, and some respective data can be found in [1.1–9] (Table 1.1). In region II the magnitude of Q is usually close to the activation energy for grain boundary diffusion, but increases with an increasing or decreasing strain rate; in region III it usually approaches the activation energy for volume diffusion.

Grain Size

Among the structural factors affecting the SP effect, grain size is the most important since a stable microcrystalline structure ($d < 10$–$15\,\mu m$) is an essential condition for realizing structural superplasticity.

Since in the near-single phase alloys it is possible to produce equiaxial grains having a size ranging from a few microns up to several hundreds of microns, one can follow the transition from the usual deformation to the superplastic flow with increasing microstructural refinement. For example, in the case of the Zn-0.4%Al alloy, it was shown in [1.3] that at $d > 10\,\mu m$, the dependence of $\sigma_{0.2}$ on d (Fig. 1.2) obeys the Hall-Petch relationship according to which $\sigma \sim d^{-1/2}$. However, at $d < 10\,\mu m$, the grain size dependence of the yield stress follows a reverse trend and $\sigma_{0.2}$ decreases with decreasing d.

With increasing refinement of the structure, the ductility of the Zn-0.4%Al alloy and the strain rate dependence of its properties undergo changes. In particular, abrupt changes in the relative elongation are observed as the grain size changes from 10 to $1\,\mu m$; increases from 25 up to 400% (Fig. 1.2) and, simultaneously, the characteristic strain rate dependence of the properties of the alloy becomes distinct. The results of numerous studies of the relationship between the flow stress and the grain size are summarized in [1.1–4]. It is shown that during superplastic deformation, the following form exists:

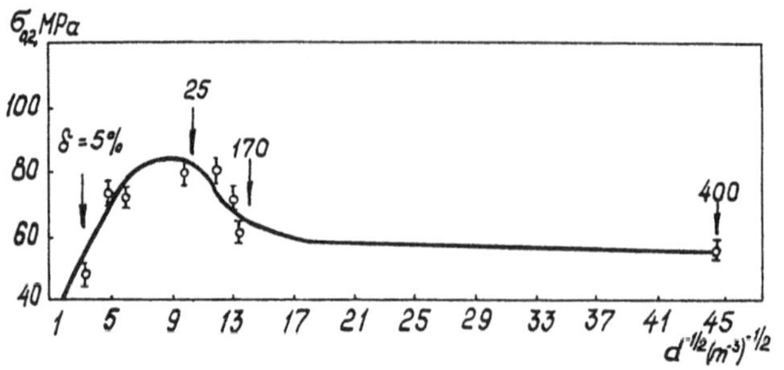

Fig. 1.2. Dependence of the yield stress of the Zn-0.4%Al alloy on the grain size d at 20°C and $\dot{\varepsilon} = 10^{-3}\,s^{-1}$ [1.3]

$$\sigma \sim d^a \,, \tag{1.7}$$

where $a = 0.7 - 2$ and, most frequently, $a = 1$.

It was established that at a constant σ the effect of grain size on the strain rate in the second region can be expressed as

$$\dot{\varepsilon} \sim d^{-b} \,, \tag{1.8}$$

where $b = 2 - 3$, most frequently with b being closer to 2.

A change in the grain size has a significant effect also on the strain rate dependence of the flow stress, i.e., decreasing the grain size makes the optimum $\dot{\varepsilon}$ II range, within which the SP effect is developed, shift towards higher strain rates. The maxima of the $m - \dot{\varepsilon}$ and $\delta - \dot{\varepsilon}$ plots shift towards higher $\dot{\varepsilon}$; moreover, the maximum value of δ increases.

The role of grain size refinement is clearly displayed in SP alloys with superfine grains. In recent years a number of ways of producing submicron-grained structures with a grain size of about 0.1 μm has been developed (Chap. 3), and data on the SP deformation of these alloys have been published in [1.14–17]. *Valiev* et al. [1.14] were the first to report an abrupt drop of a temperature at which SP flow occurs. In the Al-4%Cu-0.5%Zr alloy, a decrease in SP temperature from 500 to 220°C was observed when grain size reduced from 8 to 0.3 μm at a constant strain rate of $3 \times 10^{-4}\,s^{-1}$. The values of SP flow parameters remained almost unchanged – the coefficient m was close to 0.5, flow stress was about 20–30 MPa and relative elongation to fracture was more than 400%. The said drop in the SP temperature can also be observed in submicron-grained Ti-based and Mg-based alloys and steels (Chaps. 4–6).

However, the submicron-grained structure is unstable, and intensive grain growth occurs at temperatures higher than $0.4\,T_m$. Therefore, the studies [1.16, 17] of mechanically alloyed specimens in which a structure with a grain size of 0.2 μm can be retained up to 0.7–0.8 T_m due to disperse oxide particles are of great interest. It was established that typical SP behaviour is observed at 500°C in a strain rate range from 10^0 to $10^1\,s^{-1}$ (these strain rates are close to blow). It is evident that SP deformation of submicron-grained alloys is of great importance for practical application because of the possibility of abruptly increasing the production output and significantly prolonging the service life of the punch tackle.

In addition to the study of the SP deformation of submicron-grained alloys, the investigations of the SP of coarse-grained materials are also interesting. Typical features of superplasticity (SP) in coarse-grained alloys were observed in [1.1, 18, 19]. There are apparently several reasons underlying such unusual behaviours of these alloys. In particular, during the deformation process of coarse-grained titanium alloys a substructure forms within the grains whose behaviour is probably similar to that of the fine-grained structure of SP materials (Chap. 6).

Furthermore, in specimens of the Al-Tl and Al-Ge alloys, superplasticity is observed at $d = 100 - 200\,\mu m$ [1.19]. In this case, no substructure was observed during the deformation process. At the same time, a correlation exits between the initial porosity of the alloys (caused by phase transformations during the heat increase to test temperature) and the relative elongation under the conditions of superplastic flow. The maximum ductility was obtained in the Al-0.4%Ge alloy in which the initial porosity attained the highest level (approximately 0.8 vol.%). *Kuznetsova* et al. [1.19] showed that the SP effect in the Al-Ge alloys is due to the presence of porosity and facilitates the development of the combination of mechanisms that operate in the common superplastic alloys; and because the porosity is maintained at a constant level, it does not lead to a fracture of the material.

Grain Size Distribution

When analyzing the effect of the structural characteristics on the superplastic behaviour, one usually refers to an average grain size. However, in individual cases the size of the crystals in the experimental alloy can vary over an order of magnitude or more. This is particularly true for the industrial alloys where, during the process of obtaining fine grains by severe prior plastic deformation (working), coarse-grained zones are developed within the fine-grained structure. Naturally, the presence of different grain sizes changes the properties of the alloys under the conditions of SP flow, and the development of coarser grains expected to affect the SP effect adversely.

The behaviour of the industrial 1420 aluminium alloy (Al-5%Mg-1.8%Li-0.12%Zr) clearly illustrates this fact. The alloy was obtained in three conditions: with an equiaxial fine-grained microstructure ($d = 6\,\mu m$), with a coarse-grained microstructure ($d = 75\,\mu m$), and with a mixed structure in which equiaxial fine grains ($d = 6\,\mu m$) are present along with coarse elongated grains. In this condition, the ratio of the areas of the fractions of the fine and the coarse grains was found to be 1:1.

The dependence of δ, σ, and m on the microstructure during the deformation of the 1420 alloy is shown by the data below ($t_{def} = 450°C$ and $\dot{\varepsilon} = 10^{-3}\,s^{-1}$):

d [μm]	6	75	Mixed
δ [%]	720	210	390
σ [MPa]	5	20	13
m	0.55	0.22	0.40

These data show that the features of the SP state appear even in the presence of a mixed structure in the alloy. However, in this case, there is a certain increase in the flow stress and a decrease in the relative elongation. At the same time, the range of strain rates corresponding to that of SP shifts towards the region of lower strain rates.

The studies carried out on the Zn-22%Al and Ti-6%Al-4%V alloys and brasses [1.13, 20] showed that the location of the optimum range of strain rates and the value of m depend on the nature of grain size distribution. In [1.21] approaches that make it possible to predict the properties of the alloy are discussed. These take into account the volume fractions of the grains of different sizes, under the assumption that their contribution to the flow stress of SP is additive. A comparative study of the results calculated according to this approach shows a satisfactory agreement with the experimental values of the coefficient m, but there is a considerable difference in the values of the flow stress. The latter is probably caused by the interaction of deformation processes in the areas with coarse and fine grains.

The presence of microheterogeneities in the form of elongated particles along the prior boundaries of the grains of a high-temperature phase can be considered to be one of the types of bimodal structures arising during the process of deformation. In Zn-22%Al alloy and brasses, the presence of a network of microinhomogeneities in the boundaries of the coarse grains with the background of an ultrafine-grained structure significantly decreases the relative elongation and increases the flow stress under the conditions of SP [1.1, 20, 22]. Destruction of the network of these boundaries by prior plastic deformation leads to a decrease of σ and in increased δ.

Phase Composition

Grain size is an important, but not the only structural characteristic determining the properties of superplastic alloys. Chemical and phase compositions have a significant effect on their structure and properties under the conditions of superplastic flow. The relationship between the chemical composition and the SP effect will be examined in greater detail when the effect of alloying on the properties of concrete industrial alloys are analyzed. It is known [1.2, 3] that chemical composition has an indirect effect on SP through the microstructure, i.e., by creating the conditions required for obtaining a stable fine-grained structure. The effect of phase composition on the phenomenon of SP flow has also been examined primarily from the standpoint of microstructural stability [1.1–9]. In fact, this is the most evident aspect of the effect of phase composition on SP. It is not accidental that the very first observations and studies on SP were made on alloys having nearly an identical proportion of the phases of the eutectic or the eutectoid composition. In such materials, a fine-grained structure is most easily obtained and retained. It is readily apparent that the structural stability of the alloys increases with increasing content of the second phase; therefore, it is easier to convert multiphase alloys into the SP state as compared to single-phase alloys, and it is virtually impossible to attain the SP state in ultra-pure metals using the principles of structural superplasticity. However, microstructural stability and the grain size of alloys are not the only factors reflecting the effect of phase composition on SP; the deformation characteristics of the phases have a significant effect on the superplastic behaviour.

Let us now examine this aspect in detail. The deformation characteristics of the phases depend on their chemical composition, the type of their lattice and the homologous temperature of deformation. It might appear that knowing these characteristics, one can predict whether a multiphase alloy can be deformed superplastically. However, the actual behaviour of such alloys is much more complex, because interaction of the phases occurs during the deformation process, which must be taken into account. For instance, during the process of deformation, grain growth of the phases cannot occur without interdiffusion of these components, consequently, the diffusion characteristics of the phases can have a significant effect on SP. Furthermore, a change in the phase composition leads to a change in the structure and in the extent and the fraction of the inter-phase boundaries in the alloy. The fact that this is significant for superplastic deformation follows from the differences in sliding at the grain boundaries and the interphase boundaries [1.1]. Finally, the predominant slip systems in the phases depend on the phase composition. It was established that in the ultrafine-grained Zn-Al alloys, the active slip systems of the zinc-based beta-phase depend on the content of the alpha-phase [1.23]. Increasing the content of the alpha-phase activates the nonbasal slip systems in the beta-phase. This can be attributed to the fact that it becomes easier to generate the $(c + a)$ type dislocations at the interphase boundaries than at the grain boundaries in the beta-phase.

Detailed and systematic studies on all these aspects have not yet been carried out. However, it is already clear that, from the standpoint of phase composition, the optimum conditions for the development of SP do not simply imply obtaining a 1:1 ratio of the phases. Depending on deformation characteristics of the phases, their optimal ratio (with respect to superplastic deformation) can increase or decrease. A special investigation was made on this aspect using alloys of the Zn-Al system containing 0.4–72 wt.%Al [1.3, 22]. The homologous temperatures for the alpha and beta phases are significantly different in these alloys and, consequently, at 250°C, the possibility of plastic deformation is considerably greater in the beta-phase than in the alpha-phase. Thus, as the strain rate increases, the Zn-10%Al alloy exhibits a greater relative elongation than the alloy of the eutectoid composition. The importance of the deformation characteristics of the phases becomes particularly evident at low temperature deformation where stability of grain size is less important. In this case, the zinc-rich alloys are most ductile in spite of the fact that, from the standpoint of superplastic deformation, their structure is less favourable than that of the eutectoid alloy. The behaviour of the alpha-phase in the alloys containing more than 50 vol.% beta-phase is somewhat similar to that of the hard particles present in a ductile matrix.

Similar results were subsequently obtained during a study of the effect of the phase composition of alloys of the Cu-Zn system on their SP behaviour [1.24]. Thus, the role of the phase composition of the alloys apparently does not only reside in evolving the optimum microcrystalline microstructure having a highly stable grain size at $T > 0.4\,T_m$ in the formation of two-phase microstructures, it is also necessary to take into account the deformation characteristics of each phase and the interaction between the phases during the deformation process.

Texture

The effect of texture on the properties of alloys is traditionally related to the development of preferred grain orientation in the polycrystalline matrix. The mechanical properties of single crystal depend on the orientation of the lattice with respect to the direction of load application due to which, in the presence of preferred grain orientation, polycrystalline materials exhibit anisotropy of properties.

Under the conditions of superplastic deformation, this effect was studied for the first time on the Zn-0.4%Al alloy [1.25]. It was established that σ, δ, and the coefficient m vary considerably depending on the direction of cutting the specimens relative to the rolling direction. According to *Naziri* and *Pearce* [1.25], this effect is associated with the presence of a crystallographic texture in the Zn-0.4%Al alloy. Subsequently, anisotropy of properties under the conditions of SP was observed in a number of alloys, viz., Zn-Al [1.26, 27], aluminium bronze [1.28], brasses [1.20], Sn-Bi [1.29], etc. However, the results cannot always be considered as a consequence of the presence of preferred grain orientation. The microstructure of alloys (in particular, industrial alloys) in which superplasticity is observed shows a certain degree of structural nonuniformity (elongation of grains and directionality in the distribution of precipitates and phases) because of certain aspects of the sample preparation and the presence of impurities and inclusions; such a nonuniformity can influence the SP effect. Many investigators consider that anisotropy of properties is a result of anisotropic distribution of the particles of a second-phase in the structure. For example, it was established [1.20] that microstructural anisotropy as well as the crystallographic texture of the L59 brass cause nonuniformity of the properties in relation to the rolling direction. Apparently there is a cumulative effect of these factors on the anisotropy of properties even in other cases, but this aspect has not yet been studied in sufficient detail.

It should be noted that anisotropy of relative elongation alone is usually observed and the flow stress is isotropic [1.27]. Anisotropy of the flow stress was absent in spite of the presence of a texture, and its dependence on the direction in which specimen cutting was carried out was seen only in the case of an alloy having grains elongated in the rolling direction. The absence of anisotropy of the flow stress makes it somewhat difficult to follow the effect of textures on the properties of the alloys; therefore, some investigators believe that preferred grain orientation does not change the SP effect. However, the absence of anisotropy of the flow stress can be one of the specific features of the effect of texture on SP and this, by no means, implies the absence of such an effect. In order to study the "pure" effect of preferred grain orientation on the superplasticity of an alloy, it is necessary to carry out property measurements in a textureless condition and in the presence of different textures maintaining identical microstructures of the alloy. Such studies have been carried out on the Zn-22%Al and VT6 titanium alloys [1.3, 29, 30].

The most detailed study on the effect of texture was carried out on the Zn-22%Al alloy in which, using the specific features of the monotectoid decomposition during quenching and controlling the regimes of rolling, it is possible to obtain a wide variety of conditions ranging from the textureless state up to a sharp preferred orientation within the beta-phase alone as well as in both alpha- and beta-phases. Table 1.2 shows the processing conditions selected for the experimental study. All the processing regimes yielded equiaxial structures, and an equal grain size (approximately 0.5 μm) was obtained by varying the duration of annealing at 250°C.

A comparative study of the properties of the textures and the textureless alloy in the as-quenched state makes it possible to reveal the effect of texture on the mechanical properties in the "pure form", because as a final treatment, quenching makes preferred grain orientation the main structural parameter distinguishing between state 1 and 2 (Table 1.2). As seen from Table 1.2, the presence of a texture leads to a decrease in the flow stress over the entire experimental range of temperature and strain rate.

An analysis of the true stress-strain curves shows that the difference in their magnitudes retained up to ε = 150% and more. The relative elongation of the quenched textured alloy is significantly greater than that of a texture-free alloy (Fig. 1.3).

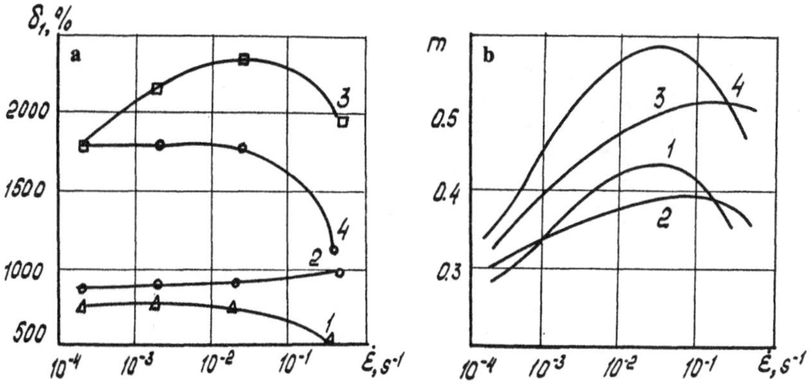

Fig. 1.3. Dependence of relative elongation (a) and the coefficient m (b) on Zn-22%Al at 250°C on the strain rate [1.1]. The curves *1–4* correspond to the states of the alloy indicated in Table 1.2

Table 1.2. Effect of the texture on the flow stress σ_{40} at $\dot{\varepsilon} = s^{-1}$ of the Zn-22% Al alloy. From [1.27]

Number of the state of the alloy	Mode of producing structure	Type of texture		σ_{40} [MPa]								
		α-phase	β-phase	$2.8 \times 10^{-4}\,s^{-1}$			$2.8 \times 10^{-2}\,s^{-1}$			$2.8 \times 10^{-1}\,s^{-1}$		
				20°C	150°C	250°C	20°C	150°C	250°C	20°C	150°C	250°C
1	Quenching, rolling at 250°C, annealing at 350°C for 5 h, and quenching	none	none	183	21	2.5	261	118	16.6	–	182	44
2	Quenching, rolling at 20°C, annealing at 350°C for 5 h and quenching	$(001)[\bar{1}00]+$ $+(110)(11\bar{2}$	$(10\bar{1}0$ $[0001]+$ $+(1\bar{2}10)$ $[10\bar{1}0]$.	114.5	15.3	2.2	295	79	13.4	–	110	26.4
3	Quenching and rolling at 20°C	$(001)[\bar{1}00]+$ $+(120)[21\bar{2}]$	$(10\bar{1}0$ $[0001]+$ $+(0001)+$ $+24°$ TD–RD$[10\bar{1}0]$	88.6	3.2	1.04	270	40	9.4	–	71	25
4	Quenching and rolling at 250°C	none	(0001)	101.0	9.9	1.26	304	96	11.7	–	163	32

Texture influences the location of the optimum range of strain rates for realizing the SP effect. It can be seen from Fig. 1.3 that the maxima of the parameters m and δ of the alloy existing in state 2 shift towards higher strain rates as compared to those of the alloy corresponding to state 1. Thus, at a given microstructure, the presence of a texture in the Zn-22%Al alloy leads to a decreased flow stress, an increased relative elongation, and a shift of the optimum strain rate interval towards higher strain rates. With a decreasing test temperature, the effects of textures are not only retained, but are intensified (Table 1.2). Thus, after quenching, at 205°C and $\dot{\varepsilon} = 2.8 \times 10^{-4}\,s^{-1}$, the value of δ of the textured alloy is 1.3 times higher and σ is 1.25 times less than those of the textureless alloy. At 150°C, these values increase to 2.1 and 1.5 times, respectively. The optimum strain rates corresponding to states 1 and 2 differ by 2–3 times at 250°C and by 15 times at 150°C. The difference in the room temperature properties of the alloy in states 1 and 2 is still greater, but its quantitative evaluation is difficult because the optimum range of strain rates corresponds to $10^{-4}\,s^{-1}$.

As a result of prestraining of a quenched texture-free alloy, the SP effect in it increases significantly [1.1, 3]. *Portnoi* et al. [1.20] attribute such a change in the properties to the destruction of the network of the elongated particles along the grain boundaries of the high temperature phase and not to the development of textures during the course of deformation. Our studies [1.27] show that structural inhomogeneity affects the flow stress and the relative elongation during the initial states of deformation but has virtually no effect on the strain rate dependence of these mechanical properties. The results in Table 1.2 and Fig. 1.3 show that in the rolled alloys (states 3 and 4), the relative elongation increases and the flow stress decreases significantly as compared to the properties obtained in the as-quenched alloys. Even in this case one observes a shift of the strain rate range corresponding to the SP effect in the alloy having preferred grain orientation, i.e., the observed change in the properties of the alloy after rolling is not only due to the destruction of the microstructural heterogeneities, but is also due to the texture development during the process of rolling.

The presence of a texture causes a significant difference between the properties of the alloy existing in states 3 and 4 that are distinguished only by the nature of the rolling textures. The flow stress of the cold rolled alloy and the relative elongation of the former is higher. In this case, the optimum strain rate range of the alloy existing in state 3 is shifted towards the region of higher strain rates. As in the case of alloys in the as-quenched state, the effect of texture on the properties of the rolled alloys increases significantly with a decreasing test temperature.

The universal nature of the effects of preferred grain orientation on the properties of the alloys under the conditions of superplastic flow is also confirmed by the studies carried out on the titanium based VT6 alloy [1.29, 30], which was obtained in two states, viz., without any texture and with a strong texture, using different combinations of the regimes of compressing and rolling at 800–900°C. The microstructure was equiaxial; an identical grain size (approximately 4 μm) was obtained in both states by varying the duration of annealing. Thus, the basic requirement of identical microstructures and different textures could be met in these two states.

Tensile tests showed that the flow stress of the textured alloy is less than that of the textureless alloy. Such an effect of texture was noticeable at the test temperatures ranging from 700 to 900°C, because the difference in the flow stress increases with decreasing temperature of deformation. The plasticity of the alloy also shows a significant dependence on the texture. The relative elongation of the textured alloy is greater than that of the texture-free alloy and these differences also increase with decreasing test temperature. Thus, at 900°C the relative elongation of the textured alloy is only 1.06 times higher than that of the textureless alloy, whereas at 700°C the plasticity of the textured alloy is three times higher. A study of the strain rate dependence of m showed that the optimum strain rates of the textured alloy were displaced towards the region of higher strain rates and that the extent of this displacement increased with decreasing test temperature.

The established effects of preferred orientation on the superplastic flow are most unusual and were not reported earlier. However, there are some published experimental data which can also be attributed to the effect of texture on the properties. For example, the studies of *Benedek* and *Doherty* [1.31] showed that with an increasing degree of prestraining, the grain size increases instead of decreasing. In this case, the relative elongation increased and the flow stress decreased. Similar results concerning the effect of prestraining on the properties of brasses under the conditions of SP were reported in [1.5].

The observed variations of the characteristics of SP flow can be attributed to the changes occurring in the structure of grain boundaries of a polycrystalline material in the presence of crystallographic textures. There undoubtedly exists a relationship between the texture and structure of grain boundaries, because, based on the orientation distribution function obtained from the texture data, it is possible to evaluate the distribution function of the boundaries with respect to the characteristic geometric parameters (in particular, with respect to the angles of misorientation between the grains). The structure of grain boundaries depends on its geometric parameters. The grain boundary processes playing a specific role in the phenomenon of superplasticity will be examined in Sect. 2.2.2.

The State of Grain Boundaries

In the structure of ultrafine-grained superplastic materials, the total grain boundary area is extremely high; the grain size essentially characterizes the reciprocal of the grain boundary area per unit volume. In this case, grain boundaries act not only as the geometric surfaces separating grains of different orientations, but also as important structural elements determining the properties of the polycrystalline material under consideration. Unfortunately, this aspect has not been analyzed in detail in studies on the phenomenon of structural superplasticity, mainly because of the nonavailability of adequate data on the behaviour of grain boundaries during plastic deformation processes. However, the progress made in recent years in understanding the structure and properties of high-angle grain boundaries (Sect. 2.2) allows a new direction of experimental design and analysis in revealing the role of the state of grain boundaries in the development of SP deformation.

In this context, the results of investigations carried out on the superplastic behaviour of the magnesium alloy MA8 containing twins are of interest [1.32]. After being rolled and annealed, this alloy exhibits superplasticity at 400°C and $d \simeq 10\,\mu m$. Twins were introduced into the structure of the alloy by cold working. After treatment, more than 75% of the grains showed deformation twins and their number remained unaltered duirng the process of heating up to the temperature of SP flow. The presence of twinned regions can be considered as an additional refinement of the structure; taking the twin boundaries into account, the grain size of the alloy is approximately 3 μm. As mentioned above, a decrease in the grain size often makes it easy to realize the SP effect and facilitates a shift of the range of strain rates corresponding to its development towards higher strain rates. However, as shown by results of mechanical testing, a different picture emerges in the alloy containing twins, i.e., the flow stress increases and the maximum values of the parameters of plasticity, viz., δ and the coefficient m are displaced towards lower strain rates. Structural studies make it possible to delineate the reasons underlying the observed variation of mechanical properties. Sliding does not occur along the twin boundaries and, at the same time, twins facilitate accumulation of dislocations in the structure. Twin boundaries are high-angle boundaries but belong to a special type of grain boundaries (Sect. 2.2.2) exhibiting peculiar properties, i.e., they do not act as dislocation sinks. Thus, the results directly indicate the important role of the structure of boundaries in the development of superplasticity.

In addition to the structure of grain boundaries, the type and concentration of impurities and segregation have a decisive effect on the grain boundary processes. Thus, it can be expected that they have a significant effect on the SP flow also. This hypothesis is supported by a number of recent experimental data.

In the sintered aluminium alloy SAS-I-50 (Al-27.1%Si-5.5%Ni-0.46%Fe-1.26%Al_2O_3), it is easy to obtain a microstructure having a grain size of 1.5–4 μm. This microstructure is stable up to 550°C (0.95 T_m). However, superplastic flow is not observed. Its deformation is characterized by $\sigma > 50$ MPa, $\delta < 30\%$, and $m < 0.3$. We emphasize that the alloys having a composition close to that of the aforementioned alloy and an analogous microstructure exhibit all the features of SP under identical conditions of testing. Probably, the absence of SP in the SAS alloy is caused by the obstructed movement of the boundaries that is caused either by the oxide film formed along the grain boundaries during the preparation of the alloy or by the Al_2O_3 particles. As a consequence, the operation of the characteristic deformation mechanisms of superplastic flow becomes impossible.

Myshlyaev et al. as indicated in [1.1] observed the disappearance of the SP effect in the Zn-0.4%Al alloy after aging. It was established that aging the alloy for a period of 6 months leads to a state of low plasticity ($\delta = 30\%$ and $m < 0.2$), although it has virtually no effect on the grain size (approximately 1 μm). It can be concluded that a change in the impurity concentration occurs at the grain boundaries. As a result, the grain boundaries can no longer act as effective dislocation sinks and, consequently, the progress of grain boundary sliding (GBS) becomes extremely difficult. In this context, the results on the model aluminium alloys Al-2 at.%Cu-0.16 at.%Zr and Al-2 at.%Mg-0.16 at.%Zr [1.33] are of interest. There are quasi-single phase alloys. Copper and magnesium enter the solid solution and zirconium is present mainly in the intermetallic compound Al_3Zr. A microstructure having a grain size of 7 µm and a uniform distribution of the particles of the second phase Al_3Zr was obtained using a special treatment.

The data in Table 1.3 show that for a given microstructural state and under identical test conditions the flow stress of the copper containing alloy is significantly less than that of the magnesium containing alloy, but the relative elongation is considerably higher.

Table 1.3. Structural data and superplasticity parameters of aluminium alloys Al-2 at.% Cu-0.16 at.% Zr (1) and Al-2 at.% Mg -0.16 at.% Zr (2) at a strain rate of $1.7 \times 10^{-3}\,s^{-1}$ (d_0 = 7 µm). The numerator and the denominator indicate the data obtained at T_{def} = 773 K and T_{def} = 723 K, respectively

Alloy	l [µm][a]	d_{50} [µm]	m	δ_{max} [%]	σ_{50} [MPa]
1	0.25	8/7	0.53/0.40	900/400	4/9
2	0.23	9/7.5	0.33/0.27	360/190	10/21

[a]Represents the interparticle spacing

An optical microscopic study of the structure arising during the process of deformation did not reveal significiant differences between the alloys. The obtained differences in the mechanical properties could be explained only by conducting electron microscopy of the structure of the grain boundaries. It was established that the boundaries present in the copper-containing alloy possess a greater capacity to absorb the lattice dislocations than those existing in the magnesium-containing alloy (Sect. 2.2).

Karpov et al. [1.34] studied the SP effect in berillium-yttrium alloys. Due to low solubility of yttrium in beryllium, its addition leads to the precipitation of the YBe_{13} compound along the grain boundaries which, in turn, stabilizes the fine-grained structure. According to existing concepts, this stabilization must increase the SP effect. However, in spite of the increased micro-structural stability, superplasticity of the berillium-yttrium alloys is not intensified; in fact, upon increasing the yttrium content beyond 0.15%, the magnitude of m abruptly decreases to 0.1–0.2 and the relative elongation decreases correspondingly. Such an effect of alloying is related to the changes occurring in the properties of grain boundaries due to precipitation [1.34].

The data indicate the importance of the state of the grain boundaries in the development of superplastic flow and show that the presence of an ultrafine-grained structure is a necessary, but not a sufficient condition for inducing the SP effect. Furthermore, it is important to consider the specific features of the structure and the state of the grain boundaries and their behaviour (sliding, migration) in the microstructure of the superplastic materials which facilitate the development of the processes determining SP deformation. However, it is necessary to conduct further studies for establishing such parameters of the structure and the state of the grain boundaries that would ensure the best indices of the effect of structural superplasticity.

1.2 Structure Evolution and Fracture Modes

A study of the structural changes occurring during SP flow is important for two reasons. On the one hand, it ensures evaluation of the specific features and nature of the process of SP flow, and, on the other, it permits one to predict the properties obtainable after SP deformation. Up to now, there have been numerous studies on the microstructure, the dislocation structure, the texture evolution, the deformation relief, and the porosity appearing in various metals and alloys in the superplastic state [1.1–9]. Let us examine the basic results in order to evaluate the present state of the problem.

Microstructure

Retention of the equiaxial grain shape after SP deformation is one of the most important features established thus far. Even after attaining several hundred percent elongation, the specimens deformed in the region of the maximum strain rate sensitivity of the flow stress retain to a large extent an equiaxial grain structure. This result has been confirmed on virtually all superplastic materials. In addition, a metallographic texture (grain elongation/fibrous structure) initially present in an alloy usually decreases during deformation. There are two points of view on the observed retention of equiaxial grain shape. According to the first theory, elongation of the grains occurs during deformation, but it is eliminated by the migration of grain boundaries [1.2, 3, 35]. According to the second theory, the grains remain equiaxial due to their unusual movement during which they are redistributed as a whole with a replacement of their neighbours (Sect. 2.1.1). Evidently, both these processes can occur during superplastic deformation.

Grains remain equiaxial not only when carrying out deformation in region II, but also when the strain rate is changed. However, a signficiant increase in the strain rate usually leads to elongation of grains and sometimes to grain refinement because of the loss of superplasticity in the material. *Lee* [1.35] and *Valiev* and *Kaibyshev* [1.36] observed the development of grain elongation even when deforming at low strain rates. This effect is most pronounced in alloys having an hcp structure, and its nature is controlled, evidently, by the specific features of the mechanisms operating in region I [1.37].

Grain growth is another characteristic feature of the microstructural change. It is related not only to the effect of time and temperature, but also to the specific nature of SP flow since the grain size in the deformed part of the specimen is usually higher than that existing in the heads of the specimen. As a rule, in two-phase alloys, grain growth of both phases occurs and, in the alloys having a near-single phase structure, rapid grain growth of the matrix occurs and the size of the second phase does not undergo significant changes.

A number of detailed studies have been carried out on the kinetics of grain growth as a function of the strain rate. It was established that the variation of the average grain size during the process of deformation often depends on the time of deformation by the following equation [1.1–5]:

$$\Delta d \propto \sqrt{t} \,. \tag{1.9}$$

A the same time, another parameter of the process, namely, the rate of grain growth, increases with increasing strain rate according to the empirical relationship

$$d/t = k\dot{\varepsilon}^{2/3} \,, \tag{1.10}$$

where k is the constant of proportionality.

It is of considerable interest to evaluate the increase of grain size due to deformation. This is given by

$$\Delta d_{\varepsilon} = d - d_0 \,, \tag{1.11}$$

where d is the average grain size after certain strain; and d_0 is the average grain size after annealing for a period equal to the duration of deformation up to the given strain.

It was shown [1.3, 36, 38] that in the two-phase Zn-22%Al alloy and in the near-single phase alloys like Zn-0.4%Al and Zn-5%Bi, the magnitude of the increment grain growth due to deformation increases with an increasing degree of deformation and decreasing strain rate (Fig. 1.4) and, at the same time, the rate of grain growth ($\Delta d_{\varepsilon}/t$) increased continuously with increasing $\dot{\varepsilon}$.

The aspects concerning grain growth have been studied extensively. The theoretical analysis of *Clark* and *Alden* [1.38] shows that during SP, grain growth cannot occur due to the coalescence of grains but by grain boundary migration, where the mobility of the grain boundaries increases during the process of deformation. Further development of these concepts [1.36, 39] led to the assumption that in superplastic materials, grain growth is related to the increased mobility of the grain boundaries because of the accelerated grain boundary diffusion induced by the formation of a nonequilibrium structure of the grain boundaries during deformation (Sect. 2.2.3) and that the tendency to decrease the surface energy of the grain boundaries forms the driving force for this process (just as in the case of normal grain growth). Thus, it is possible to explain the main regularities of the experimentally observed kinetics of grain growth. These concepts are also supported by recent studies in which identical effects of the initial texture [1.27]

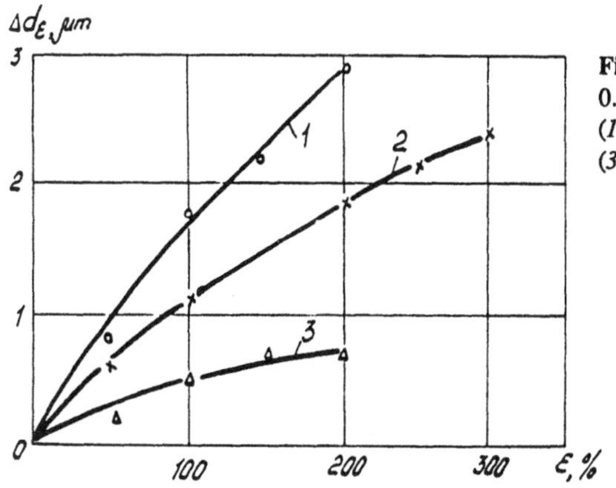

Fig. 1.4. Grain growth Δd_ε of the Zn-0.4%Al alloy at different strain rates [1.36]: (*1*) $\dot{\varepsilon} = 4 \times 10^{-5}\,s^{-1}$; (*2*) $\dot{\varepsilon} = 4 \times 10^{-4}\,s^{-1}$; (*3*) $\dot{\varepsilon} = 4 \times 10^{-3}\,s^{-1}$

and alloying [1.40] on the grain growth were observed during annealing and superplastic deformation.

Dislocation Structure

The results of numerous electron microscopic studies conducted on the dislocation structure using the thin films obtained from the specimens subjected to SP deformation showed that an absence of a substructure is a characteristic feature of the superplastic materials even at large strains [1.9–40]. As a rule, the grains are almost free from dislocations. Only in alloys containing dispersed second phase particles is it possible to observe individual dislocations (Fig. 1.5a, b). According to several investigators, single dislocation walls can also be observed in these alloys, and their number increases to some extent with increasing strain rate.

Such situations are typical for the materials deformed in regions I and II. At high strain rates (region III), where the characteristics of superplasticity decrease, a rapid increase of the dislocation density is observed in the structure and intersection of dislocations, and subboundaries are seen in several grains (Fig. 1.5c).

It should be noted that the absence of a developed substructure after SP deformation does not clearly indicate a low activity of dislocations, since their disappearance is possible during the processes of deformation, unloading, and slow cooling from the deformation temperature of the specimens (Sect. 2.1.2).

Worth a special mention are the individual publications [1.1, 2, 23, 41] concerning the observation of defects at the grain boundaries after SP (Fig. 1.6). It was concluded [1.41] that in the magnesium alloy MA8 the defect density increases with an increasing strain rate and that these defects are the lattice dislocations trapped by the grain boundaries. However, in many materials, the grain boundaries are defect free because of their accelerated relaxation at high temperatures (Sect. 2.2.2).

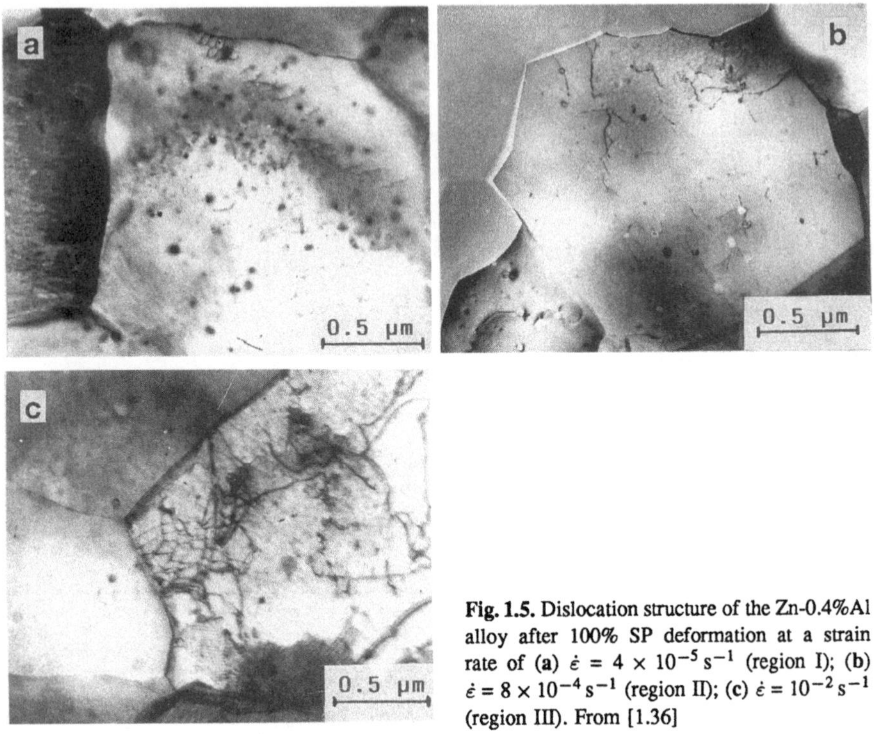

Fig. 1.5. Dislocation structure of the Zn-0.4%Al alloy after 100% SP deformation at a strain rate of **(a)** $\dot{\varepsilon} = 4 \times 10^{-5}\,s^{-1}$ (region I); **(b)** $\dot{\varepsilon} = 8 \times 10^{-4}\,s^{-1}$ (region II); **(c)** $\dot{\varepsilon} = 10^{-2}\,s^{-1}$ (region III). From [1.36]

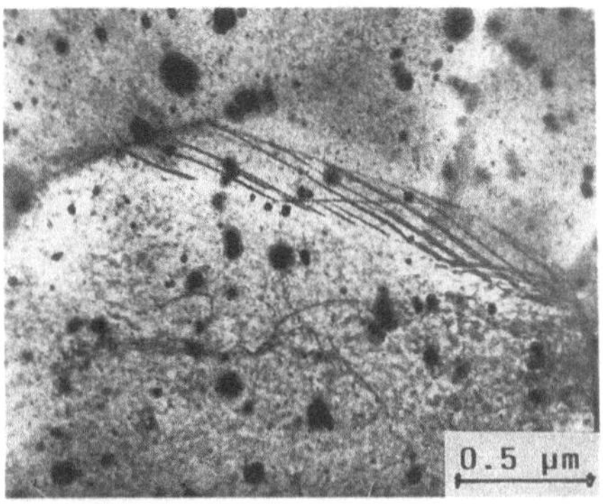

Fig. 1.6. Grain boundaries of the magnesium MA8 alloy after 50% SP deformation at a strain rate of $10^{-3}\,s^{-1}$ at 400°C [1.37]

Textures

A study of the textures is also important for controlling the properties of the superplastic materials and for understanding the operating deformation mechanisms. Most frequently, the original material exhibits a texture developed during its prior working carried out for transforming the alloys into the SP state. Even the first investigations on the texture variations [1.3, 4] established that, with an increasing degree of deformation, the original texture is significantly blurred in that it occurs most rapidly at the strain rates corresponding to region II which sometimes leads to a completely textureless state. Such a disappearance of textures is usually attributed to chaotic rotations of the grains during their relative displacements as a whole unit. At the same time, systematic studies [1.3, 4] showed that in addition to a spreading of the textures, the development of new peaks in the pole figures and the retention of the intensity of the original peaks are recorded. These data are important for studying the phenomenon of intragranular slip under the conditions of superplastic deformation. A detailed analysis of these data is given in Chap. 2.

Topography

In order to carry out topographic studies on the ultra-fine-grained superplastic materials, electron microscopy (scanning, as well as the replica methods) is widely used. For this purpose, optical microscopy also can be used if the material has relatively coarse grains (approximately 10-15 μm). Figure 1.7a shows a typical micrograph of a replica obtained from the surface of a deformed specimen of the Zn-0.4%Al alloy, which was polished before straining. We shall identify certain features of development of the deformation relief that are characteristic for other superplastic materials also.

It has been reliably established [1.5] that slip lines are absent. These are usually found only after deforming at high strain rates. However, slip lines can be observed in some alloys with an hcp lattice after deforming it in region II [1.42]. This was attributed to the absence of grain growth during deformation [1.42]. It is known that for the formation of a visible slip line, it is essential that a relatively large number of dislocations (more than 100) must move along one plane. Obviously, during SP deformation, grain boundaries act as sources of dislocations (Sect. 2.1.2) and, consequently, slip lines cannot form, because grain growth occurs during the process of SP deformation which leads to a continuous change in the locations of the dislocation sources during deformation.

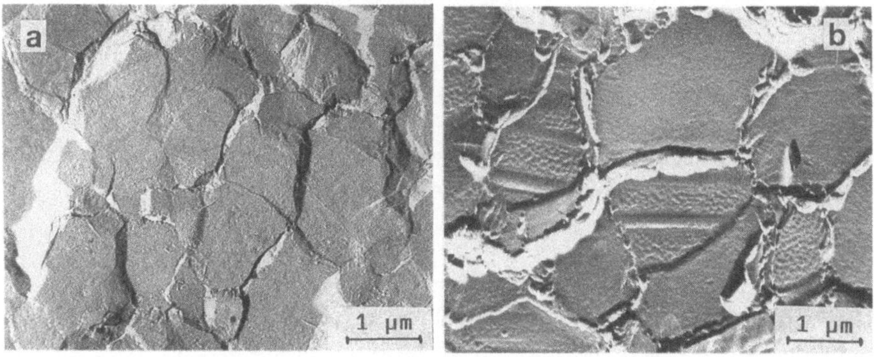

Fig. 1.7. Surface replica of the superplatic specimens of the Zn-0.4%Al (**a**) and Zn-22%Al (**b**) alloys after deforming up to $\dot{\varepsilon}$ = 20% at a strain rate of $4 \times 10^{-4}\,s^{-1}$ at 20°C [1.36]

The second specific feature of the relief being developed is that the grain boundaries are revealed even at low strain because of mutual displacement of grains. Discontinuities and displacements of the scratches previously made at the surface have been observed at the grain boundaries in numerous studies.

When superplastically deforming the Pb-Sn eutectic alloy "in situ" in a transmission electron microscope [1.43], it was established that a significant displacement and mutual arrangement of the entire grains, their exchange of sites, and their rotation occur. Those results were latter confirmed in [1.44–49]. All these data clearly prove that grain boundary sliding (GBS) occurs intensively in the process of SP flow.

Observation of the formation of a deformation zone in the vicinity of the grain boundaries (Fig. 1.7b) is yet another important result of the topographic studies. This structural phenomenon is extremely interesting, because the developed zones carry important information regarding the operating deformation mechanisms. *Backofen* et al. [1.47] were the first to identify these zones. Subsequently, they were observed in diverse superplastic materials [1.1, 5, 48].

When studying the deformation zones using the replica technique, a specific striated relief is frequently observed in most of them. Therefore, they are often referred to as "striated" zones. However, this definition is not quite correct, since striation is related to the quality of preparation of the specimen surface prior to deformation, and it can be absent in many cases [1.5]. As indicated by the studies, the width of the zones increases with an increasing degree of deformation and with a decreasing strain rate during tensile loading. Careful studies on the deformation zones developed in the two-phase Zn-22%Al alloy [1.5, 49] showed that these zones form preferentially at the interphase boundaries and make an angle of 10–20° relative to the specimen surface in that the average angle decreases with decreasing strain rate.

The nature of the observed deformation zones is complex. Backofen's studies suggested that the zones form as a result of diffusional creep. Subsequent investigations [1.38] made it possible to conclude that formation of these zones occurs due to grain boundary sliding (GBS), during which the internal surfaces of the boundaries are revealed. However, formation of new surfaces can also occur due to diffusional mass transport [1.5] and, therefore, it is justifiable to consider the formation of the deformation zones as a result of GBS and diffusion. Evidently, grain boundary migration also participated in their formation, in particular, in the case of single-phase materials [1.1, 3, 37]. It is this complex nature of the formation of the deformation zones that is responsible for their diverse manifestations observed in the topographic investigations.

Void Formation

The increased porosity observed after SP is a very important structural effect, and a number of papers have been devoted to this problem [1.50, 51]. Initially, this effect was observed in alloys of copper, but thereafter, it was revealed in a wide variety of materials. It was found that, just as in the case of creep, the alloys based on copper, iron, aluminium, nickel and magnesium are the ones most prone to void formation; in these alloys, voids can be observed after superplastically deforming up to a relatively small strain (50–100%). In the zinc-based alloys, voids are usually observed only at large strains [1.52]. As a rule, in titanium, lead, and zirconium alloys, formation of a significant porosity is not observed. In addition, the development of porosity is influenced by the load scheme, i.e., the density of voids obtained after compression loading is significantly less than that obtained after tensile deformation under identical test conditions (i.e., at the same temperature and strain rate).

The volume fraction of voids increases monotonically with an increasing degree of deformation and frequently attains a value of 3–5% at the instant of fracture. The quantitative relationship between the volume fraction of the pores f and the strain can be expressed in the following form:

$$f = a\varepsilon^n , \tag{1.12}$$

where a and n are constants.

It was established [1.53] that, in the case of steels, n varies from 1.5 up to 2.5 with an increasing strain rate (transition from region I to regions II and III). On the other hand, we found that n increases from 1.3 to 2.2 in the MA8 alloy [1.54]. At a given strain, the volume fraction of the voids and their size usually increase with decreasing strain rate (Fig. 1.8); however, in certain materials, the maximum porosity corresponds to the maximum degree of plasticity [1.50].

At the present time no direct data concerning void nucleation are available, but the phenomenological observations on void formation can be of use. As shown by numerous investigations, the smallest voids invariably originate at the grain boundaries. In brasses voids are preferentially nucleated at the interphase boundaries which also show the maximum sliding [1.50, 51]. Voids are often observed in the vicinity of the precipitates and porosity increases with the increasing size of the particles [1.55]. Based on direct observations made in a high-voltage

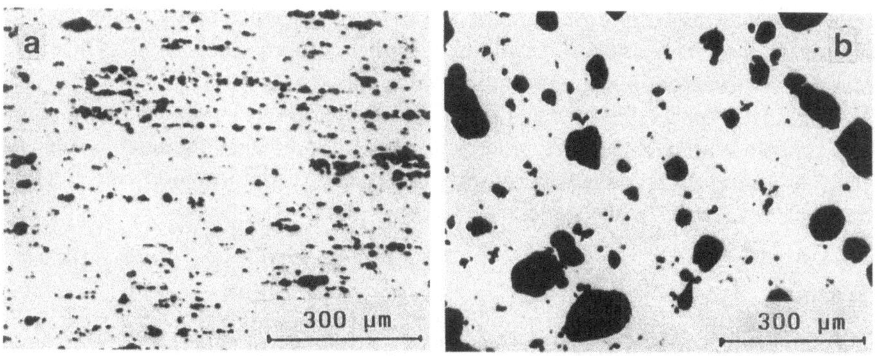

Fig. 1.8. Voids in the deformed specimens of a copper alloy after tensile loading at 827 K at a strain rate of 3×10^{-2} (a) and $2 \times 10^{-5}\,s^{-1}$ (b). The tensile axis is vertical [1.50]

electron microscope, *Fleck* et al. [1.56] proposed a model for void formation according to which the void nuclei emerge as a result of the stress concentration developed when grain boundary sliding is hindered by the particles. On the other hand, *Smith* and *Ridley* [1.57] suggested that the nuclei of voids can exist even before the commencement of deformation because of incoherency at the precipitate-matrix interfaces.

As in the creep process, the growth of voids can occur in two ways [1.51, 58], viz., by vacancy diffusion and deformation (for example, due to the opening of the void nuclei as a result of sliding along the boundaries). In the first case, voids must be spherical, and in the second case they are expected to have sharp corners. The characteristic changes occurring in the shape and the distribution of voids with increasing strain rates (Fig. 1.8) (such as their elongation and directionality) are often related to the decreased role of diffusion and the increased role of deformation-induced void growth [1.50]. Yet another standpoint exists. As was shown earlier [1.59], void shape must depend on the ratio of the velocities of the diffusional fluxes along the void surface and along the grain boundaries. If the rate of the grain boundary diffusion is less than the diffusion rate along the void surface, the void assumes an equiaxial shape, and if the rate of the grain boundary diffusion is higher than the diffusion rate along the void surface, the void becomes elongated. In view of the fact that the diffusion rate along the grain boundaries increases with increasing strain rate (Sect. 2.2.2), this factor can become decisive in changing the void shape at high strain rates.

The role of voids during SP flow has not been investigated completely. It is usually believed that void formation is a process having detrimental effects. However, at a certain level of porosity, the presence of voids can be useful for superplasticity, because it facilitates easy grain boundary sliding [1.19, 60]. It was experimentally shown [1.60] that the amount of sliding increases significantly in the vicinity of voids. Evidently, this effect can play a decisive role in the development of superplasticity in the coarse-grained materials [1.19]. However, in the typical superplastic materials (having a fine-grained structure), porosity does not appear to play such a role. Firstly, during the initial stage of deformation, porosity is rarely observed. Secondly, with increasing deformation (which is accompanied by an increased porosity), no increase in GBS is observed. Finally, when the deformation mode is changed (in particular, during compression), significantly less porosity is observed, but the plasticity is not affected.

Thus, it appears likely that during tensile deformation of the fine-grained superplastic materials, porosity does not play a significantly favourable role. At the same time, its accumulation during the process of deformation considerably accelerates fracture of the materials.

Fracture

Unfortunately, in spite of the importance of the fracture of superplastic materials, limited information is available on this aspect. Generally, two modes of fracture are observed in superplastic materials. The deformation of some materials, such

Fig. 1.9

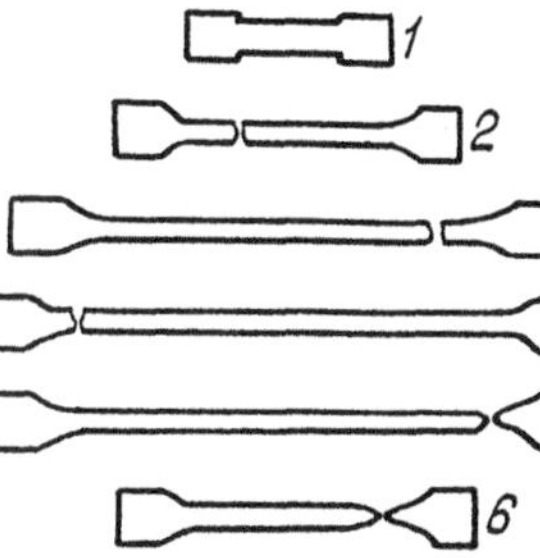

Fig. 1.10

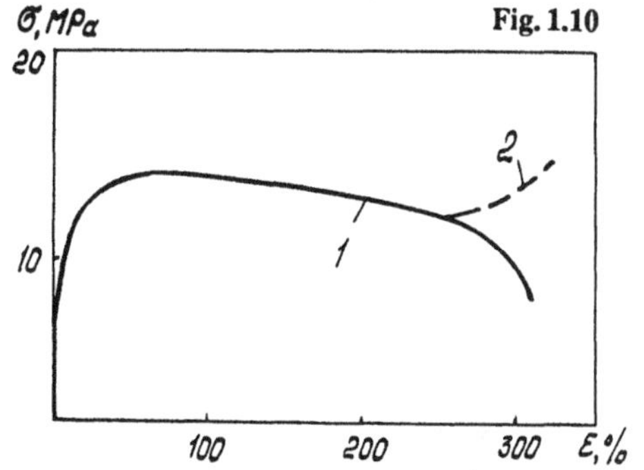

Fig. 1.9. Specimens of the 0.2%C-13%Cr steel in the prestrained state (1) and after deforming at 800°C up to fracture at a strain rate of $3 \times 10^{-5}\,s^{-1}$ (2), $3 \times 10^{-4}\,s^{-1}$ (*3*), $3 \times 10^{-3}\,s^{-1}$ (*4*), $3 \times 10^{-2}\,s^{-1}$ (*5*), and $3 \times 10^{-1}\,s^{-1}$ (*6*) [1.1]

Fig. 1.10. The true stress-strain curve of the MA8 alloy [1.63]: (*1*) at 400°C and $\dot{\varepsilon} = 8 \times 10^{-3}\,s^{-1}$; (2) same, taking void formation into account

as the Zn-Al eutectoid and the titanium alloys, is usually accompanied by a significant neck formation which continues until the cross section almost becomes a point. However, most of the materials exhibit brittle fracture without a noticeable neck formation (Fig. 1.9). In this case, the tendency to form a neck increases with increasing strain rate.

The studies conducted on alloys failing in the brittle mode show [1.50, 61] that considerable accumulation of porosity occurs in them and that the voids attain large sizes (on the order of several grains). At the same time, the role of the voids in the fracture process is not clear. Some investigators [1.4, 51] assume that crack formation and fracture occur as a result of void coalescence. In addition, an increase of porosity can lead to a change in the mechanism of SP flow [1.62], which must be reflected on the nature of the flow stress of the alloys. In this context, the $\sigma - \varepsilon$ relationship taking the true cross sectional area into account is of interest; for this purpose, it is necessary to subtract the area occupied by the voids from the total area of the specimen. Such an analysis was carried out in [1.50, 54], where it was shown that in the region of the strains corresponding to a stage just prior to fracture, a significant strain hardening of the material occurs (Fig. 1.10) which is uncharacteristic of superplastic flow. This indicates that there is a change in the nature of deformation under these conditions. These data confirm the suggestion concerning the change in the typical deformation processes of superplastic flow when a significant volume of voids accumulates in the specimens.

Apparently, the fracture of the materials in which deformation is accompanied by neck growth is also related to the specific features of the structural changes and the mechanisms operating during the final stage of deformation. In this case, rapid grain growth occurring during the deformation on process can be a possible reason for the loss of superplasticity of the material. However, further studies are necessary for a detailed understanding of this aspect.

2. The Nature of Superplastic Flow

Numerous hypotheses to explain superplastic behaviour of materials have been proposed [1.1–9, 2.1]. The earlier theories presumed that the phenomenon is unique to only certain alloys and related to their specific features. However, superplasticity is a universal state of metallics that appears during deformation of microcrystalline materials within a certain range of temperature and strain rate. The similarity of the observed phenomena and the generally observed features of the structural changes occurring in various superplastic materials indicate the universal nature of the effect. However, a universal theory has not yet been developed. It is difficult to develop such a generalized theory due to a) the difficulties involved in studying the structure of the microcrystalline materials when the structural changes are not large; b) simultaneous operation of several deformation mechanisms during SP flow make it necessary to separate their individual contributions to the total deformation and mutual interaction; and c) although grain boundaries are known to play an important role in the deformation of fine-grained superplastic materials, the structure of these defective regions and the nature of the grain boundary processes, such as sliding and migration, are not completely clear.

These reasons demanded a comprehensive experimental study of deformation mechanisms operating under the conditions of superplastic flow and special investigations of the grain boundary behaviour during deformation. Based on these studies, a new understanding of the nature of the phenomenon of superplasticity has been developed.

2.1 Experimental Investigations of Superplastic Deformation

It was established that the well-known deformation mechanisms, e.g., grain boundary sliding, intragranular dislocation slip, and diffusion creep, operate in the superplastic state, and that superplasticity is developed in the presence of a favourable combination of these mechanisms. It was necessary to show the role and the specific features of the operation of each of these mechanisms and their interrelationship during SP flow. This information can be obtained only by using a set of complementary procedures for observing the structural changes occurring during deformation. Revealing the nature of the developed deformation relief makes it possible to relate the changes in the mechanical properties of the material to the structural changes.

The objectives of the investigations are important for establishing the nature of SP deformation, since direct observations are possible only in a few materials. The results of the studies on deformation mechanisms operating during superplastic flow are given below.

2.1.1 Grain Boundary Sliding

Grain boundary sliding (GBS) is a deformation mode of polycrystalline materials that involves relative displacement of grains along the surface of their common boundary. Sliding of grains is due to their displacement along the boundary or within the near-boundary zone [2.2, 3]. The operation of GBS under the conditions of superplastic flow has been known since *Pearson* first observed it. The subsequent studies on various materials confirmed these results and made it possible to establish the basic similarities appearing during the operation of this deformation mechanism during SP flow.

Procedure of Investigation

Development of GBS leads to the formation of characteristic deformation relief on the polished surface due to relative movement of the grains, and it also causes the displacement of the marks (scratches) (previously engraved on the specimen) at the grain boundaries (Sect. 1.2, Fig. 1.7). The magnitude of sliding is geometrically described by the displacement vector in the plane of the boundary which can be resolved in three mutually perpendicular directions with the following components (Fig. 2.1): U – along the tensile axis; and V and W – perpendicular to the tensile axis. It can also be seen that two angles (φ and θ) are necessary for determining the orientation of a boundary.

In order to study the phenomenon of sliding in ultrafine-grained materials, single or two-stage replicas are usually employed. Based on the displacement of markers or specially engraved grid [2.4], only the components of GBS in the plane of the polished section (U and W) can be determined. To measure

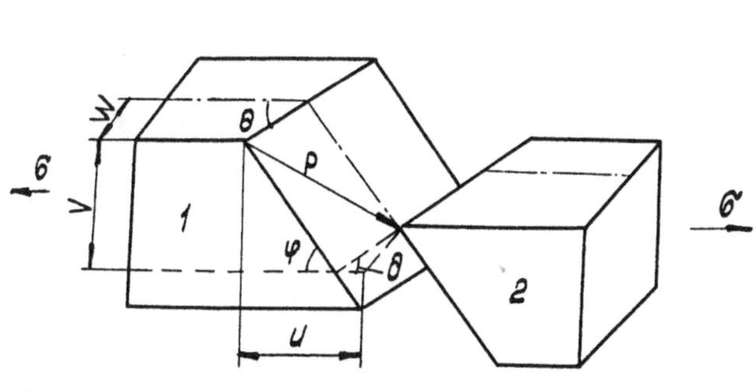

Fig. 2.1. Schematic representation of sliding along the boundary between grains (*1*) and (*2*): $\boldsymbol{P}$ is the total vector of displacement with the components U, V, and W [2.19]

the vertical component (V in Fig. 2.1), we employed chromium shading during replica preparation [2.5] which made it possible to evaluate the magnitude of V along the width of the relief on the replica.

In superplastic alloys having a relatively coarse grain size (approximately 15 μm, it is convenient to use interference light microscopy. However, we note that the aforementioned procedures can be followed only at relatively low strains (up to 20–40%), i.e., until a significant distortion of the surface relief does not occur. To understand GBS, the in-situ electron microscopic studies of the deformation of the superplastic materials are of great importance as these permit one to follow the dynamics of the movement of the grains during the deformation process [1.43, 46]. Similarly, using the method of "sighting" electron microscopy of replicas is valuable in spite of the considerable experimental difficulties [1.48, 49].

When carrying out topographic studies on GBS, it must be borne in mind that the conditions of deformation of grains at the surface and within the body of the metal are different. In view of this, a natural question arises regarding the extent to which the sliding observed at the surface of a specimen reflects GBS within its body. The results available on this aspect are due to *Matsuki* et al. [2.6]. They studied the superplastic Al-9.03%Zn-1%Mg-0.22%Zr alloy within which continuous networks of discrete aluminium oxide particles were obtained using a special treatment. GBS was evaluated on the basis of the discontinuities (ruptured sites) of these internal markers at the grain boundaries after deformation. Based on the obtained data, the authors concluded that the general picture of development of GBS and the relative movements of the grains within the body of the specimen is similar to that observed at the surface. This result agrees with the data of the experiments involving tensile deformation of massive (thickness up to 4 μm) foils of the superplastic materials in the column of a 1000 kV transmission electron microscope [1.44-6]. In view of this, it can be assumed that the basic features of GBS observed during topographic studies are similar to those of GBS within the body of the specimens. However, the quantitative data can show some differences.

Specific Features of Grain Boundary Sliding During Superplastic Deformation

One of the typical results reported by many investigators is the absence of bending of the scratches during their offset at the grain boundaries due to the displacement of grains under the conditions of SP flow. This fact indicates that sliding of grains occurs directly along the surface of the boundary. Only in rare cases, can one observe distortion of the scratches at the boundaries which is indicative of the significant deformation of the grains during the occurrence of GBS. It should be noted that in contrast to the deformation of the coarse-grained materials [2.7], the magnitude of sliding taking place during superplastic flow differs marginally at different regions of a boundary. Furthermore, the extent of GBS does not decrease at the triple junctions of grains. At the same time, one observes a significant

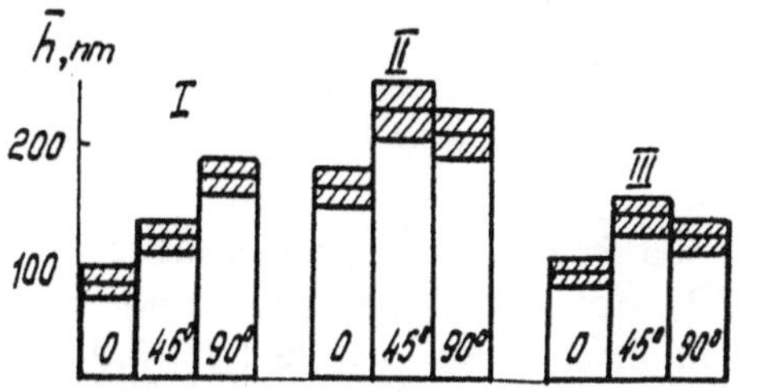

Fig. 2.2. The vertical component of GB sliding in the MA8 alloy at the boundaries whose trace at the specimen surface makes it inclined at 0°, 45°, and 90° with respect to the tensile axis ($\varepsilon = 6\%$) [1.37]: I – $\dot{\varepsilon} = 2 \times 10^{-5}\,s^{-1}$; II – $\dot{\varepsilon} = 8 \times 10^{-3}\,s^{-1}$; III – $\dot{\varepsilon} = 10^{-1}\,s^{-1}$

difference in the extent of displacements at individual grain boundaries and, therefore, when evaluating the extent of sliding, its statistical average values are used.

Anisotropy of grain boundary sliding has been observed in a number of alloys having a matrix structure [1.36, 2.5] (Fig. 2.2). At low strain rates (region I), the vertical component of the displacement of grains at the boundaries perpendicular to the tensile axis is greater than that at the boundaries parallel to the tensile axis and at the boundaries lying at an angle of 45°. At the optimum and high strain rates, this difference decreases and the maximum displacement occurs along the boundaries inclined at an angle of 45°. In this case, sliding occurs most rapidly in the second strain rate interval (region II). Unfortunately, such data are not available for the alloys having a microduplex structure.

In two-phase alloys, the extent of GBS is different at different types of boundaries. In the Zn-22%Al eutectoid alloy [1.49] and the two-phase brasses [2.5], the extent of sliding along the interphase boundaries is greater than that occurring between the grains of one phase. At the same time, a different situation is observed in the Sn-38%Pb alloy. The following table shows the magnitude of sliding along different types of boundaries during superplastic deformation of the Sn-38%Pb alloy in the I, II, and III regions at 425 K [2.8]:

Type of boundary	Sn-Sn	Pb-Sn	Pb-Pb
Sliding, nm, in the region:			
I	330	270	0
II	870	430	0
III	310	220	0

These data show that the maximum sliding occurs along the Sn-Sn grain boundaries and that no sliding takes place along the Pb-Pb boundaries; an intermediate situation exists at the Pb-Sn interfaces. The published data concerning GBS in the Zn-22%Al eutectoid alloy are contradictory. *Novikov* et al. [1.49] and *Naziri* et al. [1.44] observed that maximum sliding occurs at the interphase boundaries; whereas *Shariat* et al. [2.9] found that maximum sliding takes place along the Zn-Zn boundaries, whereas minimum GB sliding occurs at the Al-Al boundaries.

Different grain boundary structures and, consequently, different diffusional parameters are, apparently, the reason underlying the differences in GB sliding at different types of boundaries. However, to prove this absolutely, additional studies are required.

Relative displacement of the grains and their rotation appear as a result of a significant amount of GBS during SP deformation. This problem has been examined during hot deformation of usual, coarse-grained materials [1.3, 5, 2.10]. Such a study is particularly important in the case of SP deformation. During usual deformation, the elongation of the specimens is determined by the average elongation of each grain, and in this case the number of grains in the cross section of the specimen remains unaltered. Under the conditions of SP deformation, even at extremely large strains, the grain shape changes insignificantly (Sect. 1.2). Therefore, the displacement of the grains leading to an increased number of grains along the tensile axis and a decrease in the perpendicular direction forms the main factor determining the deformation of the specimen.

Experimental data on the nature of grain displacement were obtained by means of the "sighting" techniques, i.e., by studying the structural changes in one and the same sample area during deformation. By means of this technique, it was shown that during SP deformation, the shape of grains does not undergo significant changes in the process of deformation. While moving, they are separated from each other and change their neighbours [1.1, 5, 43, 44, 2.11].

Figure 2.3 shows an example of structural changes in one and the same area of the Zn-22%Al sample at different strains [2.11]. The marks were scratched on the specimen surface perpendicular to the tensile axis before deformation. In the course of deformation grains move against one another. Initially single scratches (Fig. 2.3a) are divided into segments bounded at grain boundaries. The scratches do not bend within individual grains, while at the boundaries of neighbouring

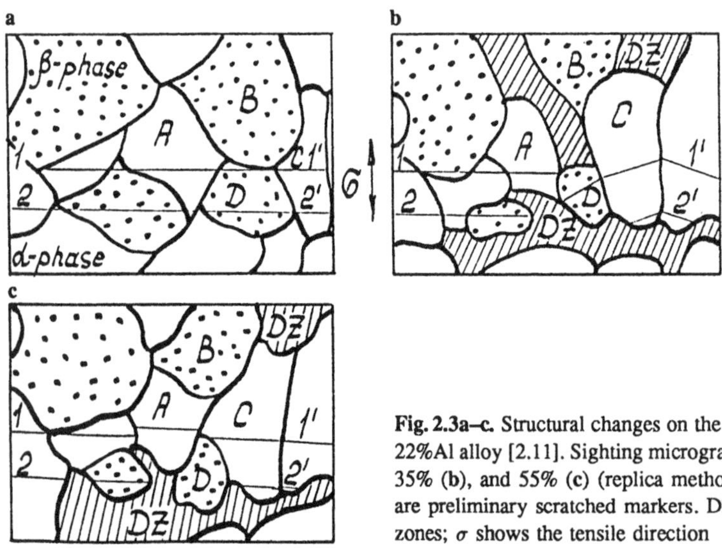

Fig. 2.3a–c. Structural changes on the surface of the Zn-22%Al alloy [2.11]. Sighting micrographs at $\varepsilon = 0\%$ (**a**), 35% (**b**), and 55% (**c**) (replica method). 1–1′ and 2–2′ are preliminary scratched markers. DZ are deformation zones; σ shows the tensile direction

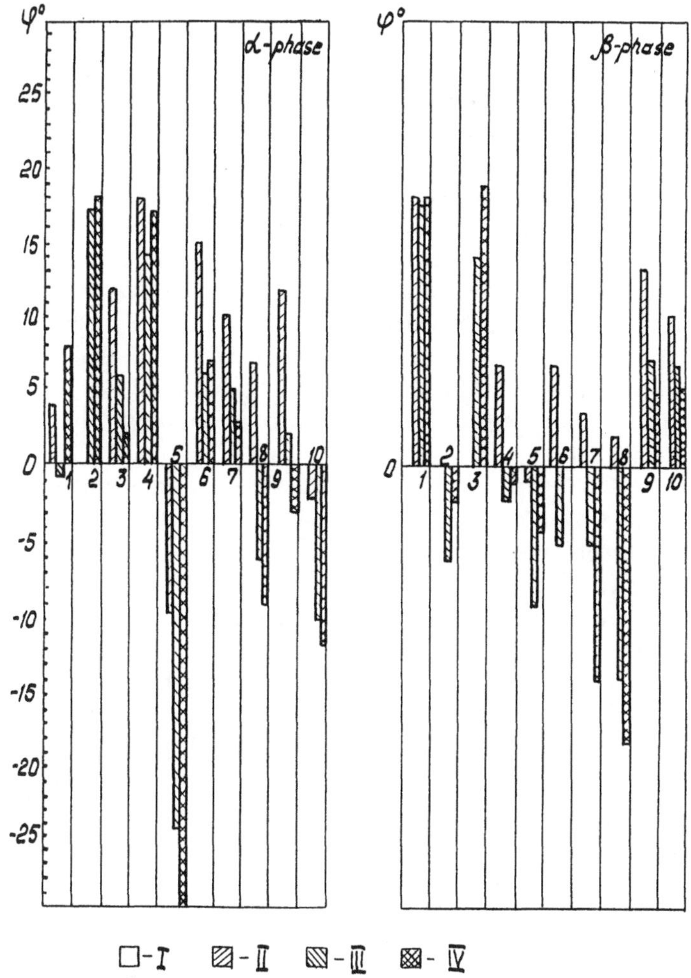

Fig. 2.4. Rotations of individual grains of the α- and β-phases during SP deformation: Curve (I) $\epsilon = 0\%$; (II) $\epsilon = 24\%$; (III) $\epsilon = 35\%$; (IV) $\epsilon = 55\%$; [2.11]

grains, the scratches shift to substantial distances (Fig. 2.3b, c). Such structural peculiarities result from intensive GB sliding. The scratch orientation with reference to the applied stresses also changes due to grain rotations. Figure 2.4 shows the peculiarities of grain rotations at different stages of deformation, taking as an example twenty typical grains of α- and β-phases of the alloys under study and presents quantitative data on the grain rotation angles measured by the angles of scratch rotations around the perpendicular of the tensile axis. It can be seen that the magnitude and direction of rotation are essentially different in different grains. In every individual grain these values change nonmonotonically and cyclically with the strain level. Though individual grain rotations can attain significant angles (up to 20°), the average grain rotation amounts to 5–7° and is practically independent of strain.

To illustrate the nature of grain rotations let us consider Fig. 2.3 again. Comparing microstructural changes shown in Fig. 2.3a–c, we can draw a conclusion about the regularities of the motion of the four specified grains (A, B, C, and D). It can be seen from Fig. 2.3c that at 55% strain the grains are located in such a way that the scratches have shifted at the grain boundaries, remaining almost parallel to their initial positions within each grain.

At the same time, it follows from Fig. 2.3b that the sliding of grain D along the boundaries A–D and C–D already occurred at 35% strain. Both the boundaries are of the $\alpha - \beta$ type; however, sliding along the A–D boundary is considerably larger than along the C–D boundary. As a result, grain D rotates at an angle of about 20°. At the next stage of deformation (55%), GBS along the C–D boundary becomes stronger, while almost no shift takes place along the A–D boundary. Grain D rotates again, but in the opposite direction. The net rotation angle approaches zero.

It is clear from these observations that the peculiarities of grain rotations (Fig. 2.4) are determined, first of all, by the nonuniform development of GBS at different boundaries. The observed rotations are accommodating in nature and provide for the arrangement of grains during their movement. It should be noted that these observations are in good agreement with the model of the origin of grain rotations developed in [2.12].

Another interesting observation in Fig. 2.3 is the change of neighbours (grains A, B, C, and D). It can be seen that the rearrangement of the grains takes place according to the well-known *Ashby-Verrall* scheme [2.13] (Fig. 2.5). Similar observations were made elsewhere [1.44, 48]. However, it is evident that such a scheme of rearrangement can be related only to a few specially selected grains, and it is difficult to correlate it with the movement of the grains within the entire specimen. We can agree with the analysis of the shortcomings of the geometrical Ashby-Verral model [2.1] in that this scheme fails to explain further rearrangements of grains at $\varepsilon > 55\%$ and is not applicable to two-phase alloys.

Apparently, a hypothesis offered in [2.14] is in better agreement with the experimental data on displacements of grains within the entire sample. According to this hypothesis, during SP deformation, grains move to significant distances and remain on the average equi-axial, though their shape can change (grains undergo accommodating straining and are extruded from underneath).

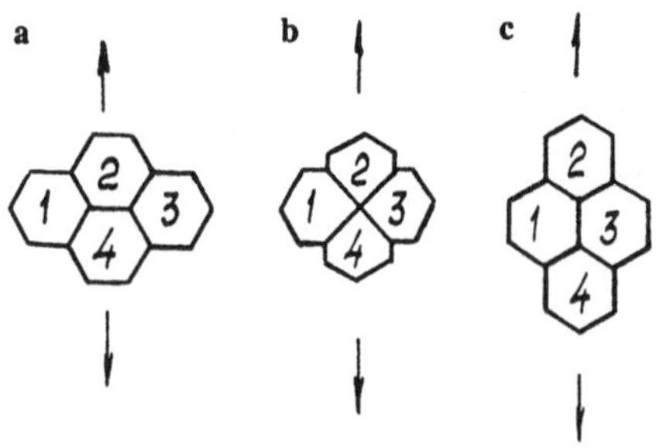

Fig. 2.5. Schematic representation of the sliding of grains during the process of SP deformation. The *arrows* show the direction of tensile loading [2.13]

Evidently, in a real polycrystal, where grains are not exactly identical, their rearrangement and intragranular accommodation strain are determined by the minimum of energy required for the realization of a specific deformation mechanism in every local area of material, and no precise scheme of the grain motion in the sample can be outlined [2.11]. The problem of grain displacement during SP deformation is closely related to the peculiarities of the development and contribution of GBS to total deformation.

Contribution to Deformation

The contribution of GBS to the total deformation has been experimentally evaluated in numerous investigations [1.3, 5, 49, 2.5, 6, 8, 9, 15, 16, 17]. Various methods are used for evaluating the contribution of GBS. We shall briefly review their merits and shortcomings [1.1, 5, 2.17].

There are two main groups of methods for evaluating the contribution of GBS to total deformation [2.4, 17, 18]. The first one is based on the determination of the components of the total displacement vector in the boundary (Fig. 2.1), which is usually done by assessing the shifts (offsets) of linear markers (scratches) or by measurements in the interference light microscope. The second group is based on the evaluation of the changes of the grain shape.

Deformation due to GBS can be most easily determined along the longitudinal displacements of the scratches marked parallel to the tensile axis U_l (the subscript $_l$ indicates the measurement along the longitudinal scratches) (Fig. 2.1). However, the magnitude of the longitudinal discontinuity of the scratches is significantly distorted by the specific features of the deformation relief developed at the surface (in particular, the deformation zones) (Sect. 1.2), and this does not permit one to measure the magnitude of displacement of the grains in the "pure" form. However, if the values of W and θ are measured for each of the boundaries interested by the scratch, the deformation due to GBS can be evaluated using the following relationship:

$$\varepsilon_{\mathrm{GBS}} = 2(\overline{W/\tan\theta})_l n_l \,, \tag{2.1}$$

where n_l is the average number of displacements per unit length; and the expression within brackets represents the average longitudinal displacement calculated from the transverse displacements of the longitudinal scratches.

Equation (2.1) was derived under the assumption that during uniaxial tensile deformation, $\bar{W} = \bar{V}$ and $\bar{\theta} = \bar{\varphi}$ for a fresh surface. In view of the fact that grain growth occurs during SP deformation, (2.1) can be used only when the strains are small.

In the case of superplastic materials, (2.1) can be simplified [2.4, 18, 19], because of the large variation in the grain shape and size

$$\varepsilon_{\mathrm{GBS}} = k_1 \bar{W}_l n_l \tag{2.2}$$

and

$$\varepsilon_{\text{GBS}} = k_2 \bar{V}_{\text{r}} n_{\text{l}} \,, \tag{2.3}$$

where k_1 and k_2 are the experimentally determined constants; and the subscript $_{\text{r}}$ indicates that the measurement is carried out on randomly selected boundaries.

The values of the coefficients k_1 and k_2 depend on the type of material, the surface condition, and the deformation conditions [2.19, 20]. In view of this, a proper justification for the choice of the values of k_1 and k_2 is necessary. The previously used values of these coefficients are conditional. For instance, k_1 was taken as 1.4 [1.37, 2.5] and k_2 was assumed to be 1.5 [2.8]. In a number of investigations, in order to evaluate the deformation due to GBS, transverse scratches were used and longitudinal displacements were measured. In this case, the following equation was used [1.3]:

$$\varepsilon_{\text{GBS}} = \bar{U}_{\text{t}} / \bar{d} \,, \tag{2.4}$$

where the subscript $_{\text{r}}$ represents the shifts perpendicular to axis of the specimen.

Novikov et al. [1.5] used the following equation:

$$\varepsilon_{\text{GBS}} = \Sigma \bar{U}_{\text{t}} / l_{\text{t}} \left(1 + \varepsilon_{\text{tot}}\right)^{1/2} \,, \tag{2.5}$$

where l_{t} is the length of a transverse scratch after deformation; and the expression $(1 + \varepsilon_{\text{tot}})$ is used for normalizing with respect to the initial length of the mark.

It should be noted that (2.4) and (2.5) were derived under the assumption that $\bar{U}_{\text{t}} = \bar{U}_{\text{l}}$, but this condition is not usually satisfied because of the features like anisotropy of GBS (see above).

Table 2.1 gives the contributions of GB sliding to SP deformation according to the data of the studies carried out using 2.1–5).

It can be seen from Table 2.1 that there is a significant difference between the obtained data, in particular when testing under the conditions corresponding

Table 2.1. Contribution of GBS to the total deformation $(e_{\text{GBS}}/e_{\text{tot}})$ [%] in various superplastic materials

Material	Contribution in the region			Equation used in the	Reference
	I	II	III	calculations	
Zn-22%Al	30	60	30	(2.2)	[2.21]
	23	49	12	(2.3)	[1.49]
Zn-0.4%Al	40	50	30	(2.3)	[2.5]
Mg-1.5%Mn-0.5%Ce	33	49	30	(2.3)	[1.37]
Mg-33%Al	12	64	29	(2.2)	[2.16]
Al-9%Zn-1%Mg	42	63	26	(2.1)	[2.6]
Al-11%Zn-1%Mg-0.3%Zn	60	80	50	(2.2)	[2.20]
Sn-38%Pb	21	56	20	(2.2)	[2.8]

to those of regions I and III. This is due to the differences in the experimental methods used by different investigators. In addition, the strain rates corresponding to regions I and III deviate from the optimum region for the SP state, but differently; therefore, in this case, comparison of the absolute values of the contribution of GB sliding is conditional. Furthermore, the results indicate that the maximum value of $\varepsilon_{GBS}/\varepsilon_{tot}$ appears in region II, but significantly decreases in regions I and III.

Rechinger [in 2.22] was the fist to suggest a method based on evaluating the grain shape for determining the contribution of GBS. In view of the fact that rearrangement of the grains that ensures deformation of the specimen (Fig. 2.3) occurs due to their relative sliding, the total deformation of the specimen can be defined as the sum of the deformation due to GB sliding and the deformation of the grains (ε_g)

$$\varepsilon_{tot} = \varepsilon_{GBS} + \varepsilon_g \ . \tag{2.6}$$

The following equation was suggested by Rechinger for calculating the deformation of the grains:

$$\varepsilon_g = \left(N_t/N_l\right)^{2/3} - 1 \ , \tag{2.7}$$

where N_t and N_l are the number of grains per unit length in the transverse and the longitudinal directions, respectively.

A somewhat different expression from (2.7) was suggested recently in [2.10]. In addition, intragranular markers can be used directly for the evaluation of ε_g. A more precise formulation of (2.6) was suggested in [2.23]. It was believed that the deformation of the grains must be accompanied by mutual displacement for maintaining the integrity of the specimen – by the accommodating GB sliding process; therefore, (2.6) is written more correctly as

$$\varepsilon_{tot} = \varepsilon_{GBS}(R) + \varepsilon_{GBS}\left(A_k\right) + \varepsilon_g \ , \tag{2.8}$$

where $\varepsilon_{GBS}(R)$ is the deformation due to GBS (according to Rechinger); and $\varepsilon_{GBS}(A_k)$ is the deformation due to accommodating GBS.

Determining the deformation due to GBS on the basis of an analysis of the grain shape has a number of advantages over the marker technique [2.4]. 1) Any number of measurements can be carried out depending on the requirement for ensuring statistical reliability; 2) it is equally easy to conduct the measurement at the surface and within the body of the specimen; and 3) in principle, there is no restriction on the strain at which studies are undertaken.

In addition, it was found that in the case of superplastic materials, the results obtained according to (2.6, 8) significantly exceed those obtained using the marker lines (Table 2.1). As was noted in Sect. 1.2, even after subjecting the specimen to significant levels of SP deformation, the grains virtually remain equi-axial in most cases and, therefore, the average value of deformation of the grains is close to zero and (2.8) reduces to

$$\varepsilon_{tot} \simeq \varepsilon_{GBS}(R) \ , \tag{2.9}$$

i.e., according to this procedure $\gamma_{GBS} \simeq 1$. Evidently, some elongation of the grains can be masked by the boundary migration (Sect. 1.2), in particular, in the alloys having an unstable microstructure. However, the magnitude of γ_{GBS} determined on a given specimen is usually found to be significantly greater than that determined according to the method of markers.

Hazzledine and *Newbury* [2.14] explain this contradiction as follows. In the case of SP deformation, the grains deform such that their total intragranular deformation and the accommodating GBS are, on average, equal to zero, but at each individual instance, a chaotic change in the shape of the grains occurs and the accommodating sliding induced by this can change the average magnitude of the measured displacements at the boundaries. However, this hypothesis requires further experimental and theoretical confirmation.

Thus, regardless of the reliably established fact that GB sliding is the most important and an inalienable mechanism of SP deformation, the aspects concerning its contribution to the total deformation are not yet completely clear. Thus, the objections raised in this regard [1.5] are not unfounded; it is considered [1.5] that the published data on the absolute values of $\varepsilon_{GBS}/\varepsilon_{tot}$ do not, in fact, represent the true contribution of GBS to the total deformation and the following expression was suggested for evaluating it:

$$\gamma = 1 - \gamma_{DC} - \gamma_g \, , \tag{2.10}$$

where γ_{DC} and γ_g are the contributions to the total deformation of diffusion creep and grain deformation due to intragranular slip.

The possibility of such an evaluation will be examined in greater detail in Sect. 2.1.4.

2.1.2 Intragranular Deformation

Under the conditions of SP flow, intragranular deformation due to dislocation movement occurs in addition to grain boundary sliding. This is a generally accepted mechanism operating when superplastic materials are deformed at high strain rates corresponding to region III. In order to reveal the role and the participation of dislocations during the process of SP deformation in regions I and II, it is necessary to pursue further studies.

The absence of dislocations in the grains after SP deformation can be a result of rapid disappearance of dislocations during the process of SP deformation and subsequent unloading. In view of this, it is not surprising that numerous studies have been carried out on the development and the changes of textures during the process of SP flow in order to establish the presence of dislocation slip. A study of the texture changes is an integral method that gives information regarding the process of crystallographic slip occurring due to the collective movement of dislocations. However, it was found that in the superplastic materials initially having a texture, blurring of the pole density with retention of only some components of the original texture is the most significant effect (Sect. 1.2). Such changes are due to the simultaneous occurrence of a number of processes (rotation of grains,

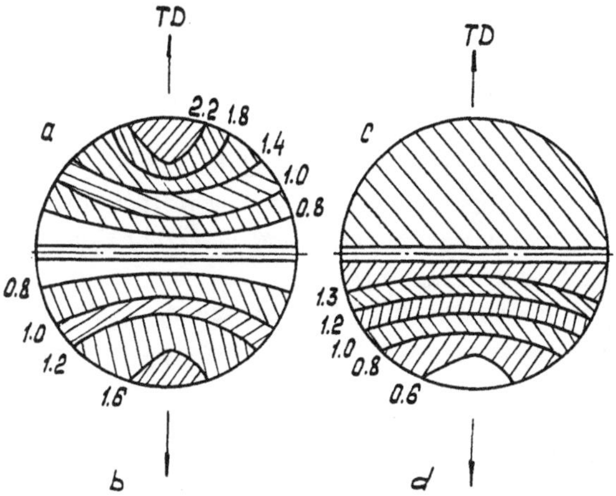

Fig. 2.6. (0001) pole figures of the *b*-phase of the Zn-22%Al alloy (d = 0.5 mm) after tensile deformation up to ε = 200% at 250°C and a strain rate of (*a*) $6\times10^{-5}\,s^{-1}$; (*b*) $1.7\times10^{-2}\,s^{-1}$; (*c*) $3\times10^{-1}\,s^{-1}$; (*d*) $5\,s^{-1}$. TD is the tensile direction. From [1.3]

their growth, etc.) and, consequently, a clear interpretation of the obtained data is difficult. In view of this, it is most appropriate to study texture formation during the process of SP flow of a material having no texture in its original state.

The first studies of this nature were carried out on the Zn-22%Al alloy having no texture in the as-quenched state [2.24] by plotting complete pole figures. It was observed that after deforming an originally texture-free specimen under the conditions of SP flow corresponding to the I and III strain rate ranges, sharply differing axial textures form in the zinc-rich phase (Fig. 2.6). The maximum of the directions of the tensile axis (LD) appears at low strain rates (region I), and a texture having a maximum (peak) in the transverse direction (TD) forms in the III strain rate range. Depending on the strain rate, the transition from one texture to another occurs gradually due to the weakening of one maximum and the evolution and intensification of another maximum. In a certain range of strain rates in region II, no significant texture is developed in the beta-phase after tensile deformation. The intensity of the maximum of the pole density increases with an increasing degree of deformation. Subsequently, it was established that there is a close correlation between the observed features of texture development and the variations of the mechanical properties of the alloy depending on the grain size [2.24] and the deformation conditions. These experimental results are of fundamental importance since they give a direct evidence for the existence of a relationship between the regularities of SP deformation and texture formation during the process of deformation (Sect. 2.4.1).

In order to understand the results obtained, computer modelling of texture formation was carried out [2.25]. The objective of modelling is to select the slip systems whose operation can lead to the experimentally observed changes in the textures. Modelling was carried out on the basis of Taylor's theory. A polycrystalline material was represented by a set of 504 crystallites that were randomly oriented in space. In each crystallite, the choice of the systems was carried out according to the principle of minimum work of deformation. Based on the specific features of SP deformation, it was assumed that slip occurs in the crystallite along different systems alternately independent of each other. In such a model, the relative contribution of the slip systems to deformation is controlled by varying the ratio of the critical shear stresses required for the operation of different slip systems.

As a result of various variants of calculations it was found that the experimentally observed types of textures were obtained when modelling the deformation process with slip along the basal slip systems (the maximum in TD) or along the pyramidal slip systems with combined operation of $(\bar{c} + \bar{a})$ and $(\bar{c} - \bar{a})$ slip (the maximum in LD). Depending on the contributions to deformation, an alternate operation of these slip systems leads to a change over from one type of textures to another.

The following quantitative results were obtained by considering GBS leading to chaotic blurring of the pole density. At a probability ratio $p = 1$ for the operation of the basal and the pyramidal slip systems, formation of a texture having a maximum in the direction of tensile loading (LD) takes place. $I_{LD} = 1.4$, at $p = 1.5$ $I_{LD} = 1.0$ and at $p = 3$ $I_{LD} = 0.6$. Thus, the results obtained from modelling explain the nature of texture evolution during superplastic deformation of the Zn-22%Al alloy. The formation of a texture maximum in the loading direction (LD) at low strain rates occurs as a result of preferential slip along the $[\bar{1}\bar{1}22]$ $\langle 11\bar{2}3\rangle$ type pyramidal systems. As the strain rate increases, more and more of [0001] $\langle 11\bar{2}0\rangle$ type basal slip systems begin to participate in the process of slip. Consequently, the maximum in LD starts to weaken and at certain strain rates, no preferred grain orientation is developed after SP deformation. Further increase of strain rate leads to the occurrence of preferential slip along the basal systems and the formation of a maximum in TD.

We observed [1.3] the strain rate dependence of texture evolution during SP deformation of various materials. This effect was attributed to intragranular dislocation slip (IDS) and to the increase in the number of operating slip systems with an increasing strain rate of SP deformation.

However, an unambiguous interpretation of the operating slip systems on the basis of the texture data is usually not possible, in particular, when studying the materials having a highly symmetric lattice (e.g., cubic one). In view of this, of great significance in texture studies is the application of the distribution function of crystallographic orientations that permits one to obtain information regarding the distribution of grains in three-dimensional space and to calculate the fraction of different ideal orientations.

Such detailed studies have been carried out and the obtained results were reviewed by *Edington* et al. [1.4]. It was established that during SP deformation, different components of the original texture are blurred at different rates in that some of them can remain without disappearing Edington came to the conclusion that in the Zn-40%Al and Al-33%Cu alloys, crystallographic slip occurs at all strain rates, but multiple slip occurs only in the III strain rate interval. However, using a similar method, it was not possible to obtain evidence for the operation of IDS in region I when studying the Al-6%Cu-0.3%Zr alloy [2.26]. Possibly, this is due to the shortcomings of the methodological approach of the authors; they studied the changes occurring in the textures, but not their formation because of which ambiguity exists in the analysis of the experimental data, in particular, in the case of materials having lattices of high symmetry.

In a critical study of the existing analysis of texture changes occurring during the process of SP deformation [1.5] it was suggested that in many alloys the appearance or the retention of the preferred crystallographic orientations is due to grain growth taking place during deformation, but is not related to dislocation slip. This conclusion is based on the study of the texture changes occurring in a brass [2.27] in which new maxima that are characteristic recrystallization textures were observed and, also, on a study of texture evolution in the Zn-22%Al alloy in the field of temperature gradient [2.28]. According to the authors, the maximum in the axial direction in the β-phase of the alloy subjected to SP deformation in region I appears, due to preferential grain growth in this direction. Such a conclusion is based on the fact that in zinc diffusion coefficient along the $\bar{c}$ axis is several orders of magnitude higher than that in the other directions. However, the hypothesis of the authors contradicts the following experimental data.

Firstly, if the texture formation along the $\bar{c}$ axis is attributed to preferential grain growth of the β-phase in the same direction, the increased content of the α-phase in the alloy must suppress texture formation. In view of this, during SP deformation, texture formation along the $\bar{c}$ axis would be expected to occur more easily in the zinc-rich alloys and to be hindered in the aluminium-rich alloys. The actual situation contradicts this argument. In fact, alloying zinc with aluminium facilitates the appearance of a maximum in the β-phase along the $\bar{c}$ axis when tensile deformation is carried out at a low strain rate [1.3, 23].

Secondly, the diffusion coefficient decreases exponentially with decreasing temperature owing to which the contribution of the diffusional processes to deformation must decrease. Consequently, the contribution of grain growth to texture formation would be expected to decrease rapidly with decreasing temperature of deformation. However, the specially conducted studies showed that in the direction of tensile loading in the β-phase of an originally textureless Zn-22%Al alloy, a maximum appears not only at 250°C, but also after decreasing the temperature up to 150°C (and even at 20°C).

Thirdly, the texture appearing in the β-phase during tensile deformation has a clearly defined axial nature corresponding to the deformation scheme of the specimen.

Thus, although a study of the textures gives useful information on the development of intragranular slip during SP deformation, the interpretation of the obtained data is not always unambiguous and additional experimental methods are required for studying the specific features of intragranular deformation. In view of this, the interest in transmission electron microscopy has been revived in recent years.

Various methods were used to fix dislocations. The possible disappearance of dislocations during the period of unloading and cooling the specimen can be avoided by the precipitate particles dispersed within the grains and rapid cooling of the specimen from the temperature of SP deformation. Annealing twins were also used to "trap" dislocations.

Several investigators [1.1, 3, 4, 36] studied thin foils prepared from alloys containing dispersed precipitates. It was found that in these cases, it is often possible to observe a dislocation structure in the grains (Fig. 1.5). As a rule, at low strain rates (in region I), individual dislocations are observed and during the transition into the II strain rate range, the dislocation density increases slightly, but a well-developed substructure does not appear; only after subjecting the specimen to tensile deformation at high strain rates, is it possible to observe dislocation intersections and subboundaries. Based on an analysis of the dislocation structures, we identified the slip systems operating in the magnesium-base MA8 alloy [1.41]. It was shown that as the strain rate increases (during the transition from region I into region II), the pyramidal slip systems begin to operate in addition to the basal systems. This conclusion agrees with the data obtained from the texture studies. In view of this, the results of the aforementioned investigation [1.41] unambiguously prove the importance of intragranular sliding during superplastic deformation of the MA8 alloy. However, we note that these experiments were carried out using a relatively coarse-grained (approximately 10 μm) material and the obtained data could reflect the specific features of its behaviour. A study of the dislocation structure of the grains of the aluminium-rich phase in the typical superplastic alloy Zn-40%Al [1.4] confirmed our results [1.41] to a large extent. It was established that although the Burgers vectors of all the dislocations belong to the $a/2\ \langle 1\bar{1}0\rangle$ family, the number of operating slip directions increase with increasing strain rate. Unfortunately, a shortcoming of the experimental Zn-40%Al specimen [1.4] is that it contains a high dislocation density in its initial state. In view of this, it is not completely clear to what extent dislocation structure observed after deformation is related to the process of SP flow.

Retention of a dislocation structure due to the presence of dispersed particles is of limited significance because only a few superplastic materials contain appropriate precipitates that are stable at the temperature of deformation and are capable of effectively stopping the dislocation. The application of "instantaneous" quenching of the specimens immediately after completing the deformation process is of greater importance. For this purpose, we used a special device [1.23, 2.29] ensuring extremely fast cooling of the specimens within 0.2 s after unloading. The studies conducted on the ultrafine-grained alloys of the Zn-Al system using this device showed [1.23, 2.29] that direct quenching of undeformed specimens does not lead to the formation of a dislocation structure. On the other hand, after superplastic deformation, it was possible to observe dislocations in many grains (Fig. 2.7) in spite of the fact that they were not found in the original material. On increasing the strain rate, we observed some increase in the total dislocation density in that the zinc-rich phase contains dislocations having other Burgers vectors than those observed at low strain rates. These data indicate the operation of new slip systems as the strain rate increases, and this fact corresponds to the nature of the texture changes observed in the experimental alloys. In spite of the fact that in the aforementioned investigations a complete statistical analysis of the Burgers vectors of the observed dislocations could not be performed because of experimental difficulties, the obtained data (and the results

Fig. 2.7

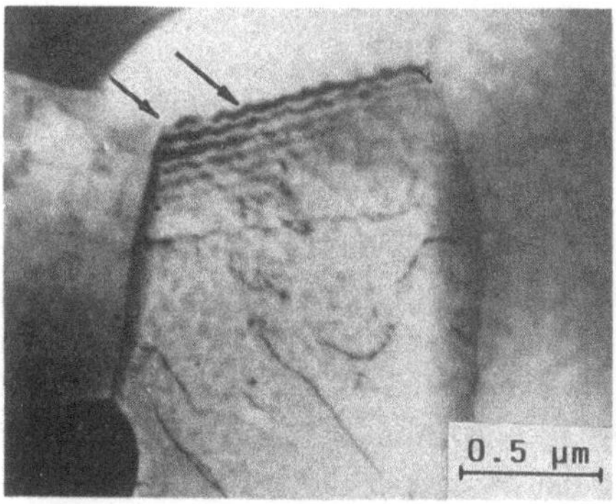

Fig. 2.8

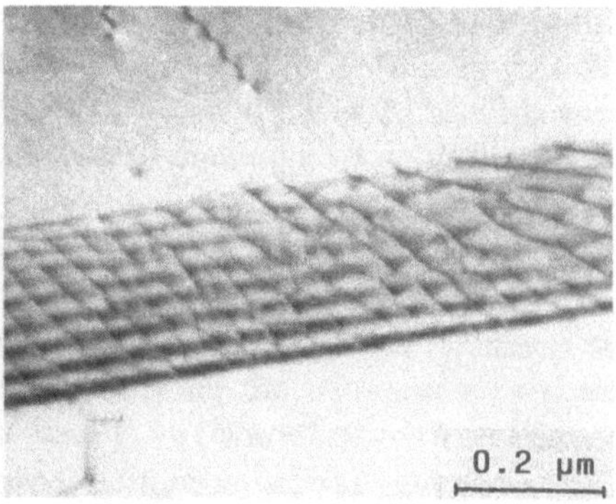

Fig. 2.7. Dislocations in the α-phase grain and its boundaries of the Zn-50%Al alloy after 100% tensile deformation and "instantaneous" quenching [1.42]

Fig. 2.8. Twin boundary in the structure of an austenitic steel after 100% SP deformation. Typical bends and the line bending of GBD indicating their mobility in the boundaries are clearly observed

of our earlier studies [1.36, 41]) confirm the conclusion of the texture studies regarding intragranular slip in all three strain rate ranges and the increase in the number of operating slip systems with increasing strain rate which is particularly distinct in the materials having an hcp structure.

In addition to the investigations revealing the dislocations formed during the process of superplastic flow, the studies of *Falk* et al. [2.30] on certain fine-grained alloys containing twins within their grains are of special mention. It is known [2.31] that at the twin boundaries absorption of the lattice dislocations does not occur almost up to temperatures corresponding to the melting point and, consequently, twins can act as "traps" for dislocations. In a study of the thin foils prepared from the superplastically deformed specimens [2.30] no dislocations were observed at the twin boundaries after deforming in region I, but they were found after deforming at the optimum strain rates, and a particularly large number of dislocations were seen at high strain rates (in region III). Similar studies were recently carried out[1] on austenitic steels having a grain size of approximately 2 μm. It was established that at the twin boundaries the lattice dislocations dissociate into grain boundary dislocations. Figure 2.8 shows the lattice dislocations at a twin boundary that are dissociated into two grain boundary dislocations. The characteristic bending and convolution of the dislocation lines can be seen. This indicates the mobility of the dislocations at the twin boundaries. Twin boundaries are special boundaries, and it is well-known (Sect. 2.2.1) that the absorption of lattice dislocations in them proceeds via their dissociation into grain boundary dislocations and subsequent mutual annihilation of these dislocations or through their sinking into triple junctions. The rate of this process

[1] The investigations were conducted by R.Z. Valiev and M.W. Grabski.

is less than the rate of the disappearance of dislocations directly in the grain boundaries, therefore, it is not surprising that dislocations could be observed in the II strain rate interval after SP deformation. At the same time, at a low strain rate, the rate of absorption can be too high to make lattice dislocations visible, i.e., in this case twin boundaries cease to be "traps" for dislocations. Therefore, a conclusion made in [2.30] that there is no dislocation activity in region I seems to be disputable.

The experimental studies in which thin foils prepared from the superplastic materials are subjected to tensile deformation directly in the column of an electron microscope are particularly important for revealing intragranular deformation. *Fike* and *Rack* [2.32] were the first to conduct such an experiment. They observed dislocation motion during the process of deforming a specimen of the eutectoid Zn-22%Al alloy. However, the obtained results are of limited value since the work was carried out using a 100 kV microscope on very thin foils whose behaviour can differ significantly from that of a bulk specimen. In view of this, subsequent experimental studies (intended for direct observations during the process of tensile deformation) were carried out using high-voltage (500–1000 kV) microscopes where the thickness of the electron-transparent foils amounts to approximately 4 μm. Such studies were conducted on the Zn-22%Al alloy at 100°C [1.44]; however, a significant dislocation activity was not found in the β-grains, and only individual dislocations could be observed in the α-grains. According to *Bricknell* and *Edington* [2.26], these data are not characteristic for SP deformation under the usual conditions because under the action of an electron beam, the vacancy concentration in the object can increase by a few orders of magnitude. It is necessary to carry out deformation of the foil at a temperature exceeding a specific critical temperature (T_c), where the concentration of the thermal vacancies is greater than that of the vacancies formed during electron bombardment, but this condition was not maintained in the investigation of *Naziri* et al. [1.44]. Such objections can not be raised against other experimental studies [1.45, 2.33], but these studies also revealed a low dislocation activity during SP deformation.

A number of experimental difficulties are encountered when one aims at a direct observation of the movement of individual dislocations during the tensile deformation of the foils obtained from superplastic materials in an electron microscope [1.46]. If dislocations are generated and disappear easily at the grain boundaries, then the possibility of observing them is restricted by the duration of their motion through a grain. Since a dislocation cannot be observed when this duration is less than 0.1 s, the rate of movement of the dislocations must not exceed 10^{-5} m/s for observing them directly. For example, in the single crystals of zinc the rate of dislocation movement at room temperature at a stress of the order of 10 MPa amounts to approximately 1 m/s. In addition, during superplastic deformation of fine-grained alloys, the number of mobile dislocations in the grains is very small. Thus, in order to ensure deformation of a grain up to a strain degree equal to half the strain of the entire specimen, it would be sufficient if one dislocation passes through each grain only after 6 s [1.46].

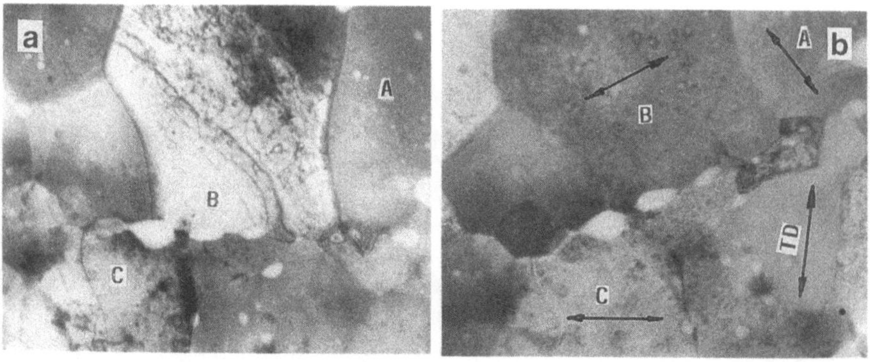

Fig. 2.9. Structure of the Zn-0.4%Al superplastic alloy in the original state (a) and after deforming up to $\varepsilon = 30\%$ in situ in an electron microscope (b). The *arrows* indicate the direction of deformation of the grains

During direct observation of dislocations in a microscope, a significant change in the diffraction contrast can occur in the grains due to the movement of the specimen and the development of GBS. Thus a special study of the intragranular slip occurring during tensile deformation of the superplastic Zn-0.4%Al alloy [1.46] was conducted. With respect to the structure, this alloy is close to the single-phase materials; it contains only aluminium based phase dispersed particles which were treated as markers in this experiment for determining the intragranular deformation. it was found that during tensile deformation of the films, the distance between the particles located within the limits of a single grain changes significantly (Fig. 2.9) although a dislocation movement could not also be observed in this case. This effect cannot be attributed to the diffusional processes (Sect. 2.1.3). Similarly, the effect is not related to the rotation of grains since rotations occurring during SP deformation have a nondirectional nature. In addition, at the strain levels used in the experiment (30–40%), their average value does not exceed 5–10° [1.5]. At such levels of rotations, the displacement of the markers must be so small in magnitude that it cannot be recorded experimentally. Moreover, grain rotation can never result in the growth of interparticle spacing which was frequently observed. Thus, the conducted experiment gives a clear proof for the operation of intragranular dislocation slip during the process of SP flow. In this connection, it should be noted that *Papirov* et al. [2.34] made measurements of acoustic emission during the SP deformation of the Zn-0.4%Al alloy and observed no peaks of acoustic emission. However, this fact is not evidence of the low activity of the dislocations, but can be related to the specific features of their movement (the number of simultaneously moving dislocations and the distance of their slip).

Summarizing the aforementioned results, we note that, in spite of the experimental difficulties, at the present time there is a large number of data suggesting that dislocation movement occurs in all three regions of SP flow. With increasing strain rate, the nature of intragranular slip undergoes a change caused by the increased number of the operating systems. These conclusions are valid for different superplastic materials.

Contribution of Intragranular Dislocation Slip to the Total Deformation

At the present time not much data are available with regard to the contribution of intragranular deformation due to dislocation motion towards the total deformation during SP deformation. For this purpose, two methods of evaluation can be used. The first is based on the shape changes of the grains and the second on the shift of marker within the grains. In the first case, Rechinger's method (2.7) or its modifications are used. However, during SP deformation, a change in the shape of the grains is caused not only by the intragranular slip, but also by other processes (in particular, grain boundary migration) because of which the results can be significantly affected [1.2, 4, 2.4].

Intragranular deformation was evaluated in [1.37, 2.15, 16] on the basis of the displacement of marker lines located within a single grain and perpendicular to the tensile axis. In this case, the contribution of the intragranular slip can be determined according to the following equation:

$$\gamma_{\mathrm{IDS}} = \varepsilon_{\mathrm{IDS}}/\varepsilon_{\mathrm{tot}} = (l_1 - l_0)/l_0\varepsilon_{\mathrm{tot}} , \qquad (2.11)$$

where l_0 and l_1 represent the distance between the lines within a given grain before and after deformation, respectively. In view of the technical difficulties of making two scratches within a single grain and remeasuring, the proposed procedure yields the most reliable results for superplastic materials with a relatively large grain size (about 10–15 μm) which can be studied by more reliable methods of optical microscopy. However, a recent study of the Zn-22%Al alloy with a grain size of 1.5 μm [2.35] should be mentioned in which a rather labourious technique involving "sighting" of one and the same area of the specimen and two-stage replicas were used. The magnitudes of IDS evaluated according to the marker method in some alloys existing in the SP conditions are given below:

	Mg-33%Al	MA8	AMg6	Zn-22%Al
$\varepsilon_{\mathrm{tot}}$	0.3	0.25	0.25	0.3
Contribution in the region:				
I	30	43	39	–
II	21	24	34	9
III	–	–	66	–
Reference	[2.16]	[1.37]	[2.15, 36]	[2.35]

Evaluation of the magnitude of γ_{IDS} on the basis of the measurements of the displacement of the markers during SP deformation of the Zn-0.4%Al alloy in situ in an electron microscope was done in [1.46]. The data obtained (~ 12%) for region II confirmed the validity of the quantitative evaluation.

It can be seen from the aforementioned data that in all experimental alloys, the magnitude of γ_{IDS} is minimum when deformation is carried out at the optimum strain rate, and it tends to increase either with increasing or decreasing strain rate. We note that the obtained results do not contradict the fact that the density of the mobile dislocations continuously increases with an increasing strain rate and can be explained by the specific features of intragranular slip during SP deformation. This point is considered in detail below.

Specific Features of Intragranular Slip During Superplastic Deformation

As indicated by the experimental studies, intragranular slip occurs in a wide range of superplastic flow rates. However, a few more questions require answers. What are the specific features of the dislocation movement? Why don't an accumulation of dislocations and a significant elongation of the grains occur during deformation due to the intragranular deformation? How does one explain the observed contribution of IDS? etc. All the aspects are associated with the specific features of the effect of IDS during SP deformation, i.e., with the nucleation, motion, and absorption of the lattice dislocations. Unfortunately, despite their importance for understanding the nature of superplastic deformation, these aspects have not yet been studied in detail.

Among the three separate stages of intragranular dislocation slip, viz., nucleation, motion, and absorption, the first stage is evidently the most difficult one to prove. Otherwise, accumulation of dislocations within the body of the grains or in the vicinity of the grain boundaries would have been the typical situation. In the fine-grained superplastic materials, nucleation of dislocations within the grains due to the Frank-Reed sources is not possible, because, as was shown by the calculations, the stress required for their operation significantly exceeds the stress at which SP flow occurs [1.1, 2, 3]. Grain boundaries are the most probable sites at which dislocations can nucleate since a high stress concentration can always be expected [1.2]. This conclusion is supported by the experimental studies on the deformation behaviour of the superplastic Zn-0.4%Al alloy sample "in situ" in a high voltage electron microscope [1.46]. Observation of the dislocations already present in the initial state (for example, the grain C in Fig. 2.9) showed that during tensile loading, most of them are retained with only a slight change in their configuration and, at the same time, there was a significant intragranular deformation. Therefore, observed dislocations did not participate in the deformation process, evidently because of the low stresses acting on them. The deformation of the grain was due to the dislocations generated at the grain boundaries under the action of high local stresses which also determine the nature of their movement.

Dislocation motion within the grains due to their small sizes occurs easily from boundary to boundary and is accomplished essentially in a conservative manner. This assumption is confirmed by a number of experimental observations.

1) The defects forming during the interaction of grain boundaries with the lattice dislocations and frequently observed after SP deformation (Figs. 1.6, 2.8)

usually reveal themselves in the form of straight lines corresponding to the trace of the intersection of the slip plane with the surface of the grain. In the case of climbing of dislocations, one would expect a bending of their lines.

2) The analysis [1.4] of the dislocations observed after SP deformation showed that in regions I and II, the dislocations remained essentially on their slip planes, and that the number of dislocations abandoning the slip planes (as a result of sliding) increased with increasing strain rate.

3) As a rule, when using the procedure of "instantaneous" quenching, the dislocations dispositioned on a single plane (but not subboundaries) could be recorded [1.23, 2.29]. It is known that intersections and subboundaries are the characteristic elements of a dislocation structure only after deforming at high strain rates wherein their movement is hindered by the accumulation of dislocations in the structure.

Under the conditions of SP deformation, the dislocations generated at the grain boundaries are absorbed by them. Thus, it can be confirmed that during SP deformation, the boundaries are the main sinks for the lattice dislocations whose operational effectiveness determines the rate of their accumulation. This is also confirmed by the data obtained using the method of "instantaneous" quenching that makes it possible to observe the presence of linear defects at the boundaries [1.23, 2.29]. Continuation of some lines in the boundary (as the dislocations within the body of the grains) (Fig. 2.7) form a confirmative evidence that these lines are the "run-in" lattice dislocations. The observed weakening and spreading of the images of some of the dislocations (indicated by arrows) show their partial absorption by the grain boundaries. The role of grain boundaries as sinks of dislocations is discussed in greater detail in Sect. 2.2.2.

As was shown by the experiments in which specimen deformation was carried out in situ in an electron microscope, the direction of elongation of the grains (for example, A and B in Fig. 2.9) corresponds to the direction of slip in these grains but does not coincide with the direction of tensile loading of the specimen. This gives direct evidence that the dislocation movement occurs under local stresses. The nature of intragranular slip was given a more detailed consideration in the aforementioned [2.35], where the development of IDS in the Zn-22%Al alloy was determined by means of measuring spacings between the marker lines within individual grains. Figure 2.10 shows the values of the intragranular strain after tensile deformation up to ε = 24.35, and 55% in ten typical grains of α- and β-phases (Fig. 2.10a, b). It can be seen that deformation in different grains is rather nonhomogeneous and its average value is rather small and almost independent on the specimen strain level. Intragranular strain is even negative in some grains. Such a characteristic of IDS under superplastic conditions essentially differs from the development of IDS during conventional deformation. Under superplastic conditions IDS is found under specific conditions. Slip occurs under the action of the stresses that vary continuously both in magnitude and in direction. In this case, the shape change of the grains has

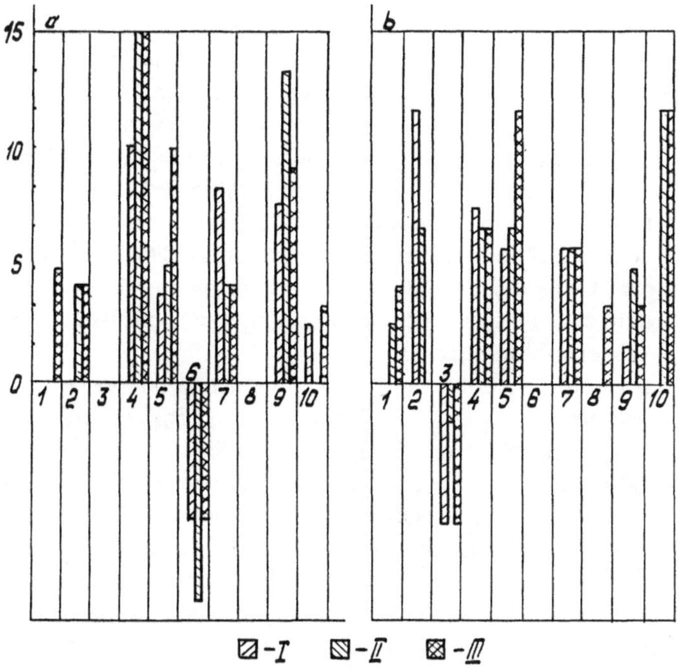

Fig. 2.10. Deformation (ε_{BD}) of individual grains of the α-phase (a) and β-phase (b) of the Zn-22%Al alloy due to the motion of dislocations at different levels of the total deformation of the specimen. Curve (I) ε_{tot} = 24%; (II) ε_{tot} = 35%; (III) ε_{tot} = 55%. From [2.35]

an oscillating character that is not related to the direction of the tensile axis, therefore, during deformation, the grains do not become elongated, but remain equiaxial. The physical situation existing during the development of IDS is close to that observed during cyclic deformation. In the latter case, the magnitude as well as the sign of the external stresses vary continuously and, as a result, intragranular deformation occurs, but the shape of the grains does not undergo significant changes. Finally, the specific features for producing IDS during SP deformation are caused by the generation of a low number of dislocations by each source. It is easy to understand this by remembering that grain boundaries form the sources of dislocations and are mobile during SP deformation since sliding and migration of the grains occur simultaneously and the magnitude and the sign of the local stresses undergo changes. Thus, visible traces of slip are not observed during superplastic deformation.

The aforementioned concepts concerning the specific features of intragranular deformation are essentially valid at the optimum strain rates (region II); the characteristic features of the operation of IDS may differ during the I and III strain rate ranges where the proportion and the role of the operating mechanisms are different.

In region III, intragranular slip becomes more and more independent on grain boundary sliding (just as in the case of usual deformation), and this is the reason for the significant contribution of IDS to the total deformation.

In region I the tendency towards the increased contribution of IDS (that has been established in a number of alloys [Sect. 2.1.2] can be understood only from the standpoint of the changes occurring in the nature of slip. In fact, as was shown in Sect. 2.1.2, there is a decrease in the number of slip systems in region I, but this is expected to lead to an increased directionality in the movement of dislocations within the individual grains which become particularly significant in the presence of a texture and during the deformation of the alloys having the hcp structure. Thus, not only an increase in the contribution of IDS to the total deformation occurs, but also an elongation of the grains and the development of a crystallographic texture in the originally textureless alloy can occur which can also be observed experimentally in region I.

Thus, IDS occurring under the conditions of SP deformation is qualitatively different from that developing during the conventional deformation. The most important difference is that during SP deformation, grain boundaries act as sources as well as sinks of dislocations, and the dislocations movement occurs, apparently, under the influence of local stresses developed at the boundaries during the process of grain boundary sliding. Consideration of these facts makes it possible to understand specific features observed during the occurrence of intragranular slip during SP deformation, viz., the structural changes and the nature of deformation relief an, also, to explain the experimental results concerning the evaluation of the contribution of IDS to the total deformation.

2.1.3 Diffusion Creep

Superplastic deformation is associated with the activation of diffusion processes. This is indicated by the data of direct measurements of the parameters of diffusion [1.3, 2.37] and by the results of the studies on the structural changes (in particular, the kinetics of grain growth) [1.36, 38].

At the present time there is no doubt regarding the importance of diffusion in ensuring superplastic behaviour. The role of directed diffusional mass transport (diffusion creep) in the development of SP flow is a more complex aspect. There have been a lot of experimental difficulties in studying diffusion creep (DC) during SP deformation; thus, a unique conclusion regarding this aspect could not be drawn until recently.

According to the classic concepts, e.g., [1.1–4] directed diffusion mass transport is caused by the concentration gradient of vacancies that develops between the differently oriented (in relation to the tensile axis) boundaries under the action of externally applied stresses. There are two types of DC, viz., the Nabarro-Herring creep during which diffusion fluxes are directed through the grain interior and the Coble creep which shows a more pronounced dependence on the grain size. It is expected that this process is preferred in an ultrafine-grained superplastic material.

Essentially, two methods have been hitherto employed for studying DC during SP deformation. The first method uses inert markers such as the precipitate

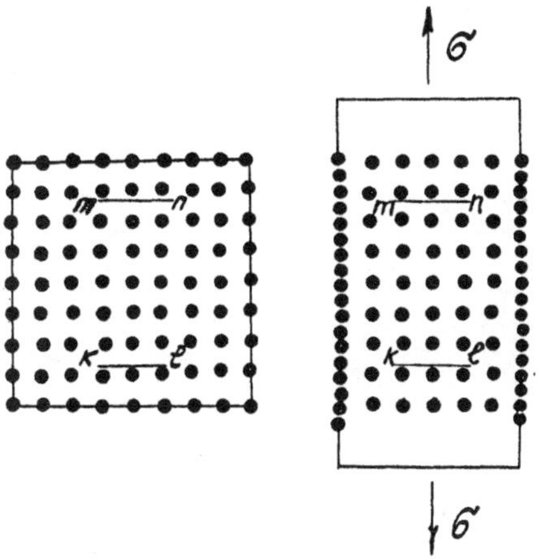

Fig. 2.11. Schematic representation of the formation of precipitate-free zones during diffusional creep

particles and the second is based on the deformation bands (striated zones) forming at the grain boundaries during SP deformation.

1. In the alloys containing dispersed particles, formation of precipitate-free zones (PFZ) occurs during the development of DC. These zones form near the transverse (with respect to the tensile axis) grain boundaries and, at the same time, an increase in the density of the precipitates takes place at the longitudinal boundaries (Fig. 2.11). Such a redistribution of the precipitates was reliably established during the occurrence of diffusional creep in coarse-grained materials [2.38].

The first observations of PFZ after SP deformation were made in 1969 [2.39, 40] by an optical microscopic study of the redistribution of dispersed precipitates and, later on, by X-ray microspectral analysis [1.3]. The formation of the depleted zones in the vicinity of the boundaries was established; however, no enrichment near the longitudinal boundaries was observed. Apparently, such a result is due to the experimental methods possessing a relatively low resolution; thus it has not been possible to establish all the specific features of redistribution of the particles (or the alloying elements), and individual precipitates could not be observed. Consequently, an ambiguity arose in the interpretation of the obtained results. Some investigators related the formation of the depleted zones to an intensive occurrence of grain boundary migration during superplastic deformation [2.41]. It is well-known [2.2] that such zones can often form behind the migrating boundary.

In this context, the application of transmission electron microscopy is of great importance for studying the redistribution of the dispersed precipitates during SP deformation. Such investigations have been carried out at our laboratory on the magnesium alloy MA8 [2.42] and the aluminium alloy AMg6 [2.15]. These alloys contain the particles of manganese-bearing phases that are stable at the temperature of SP deformation and the atoms of manganese, being heavier, diffuse slowly as compared to the atoms of the matrix; therefore during the diffusion of the light atoms, the particles can be used as inert markers.

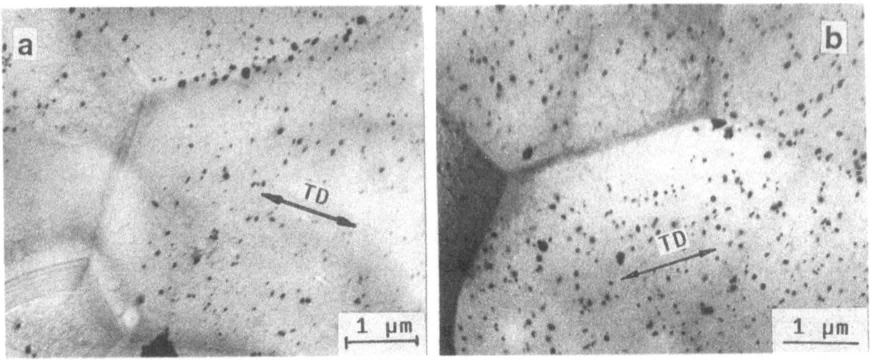

Fig. 2.12. Precipitate-free zones in the structure of the MA8 alloy after 40% superplastic deformation at a strain rate of $2 \times 10^{-5}\,s^{-1}$ (**a**) and $8 \times 10^{-3}\,s^{-1}$ (**b**). TD is the direction of tensile loading. From [2.17]

In the initial state, the precipitates were uniformly distributed in the microstructure. SP deformation leads to the formation of the zones free from the dispersed particles (Fig. 2.12a, b). It can be seen from the micrographs that the formation of PFZ at the transverse grains is accompanied by a simultaneous increase in the concentration of the particles at several longitudinal grains (see arrows in Fig. 2.12a).

Furthermore, our studies [2.42] made it possible to analyze the role of boundary migration (indicated by the grain growth during SP deformation) in the formation of the depleted zones. It is known [2.2] that a migrating boundary can drag off dispersed precipitates. In this case also a precipitate-free zone evolves. As a result, all particles "swept" by the grain boundary find themselves at the zone frontier. However, such an increased quantity of the precipitates in the vicinity of the zone could not be experimentally observed even in the case under consideration (Figs. 2.12a, b). Similarly, the mechanism of a PFZ formation due to the dissolution of the precipitates by the migrating boundary is also improbable, because, for example, PFZ did not form when annealing the specimens of the MA8 alloy at a temperature exceeding 400°C, whereas intensive grain boundary migration occurred which led to a significant grain growth. Thus, establishing the redistribution pattern of the dispersed precipitates gives a clear indication of the operation of diffusional creep under the conditions of superplastic deformation.

However, simultaneous occurrence of grain boundary migration can complicate the observed picture; therefore, it is not possible to distinctly observe an increased particle density at the longitudinal boundaries of all the grains. Apparently, the process of grain boundary migration often masks the region of increased density of the precipitates. In addition, in a number of cases, it is possible to observe the formation of individual particles in PFZ (e.g., in the zone A, Fig. 2.12a) caused by boundary migration. Such a situation can emerge, in particular, during the successive development of the processes of diffusional creep and grain boundary migration.

Thus, the rearrangement of the dispersed precipitates gives the most direct information regarding diffusion creep. However, this method can be used only for a limited number of materials such as alloys containing stable particles.

Ivanov et al. [1.48] and *Novikov* et al. [1.49] suggested another method involving measurements of the deformation zones formed at the surface of the specimens during their tensile deformation (Sect. 1.2). Assuming that the deformation zones form mainly by diffusion creep, it was proposed [1.5, 48, 49] that the regularities of the occurrence of DC can be established on the basis of the data related to the specific features of these zones. This conclusion was based on the results of *Backofen* et al. [2.39], who noted that the polished unetched surface of a deformed specimen of the hydrogenated Mg-0.5%Zn alloy shows striated zones (bands), whereas the etched sections show zones free from zirconium hydride. Both types of zones were of approximately equal thickness and were located perpendicular to the tensile axis. Backofen believes that the diffusional nature of these zones is identical.

However, we note that in the aforementioned study [2.39], the object of the investigation was a coarse-grained ($d > 30\,\mu m$) alloy. Thus, it is interesting to compare the sizes of the deformation zones and PFZ of the superplastic material.

The measured width of the deformation zones L_{DZ} and the width of the precipitate-free zones measured in the volume L_{PFZ}^{vol} and near the surface of the specimens L_{PFZ}^{surf} after superplastic deformation of the MA8 alloy up to $\varepsilon = 40\%$ [2.43] are given below.

$\dot{\varepsilon}$, s^{-1}	2×10^{-5}	2×10^{-4}	2×10^{-3}
L_{DZ}, µm	4.1	2.2	0.9
L_{PFZ}^{vol}, µm	0.8	0.4	0.2
L_{PFZ}^{surf}, µm	1.3	0.6	0.4

These data show that in the case of superplastic alloys, the width of the deformation zones is significantly greater than that of the precipitate-free zones. One of the possible reasons for such a difference is the evolution of the deformation zones at the surface of the specimen, but not in the volume because of the differences in the conditions of deformation at the surface and within the body of the specimen. Therefore, PFZ formed near the specimen surface were measured. For this purpose, after deforming the specimen up to $\varepsilon = 40\%$, it was covered on one side with lac and was thinned down from the opposite side in order to obtain thin foils. It was found that the difference in the width of the zones ($L_{DC} - L_{PFZ}^{surf}$) decreased to a certain extent, but remained fairly high, indicating that in the case of superplastic materials, formation of the deformation zones is caused not only by the DC, but also by other processes. In particular, grain boundary sliding and migration can cause the formation of these zones. This conclusion is in agreement with the results of several investigators (Sect. 1.2). Thus, although diffusion creep can evidently play a significant role in the formation of the deformation zones, it is not possible to obtain direct information regarding the occurrence of DC from the width measurement on the zones.

Specific Features of Diffusion Creep During Superplastic Deformation

In order to understand the role of DC in ensuring SP deformation, the data concerning the superplastic features of occurrence of this process during superplastic flow and a comparative study between these features and the classic concepts on the diffusional mass transport are of great significance.

Unfortunately, at the present time, the available data on the aspects are quite limited because of the difficulties arising during the experimental studies on DC. As was mentioned earlier, using the results of topographical studies on the deformation zones is of limited value. Therefore, let us examine the results obtained during a study of the nature of formation of PFZ in the magnesium-based MA8 alloy under various conditions of deformation [2.42].

It was established that at the given degree of tensile deformation of the alloy, the location and the width of the PFZ depend significantly on the strain rate. At low strain rates (in region I), they are encountered in almost half (50%) of the grains in that approximately 80% of all the zones are located in the vicinity of the transverse boundaries. On increasing the strain rate, the number of the grains containing PFZ decreases and the location of the zones becomes less ordered. In this case, they can frequently (60% of the total number) be observed even at the randomly oriented boundaries (Fig. 2.12b). Increasing the strain rate leads to a considerable decrease in the width of the PFZ also. After deforming in tension at a strain rate exceeding $8 \times 10^{-3}\,s^{-1}$ (in region III), PFZ are virtually absent. Essentially, analogous regularities in the formation of the zones were observed when studying the rearrangement of the dispersed precipitates in the AMg6 alloy [2.15].

The obtained data indicate significantly heterogeneity of the processes occurring during SP deformation within individual grains and the complex nature of the development of DC. According to the classical models of DC, when diffusional mass transport occurs under the influence of an applied stress, the depleted zones form near the transverse (with respect to the tensile axis) grain boundaries. Based on the different location of the zones which is observed during SP deformation, it can be assumed that under these conditions, diffusional mass transport is accomplished under the action of the local internal stresses. Such a model corresponds to the situation which in diffusion creep is not the main deformation mechanism, but acts as an accomodational mechanism (Sect. 1.2.4).

As a directional mass transport mechanism, DC must lead to the development of grain elongation (Fig. 2.11). However, in view of the fact that DC occurs under the action of local stresses whose magnitude and direction can undergo continuous changes during superplastic deformation, the directionality of DC decreases (similar to that of IDS) and elongation of the grains becomes insignificant. Such a situation can exist at the optimum strain rates; in region II and during the transition into the region I, the directional nature of DC and its role in developing elongated grains increase (as was shown by the experiment).

Contribution to Deformation

In order to determine the magnitude of deformation due to DC, one can measure the width of near-boundary zones – PFZ. In this case, the contribution of DC to the total deformation is calculated according to the following equation [2.44]:

$$\gamma_{DC} = (\bar{L}/\bar{S})\varepsilon_{tot} , \tag{2.12}$$

where $\bar{L}$ is the average width of PFZ in the direction of tensile loading; and $\bar{S}$ is the average distance between PFZ.

In our publications [2.15, 42], the initial grain size d_0 was taken instead of $\bar{S}$. We can agree with the earlier suggestion [1.5] that this is not quite correct, because in the given case the grain elongation caused by intragranular dislocation slip is not taken into account. However, at small strains (< 50%), the relative error in evaluating γ_{DC} (according to this method) due to the shape change of the grains does not exceed the statistical error. The results on the contribution of DC on the basis of the measurements carried out on PFZ are given below.

Contribution of DC in the region	Mg-6% Zn-0.5%Zr	Mg-1.5% Mn-0.3%Ce	Al-6.3% Mg-0.5%Mn
I	31	14	11
II	17	2	3
III	5	0	0
Reference	[2.44]	[2.42]	[2.15]

These data indicate that, in the experimental materials, the contribution of DC at the strain rates corresponding to region II is insignificant. This contribution increases with decreasing strain rate.

In [1.5, 49] it was noted that the deformation zones (bands) formed at the specimen surface can be used for determining the deformation due to DC. The following equation was proposed for evaluating the contribution of DC:

$$\gamma_{DC} = \bar{B}\,\bar{n}\cos\beta/\varepsilon_{vol} , \tag{2.13}$$

where $\bar{B}$ is the length of the bands in the striated zone; β is the angle between the bands and the tensile axis; and $\bar{n}$ is the average number of the striated zones per unit length of the longitudinal sections.

Furthermore, the authors proposed the possible application of the simpler relationship (2.12) which requires measurement of the width of the deformation surface zones instead of the average width of the precipitate-free zones. The evaluations of the contribution of DC that were carried out by *Novikov* et al.

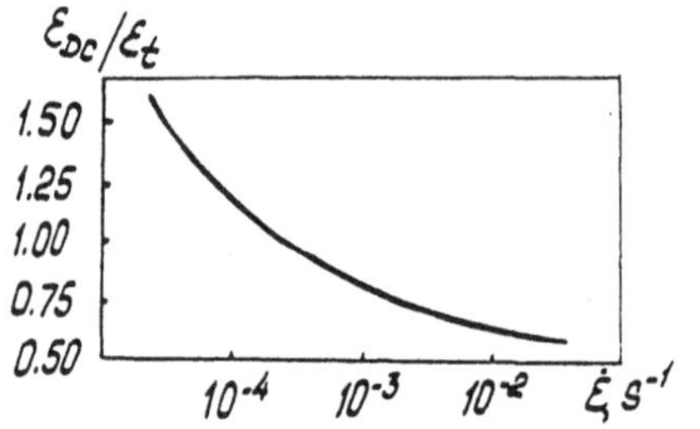

Fig. 2.13. Calculated values of deformation due to diffusional creep in the superplastic Zn-04.%Al alloy. The calculations were carried out according to (2.13), assuming that the nature of the deformation zones is diffusional

[1.49] on the basis of these concepts showed that the magnitude of $\varepsilon_{DC}/\varepsilon_{tot}$ of the Zn-22%Al alloy is quite significant. In region II, it amounts to approximately 50% and increases in the I strain rate interval up to 75–95%. At the same time, the application of (2.13) in determining the contribution of DC in the Zn-0.4%Al alloy gives a paradoxical result in that it exceeds 100% (Fig. 2.13). Thus, the procedure based on the measurement of the deformation surface zones gives significantly overestimated values which shows that it is incorrect to consider the formation of these zones due to the occurrence of diffusional mass transport only.

Summarizing the results of the studies concerning DC during SP deformation, it can be concluded that at the present time the data obtained on the alloys containing stable dispersed particles are the most reliable. According to conducted evaluations, the contribution of DC to the total deformation is insignificant and amounts to a small percent (although it increases with decreasing strain rate). Under the conditions of superplastic flow, diffusional mass transport is accomplished not under the action of the applied stresses, but at the grain boundaries whose magnitude and direction continuously change during the process of SP flow.

2.1.4 The Interrelationship and the Role of Different Deformation Mechanisms in Superplastic Flow

The experimental data obtained up to the present time convincingly prove that under the conditions of SP deformation numerous deformation mechanisms (grain boundary sliding, intragranular dislocation slip, and diffusional creep) operate simultaneously. To analyze the process which controls the rate of SP flow, it is important to determine whether the deformation mechanisms mentioned above operate independently or whether they are correlated. There are two possible approaches for analyzing this problem – microscopic and macroscopic. At the microlevel, the development of each of these mechanisms can be treated as a cooperative movement of the structural defects such as lattice dislocations, vacancies, and grain boundary dislocations, and their relationship is reflected in the interaction of these defects with the grain boundaries.

On the other hand, the macroscopic approach makes it possible to delineate the role of each of these mechanisms in ensuring SP deformation without going into the details of the microprocesses.

In contrast to the usual modes of deformation, SP deformation is distinguished by a strong strain rate dependence of the properties (the flow stress and the elongation). In this connection, it is very interesting to compare the variation of mechanical properties and actual mechanisms of deformation under the changing strain rate.

Let us compare the strain rate dependence of the properties obtained during SP deformation and the data concerning the contributions of different mechanisms to the total deformation. As mentioned earlier (Sect. 1.1), SP deformation is characterized by an strong dependence of m and δ on $\dot{\varepsilon}$. The maxima of these characteristics is attained in region II, and their magnitude decreases in region I as well as in region III. Let us consider data on deformation mechanisms operating in all three regions. At present the most complete data have been collected only for an aluminium alloy AMg6 and a magnesium alloy MA8. In these alloys independent measurements of the contributions of IDS and DC were made (Sect. 2.1). Based on these data, it is possible to determine the parameter $\gamma_{\mathrm{GBS}} = \varepsilon_{\mathrm{GBS}}/\varepsilon_{\mathrm{tot}}$ using the equation $\gamma_{\mathrm{GBS}} = 1 - \gamma_{\mathrm{IDS}} - \gamma_{\mathrm{DC}}$. The values of the contributions of all three deformation mechanisms are given in Fig. 2.14. It should be noted that the given values of γ_{GBS} are slightly higher than similar values measured in these alloys using the marker line method (Table 2.1). This difference is evidently due to the shortcomings of the estimations, which was discussed in Sect. 2.1.1. Figure 2.14 shows that the contribution of GBS is maximum in region II and can reach 70–80% of the total deformation; however, it decreases with both the increasing and decreasing strain rate. Hence, there is a clear correlation between the variation of γ_{GBS} and parameters m and δ under SP deformation. It is clear that the correlation of the variations of δ and m on the one side and the contribution of GB sliding to the total deformation on the other actually exist in spite of the technical difficulties encountered while measuring γ_{GBS}. Based on this, most of the investigators conclude that GBS is the main deformation mechanism during SP deformation.

However, the following questions remain unanswered: why is the contribution of GBS maximum in region II? Is there a correlation between the operation of other deformation mechanisms and the properties of the alloys during SP

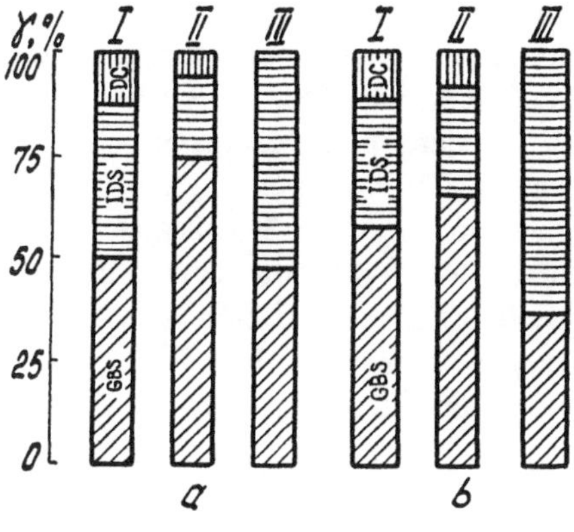

Fig. 2.14. Contributions of GBS, IDS, and DC to the deformation of the magnesium-based MA8 (a) and aluminium-based AMg6 (b) alloys in I–III regions of the curve of superplasticity. From [2.17]

deformation? Here, it is important to note that during the development of GBS in a polycrystalline material, the accommodating processes ensuring rearrangement and adjustment of the grains during deformation of the specimen must occur in order to create the conditions of simultaneous deformation and to retain the continuity of the material. IDS, DC, and grain boundary migration can act as such accommodating processes.

A comparative study of the contribution of IDS to the total deformation during superplastic flow and the variation of the mechanical properties as a function of the strain rate also makes it possible to establish some correlation between the aforementioned parameters. The contribution of IDS is minimum in region II, slightly increases in region I, and becomes noticeable in region III. One observes a correlation not only between the variation of γ_{IDS}, m, and δ, but also between the nature of IDS and the parameters of SP deformation. According to the data of texture analysis and electron microscopy (Sect. 2.1.2), the transition from region I to region II enduces both the enhancement of IDS and an increase in the number of operating slip systems (a transition from the preferential single to multiple slip occurs). Such a transition, established, for example, in the β-phase of the Zn-22%Al alloy, corresponds to the optimal superplastic conditions both with the changing grain size and with the varying temperature of deformation [2.24].

The observed correlation between the variation of IDS, the nature of IDS, and the mechanical properties during SP deformation indicates that this mechanism plays an important role in realizing SP flow. In particular, in region II, IDS can be assumed to be basically of an accommodation nature, which is accomplished under the effect of local stresses, arising during GB sliding developing alternately along numerous slip systems. Under these conditions the magnitude of IDS must be rather inhomogeneous in different grains and its dependence on the strain must be of an oscillating character. When there is pronounced strain in each of the grains (frequently with opposite signs), the average contribution of IDS to the total strain is insignificant. Such a conclusion on the role of IDS under the optimal superplastic conditions agrees totally with the experimentally observed peculiarities and the pattern of the process development (Sect. 2.1.2). It is important that, in this case, the most favourable conditions for the development of GBS proper are created. In region I, IDS becomes more directed due to the operation of the preferential slip systems. The contribution of IDS to the total deformation increases, although the intensity of dislocation slip decreases. At high strain rates (region III), IDS starts to play an independent role in the deformation process and develops under the effect of applied, not local stresses. In this case, the contribution of IDS to the total deformation increases.

In addition, during superplastic deformation, IDS does not play only an accommodating role, but also stimulates the development of GBS. This conclusion, important in understanding the nature of superplasticity, follows from a number of experimentally established facts. The stimulating effect of IDS on grain boundary sliding is well-known, e.g., from the creep experiments [2.22]. The influence of IDS on the development of GBS was established in the experiments performed on bicrystals [2.45, 46] (Sect. 2.2.2).

Special investigations were undertaken to reveal the stimulating effect of IDS on GBS during SP flow. In particular, among them are experiments on the influence of small strains, at room temperature, on the development of GB sliding [2.47], investigations in the superplastic behaviour after a short-time deformation at higher strain rates [1.1], etc. These data are significant for modelling superplasticity and will be considered in the following section.

The experimental data show that during superplastic deformation, diffusion creep is also an essentially accommodating mechanism which does not play a role in ensuring elongation of the specimen; rather, it helps in accomplishing local deformation and in eliminating the excess stresses developed during the collective deformation of the grains of the polycrystalline material. In addition, in contrast to the strain rate dependencies of the contribution of GBS and the parameters of superplasticity exhibiting an extreme nature, the contribution of DC increases monotonically with decreasing strain rate (Fig. 2.14). Based on these data, it can be assumed that during superplastic flow diffusional creep is an accommodation mechanism of deformation (Sect. 2.2.3).

The role of grain boundary migration in the accommodation processes is not completely clear, because the maximum indices of SP deformation are exhibited by the alloys having a stable microstructure that is not prone to grain growth. However, it is possible that grain boundary migration is also capable of relieving localized stress concentrations at the grain boundaries [1.1, 3], and thus contribute to high plasticity.

Thus, it can be concluded that at present several deformation mechanisms (viz., GBS, IDS, and DC), operating concurrently under the SP flow conditions, have been established. The observed correlations between the contributions of these mechanisms to the total strain and between the nature of these mechanisms and the variation of the alloy mechanical properties indicate that these mechanisms are closely related. These experimental data must be taken into account and explained to fully understand this phenomenon.

2.2 Micromechanisms and the Theory of Superplastic Deformation

A study of the operating deformation mechanisms during SP deformation is essential for developing the theory of superplasticity. The theory must explain the process and the structural changes occurring during the process of superplastic flow. In addition, the theory must be capable of predicting new and hitherto unknown effects and must form a basis for developing the methods of controlling the SP state. At the microlevel, superplastic deformation (just like usual deformation) is a cooperative process of the movement of structural defects but, in contrast to usual deformation, superplastic flow is accomplished by the cooperative simultaneous movement of the grain boundary dislocations, the lattice dislocations and vacancies. In spite of the fact that superplastic deformation involves mechanisms such as grain boundary sliding, intragranular dislocation slip, and diffusion creep that are also characteristic of usual deformation, its uniqueness emerges from a qualitatively new combination of their interaction that is achieved by the specific role of the grain boundaries during superplastic deformation, where the boundaries act as effective sources as well as sinks of the lattice defects.

Summarizing the results of these investigations, one can identify the following aspects whose understanding is essential for developing the theory of superplasticity:

1) the characteristic shape of the stress-strain curves with a high strain rate sensitivity of the flow stress; the presence of three typical regions in the $\dot{\varepsilon} = f(\sigma)$ plot;
2) the relationship between the superplastic behaviour and the structure (in particular, the dependence of the flow stress on the grain size and the structure and state of the grain boundaries);
3) the characteristic microstructural changes that depend on the degree of deformation and the strain rate (deformation induced grain growth, change of grain shape, etc.);
4) the experimentally established specific features of the operating deformation mechanisms and the data concerning their contributions to the total deformation.

Detailed reviews [1.1–7] on the earlier hypotheses proposed for explaining superplasticity are available. In this section, a brief summary of these models will be given mainly from the standpoint of the recent experimental data and the recently developed models which will be examined in a greater detail.

2.2.1 Physical Models

The experimental studies on the specific features of development of the operating deformation mechanisms and the evaluation of their contribution to the total deformation prove that GBS is the predominant mechanism of SP flow (Sect. 2.1.4). Thus, it is not surprising that most of the recent models of SP deformation are based on this fact. However, realizing the predominant role of GBS does not solve the problem completely. Other mechanisms (IDS and DC) are known to operate simultaneously along with GBS; consequently, it is necessary to take their interrelationship into account. In principle, the primary mechanism of GBS as well as the accommodating processes can control[2] superplastic deformation. Furthermore, there are various viewpoints on the nature of GBS itself (Sect. 2.2.2) which have led to a wide variety of models of SP deformation. These models can be divided into three groups in which diffusional accommodation, movement

[2] A controlling mechanism is one that determines the flow stress when deforming at a given strain rate or restricts the strain rate when loading under a given σ

of the lattice dislocations, and grain boundary sliding itself form the controlling processes.

It should be noted that in the models of the first and the second groups, it is usually assumed that the occurrence of GBS is relatively easy (as a rule, its micromechanism is not examined) and that the rate of deformation is controlled by the accommodating processes.

Most of the theoretical models aim at describing the superplastic behaviour of the alloys under the optimum strain rate conditions, i.e., in region II. For region III, it is assumed that IDS is the controlling mechanism and the behaviour of the material is described by the characteristic equations of diffusional creep. Developing the models of SP deformation corresponding to region I is of great importance from the standpoint of the physics of the processes, but the progress made in this direction has not been relatively significant. This is due to the inadequate understanding of the phenomenological regularities in this region (Chap. 1) and the fewer structural studies concerning this aspect.

In the models examining GBS with diffusion accommodation, the strain rate is described by equations that are, in principle, analogous to those used in the models based on the concepts of diffusional creep. Here, the Ashby-Verrall model [2.13] is among the most widely known models. The special feature of this model is that it takes into account the rearrangement of the grains during the deformation of the superplastic specimens (Fig. 2.5). Therefore, the diffusion paths become shorter and the stress dependence of the strain rate assumes the following form:

$$\dot{\varepsilon} = A \frac{D_{\text{eff}} G b}{kT} \left(\frac{b}{d} \right)^2 \frac{\sigma - \sigma_0}{G} , \tag{2.14}$$

where k is Boltzmann' s constant; T is the temperature; G is the shear modulus; b is the Burgers vector; d is the grain size; A is a constant; D_{eff} is the effective diffusion coefficient that takes the grain boundary diffusion and the volume diffusion into account; $\sigma_0 = 0.72\,\gamma/d$ is the threshold stress; and γ is the grain boundary energy.

The threshold stress physically appears due to the increased surface of the boundaries during the rearrangement of the grains, and its task is to decrease strain rate sensitivity of the flow stress at low strain rates, i.e., σ_0 causes the appearance of region I.

The proposed model is rather attractive primarily because it explains the experimentally observed change of neighbours and retention of the equi-axiality of grains during deformation. However, it has some essential shortcomings, too. For example, based on the experimental data reported later in [2.48, 49] it was shown that in the proposed form the model failed to give a satisfactory explanation of the sigmoidal shape of the $\sigma = f(\dot{\varepsilon})$ curves for different grain sizes. Specifically, the Ashby-Verrall model predicts that under optimal conditions of deformation the coefficient m is near unity, which is significantly larger than values usually provided by experiments. In addition, the model does not explain the experimentally observed increase in the activation energy and the decreased contribution of GBS during the transition from region II into region I. Furthermore, it should be noted that, although the model agrees with some structural data, it does not take into account the movement of dislocations within the grains and neglects the grain boundary structure without which, as shown by the experimental results, it is not possible to explain SP deformation.

Models based on the concepts of GBS controlled by the movement of lattice dislocations within the grains are more numerous.

Ball and *Hutchinson* as indicated in [1.3, 4] examined the sliding process of a group of grains. The very process of GBS occurs easily, but it is restricted by the increased stresses at obstacles such as triple joints. Accommodation is accomplished by the generation of dislocations at these obstacles and their subsequent movement through the grain, such that they are accumulated at the opposite grains. Sliding of the leading dislocations in the pile-up along the boundary towards the region of their annihilation forms the stage controlling the rate in this sequence. The following expression was derived for the strain rate:

$$\dot{\varepsilon} = A\frac{D_b G b}{kT}\left(\frac{b}{d}\right)^2\left(\frac{\sigma}{G}\right)^2, \tag{2.15}$$

where A is a constant; and D_b is the coefficient of grain boundary diffusion.

A similar analysis was carried out by *Mukherjee* as indicated in [1.3, 4] with the assumption that sliding of grains is accomplished not in groups, but individually.

Both these hypotheses predict a grain size dependence of the strain rate close to that observed experimentally and that $m = 0.5$. The main shortcoming of the hypotheses based on these models is that there is not experimental evidence for the presence of the anticipated dislocation pile-ups in the grains after SP deformation.

Modifications of the dislocation models are described in [1.4, 2.1, 49, 51, 52]; in these models, accommodation of GBS is assumed to occur due to the movement of individual lattice dislocations. *Hayden* et al. [2.51] assume that the dislocations generated at the grain boundaries move through the grains and subsequently climb along the opposite boundaries where their annihilation takes place. A new aspect of the last model of *Arieli* and *Mukherjee* [2.52] is the assumption that the climbing dislocations themselves get multiplied according to the Bardin-Herring mechanism. On the other hand, in the model of *Gifkins* [2.1, 50], accommodation of GBS is assumed to be caused by the movement of lattice dislocations in the narrow near-boundary "mantle" zone. The model specifically treats the process of sliding along the boundaries as a result of the movement of grain boundary dislocations which, after accumulating at the triple junctions, cause generation of lattice dislocations and their movement in the mantle along the boundaries of the neighbouring grains.

Thus, a common point of the dislocation models considered above is the assumption that the removal of the obstacles for the occurrence of GBS (triple junctions, steps) is related to the movement of lattice dislocations and their climb in the vicinity of the grain boundaries which forms the rate-controlling factor of the process. Although each of these models has certain merits, they are not in agreement with several experimental data obtained in recent years (Sect. 2.1). In particular, their basic concepts contradict the established peculiarities of the movement of the lattice dislocations by intragranular slip and their easy entrance into the boundaries, i.e., the concept that the movement of lattice dislocations cannot be a controlling process of superplastic flow is contradicted. There is another matter of principal importance. The suggested models take into account processes developing in the grain boundaries or near them; however, these models completely ignore the actual structure and properties of grain boundaries.

The models of the third group (which treat GB as a controlling deformation mechanism) significantly differ from the models of the first two groups. Among them, the models of *Grabski* [1.2] and *Padmanabhan* [1.8] are widely known.

The model of Grabski was the first attempt at developing the theory of superplasticity on the basis of the data related to the structure and the properties of grain boundaries. However, studies on the behaviour of grain boundaries during the deformation processes began during that period (1973). Therefore, it is not surprising that the proposed model contradicts the modern concepts on the role of grain boundaries in the nature of superplasticity (Sect. 2.2.2). For instance, it does not take the process of dissociation of the lattice dislocations in the boundaries and the effect of the nonequilibrium defects on the development of the grain boundary processes into account. In addition, the model neglects the most important aspect, namely, the interaction of different cooperative deformation mechanisms during SP deformation.

Padmanabhan [1.8] also assumes that the triple junctions and other obstacles do not stop viscous flow along the grain boundaries, i.e., the accommodation processes do not restrict the rate of GBS; therefore, shear along the boundaries ensures a high degree of ductility. He distinguishes two stages during the process of SP deformation, viz., the initial stage of unstable flow and the stage of stable flow (superplasticity proper). During the initial stage of deformation, rearrangement of atoms in the boundary leads to the occurrence of sliding which is, however, hindered by the obstacles such as steps, triple junctions, and other uneven features of the boundaries. The resistance offered by the obstacles to sliding leads to the development of reverse elastic stresses. When these stresses begin to exceed the shear stresses, local directional diffusion occurs at the boundary which leads to such a redistribution of the material of the uneven features that the hindering effect of the obstacles decreases and the development of sliding continues. The transition to stable flow is dictated by this. However, the model does not consider the micromechanism of grain boundary sliding and assumes that the nature of sliding is determined by the diffusional rearrangement of atoms at the boundaries according to concepts developed as long ago as 1949 in the pioneer work of *Ke Tensui* and used to explain the nature of the grain boundary peaks of internal friction.

The following equation was obtained [1.8] for the strain rate:

$$\dot{\varepsilon} = \left(cl/d^2\right)\sigma^n \exp(-Q/kT) \tag{2.16}$$

where c is a constant; l is the distance of diffusional transport of atoms; and Q is the activation energy.

Padmanabhan's model and its slightly more accurate variant [2.53] were discussed in detail in [1.1, 6, 7, 2.48]. The concept concerning the unstable and steady stages of SP deformation is correctly introduced in this model, but its major shortcoming is that it neglects the role of the accommodation processes without which many specific features of the structural changes occurring in the superplastic materials (grain growth, bending of the boundaries, void formation, etc.) cannot be explained. In addition, since the micromechanism of sliding was not taken into account, numerous arbitrary assumptions were made when deriving the main equation of SP flow in the form $\dot{\varepsilon} = f(\sigma)$.

Experimental data on the $\sigma - \dot{\varepsilon}$ dependence (mainly for the optimal strain rate region) in the Zn-22%Al alloy and assessments made using different models were compared in [1.6]. The conclusion was that the results provided by models based on GBS and dislocation accommodation were most comparable to the experimental data.

At the same time, it should be pointed out that the value of such comparisons is limited. The problem is that in this case state equations of the following type are used (notations are the same as above):

$$\dot{\varepsilon} = A\left(\frac{b}{d}\right)^p \left(\frac{\sigma}{G}\right)^n \exp(-Q/kT)\,. \tag{2.17}$$

Such equations look quite similar, i.e., the parameters p, n, A, and Q are almost the same in the majority of models, in spite of different (and sometimes opposite) concepts underlying these models. This is due to a number of assumptions and simplifications, introduced when these equations are derived. There is no doubt that the models must be developed and verified based on more complete and reliable microscopic parameters and regularities; the models must not only describe these regularities, but predict new ones. From this viewpoint, we can assert that not a single existing model is satisfactory enough. This is the conclusion made by most authors of the latest reviews [1.1–9, 2.54] devoted to the nature of superplasticity.

The complexity of the problem is that when regarding the structural aspects, it is necessary to consider the unusually large area of grain boundaries, and from the point of view of the deformation mechanism, it is necessary to define the role and behaviour of grain boundaries during plastic flow. Furthermore, until recently, the physical aspects of grain boundaries were studied insufficiently and were not employed in the analysis of the nature of superplasticity. Works based on the modern concepts of grain boundary structure and properties appeared only in recent years. The results of these studies will be described in Sect. 2.2.3. But before that, it would be useful to consider results obtained during the study of grain boundary structure and properties in the course of plastic and superplastic deformation.

2.2.2 Grain Boundaries in the Deformation Processes

As was mentioned in Chap. 1, the microstructure of the superplastic materials is characterized by a significant area of high angle grain boundaries. In spite of the fact that grain boundaries are one of the "oldest" microstructural defects of the polycrystalline materials, it was only during the recent past that there was significant progress in understanding their atomic structure, behaviour and role during plastic deformation [1.39, 2.2, 3, 55–58].

Perfect Grain Boundary Structure

At present, many problems concerning the structure of high-angle grain boundaries have been solved based on the achievements of the high-resolution experimental methods and computer simulation. Considerable progress has been made due to the development of the structural units model [2.55–60]. According to this model, any grain boundary may be represented as a mixture of certain low-energy structures. Coincident boundaries (special boundaries with a small reciprocal coincident site density Σ) are formed by a periodic array of structural units of one type, whereas any deviation from the coincident site lattice (CSL) results in a mixture of structural units of different types. Computer simulation was used to find out "bricks" of which any boundary can be constructed. It was established that there exists a limited number of coordinate polyhedra whose vertices are atomic sites in grain boundaries. Unlike the structure of amorphous bodies in which atomic polyhedra are in disorder, the polyhedra in grain boundaries are arranged in one layer, there being periodicity of crystals on either side of the boundary, which results in the well-ordered arrangement of atomic groups in the grain boundary structure. The ordered structure is characteristic of all grain boundaries.

It is important to note that the structural units theory does not contradict the dislocation models of high-angle boundaries, but develops them. This was confirmed by experiments, primarily by the high-resolution electron microscopy and high-resolution methods of X-ray and electron diffraction analysis, which allowed the atomic-scale experimental studies to become possible. Using these methods as well as ion microscopy, the grain boundary was found to be as thick as several atomic spacings, as predicted by the crystal-geometrical theory. The high-resolution electron microscopy gives direct evidence for the periodic structure of random as well as special boundaries. In addition, a secondary periodic structure due to periodic networks of grain boundary dislocations (GBD) that cause a deviation from coincidence misorientations was observed in the boundaries deviating from special misorientations.

At the same time, the study of the grain boundary structure refer in general to a low temperature range. At present, it is not quite clear what happens with the atomic structure of grain boundaries during heating. In the meantime, there is no strict experimental evidence for the grain boundary phases being transformed either from one crystallographic modification into another or into an amorphous state [2.55, 58, 59]. Hence, it can be assumed that the grain boundaries have an ordered crystalline structure at any temperatures at which solids are stable.

The above discussed concepts refer to the perfect grain boundary structure; grain boundaries in well-annealed materials are very similar to the perfect boundaries. The perfect grain boundaries possess minimum energy at the specified crystallographic parameters of the boundaries and are characterized by the absence of long-range elastic stress fields. Such boundaries have a stable and equilibrium structure. However, in real metals, the structure of the grain boundaries is determined not only by the crystallographic parameters, but also by the result of interaction of the boundaries with other lattice defects such as dislocations and point defects. Let us examine this aspect in greater detail.

Grain Boundary Structure and Properties During Plastic Deformation

Numerous studies on the nature of distribution of the slip lines and the data of electron microscopy indicate [2.2, 58] that grain boundaries are effective sources of lattice dislocations. Generation of dislocations has been observed at different uneven features of the grain boundaries such as ledges and steps and in the triple junctions. At the same time, as was shown by the studies on the interaction of the grain boundaries and the lattice dislocations (LD) during specimen deformation "in situ" in an electron microscope [2.62, 63], dislocation generation occurs most frequently at those regions of the grain boundaries where no visible defects are present. It should be noted that even after the generation of dislocations, visible defects are not revealed in the structure of the boundaries, although a local change in their structure does occur. Thus, the experimental data show that dislocations are generated not only at the uneven geometric features of the grain boundaries, but also at the structural elements of the boundaries themselves due to the peculiar rearrangement of atoms along their surface.

The effect of grain boundary sliding on the generation of lattice locations at the grain boundaries requires particular attention. As was shown by computer simulation [2.64], during the generation of dislocation at the steps of atomic sizes existing at the grain boundaries, sliding along the boundary leads to a decrease in the stresses required for their generation by more than two orders of magnitude. This is due to the high stress concentration observed at the steps during the development of shear stresses in the plane of the boundary.

Another important aspect of the interaction of the lattice dislocations with the grain boundaries is the capacity of the grain boundaries to absorb the dislocations, i.e., their ability to act as sinks for LD. It is well-known [2.56, 58] that during plastic deformation and recrystallization, dislocations can enter the grain boundaries. The arising defects, i.e., trapped lattice dislocations (TLD), can be clearly observed when their density is not high, for example, after cold working up to a small strain (Fig. 2.15a). As the temperature increased, the TLD become unstable and dissociate into extrinsic GBD[3]. In the case of special boundaries and the boundaries near them, this process can be observed directly in an electron microscope. As was established during the process of annealing thin

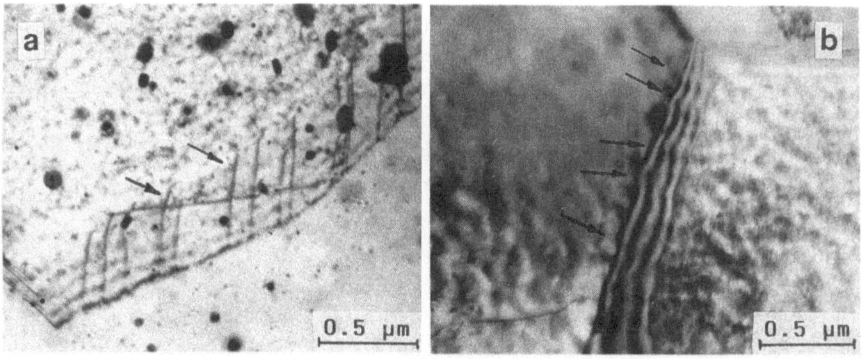

Fig. 2.15. Trapped lattice dislocations in the grain boundaries of the MA8 alloy (**a**) and magnesium (**b**) at $\varepsilon = 2\%$ and $t_{def} = 20°C$. The extinction fringes of magnesium show the typical bends developed due to relaxation of TLD. From [2.55]

foils in situ in an electron microscope, relaxation of the TLD occurs in the random boundaries which can be observed in the microscope by a weakening and spreading of their images (Fig. 2.15b). Studies conducted on the relaxation of TLD [1.41, 2.58, 65–67] indicate that in pure metals it occurs at a temperature corresponding to 0.3–0.4 T_m, but this temperature increases in the presence of impurities. For example, in magnesium, the relaxation temperature of TLD is below room temperature [2.67], but in the MA8 alloy (Mg-1.5%Mn-0.3%Ce), it is equal to 350–400°C [1.41]. It was established [2.67–69] that after the relaxation of TLD and the disappearance of their images, the structure of the grain boundaries does not become perfect and does not attain the equilibrium state. This indicates a disturbance of the periodicity in the structure of the grain boundaries because of the presence of GBD, i.e., defects of grain boundary structure.

By considering the behaviour of GBD, two stages of absorption of the lattice dislocations by the grain boundaries can be identified [2.67, 70]. Initially, dissociation of TLD into the extrinsic GBD occurs, and this is followed by a rearrangement of the grain boundary structure that leads to the elimination of GBD due to annihilation, grain boundary sliding or migration, or due to the formation of equilibrium configurations of GBD, i.e., the transformation of the extrinsic GBD into intrinsic GBD. The absorption kinetics of the lattice dislocations is of a diffusional nature [2.70], and, therefore, the process occurs only at sufficiently high temperatures.

At elevated temperatures, the grain boundaries in the polycrystalline materials also act as active sources and sinks of vacancies. This fact is indicated by several experimental observations of the precipitate-free zones formed after quenching the specimens caused by a sinking of vacancies at the grain bound-

[3] It is precisely these extrinsic GBD (EGBD) which, in contrast to the structural GBD, form distortions in the periodic structure of the grain boundaries. They will be simply referred to as GBD in our subsequent discussions.

aries [2.54]. Grain boundary changes during the interaction of grain boundaries with lattice defects. These changes are related to the disturbance in the periodicity of the structure of the boundaries, i.e., to the development of the grain boundary defects and the long-range stresses around the grain boundaries and lead to a situation, wherein, at the specified crystallogeometric parameters, the grain boundaries would possess higher energy [1.39, 2.70]:

$$E_n = E_0 + \Delta E$$
$$\Delta E = \frac{Gb^2}{4\pi(1-\nu)2h}\ln\frac{e\alpha h}{\pi b}\,, \tag{2.18}$$

where E_0 is the energy of the boundary having a perfect equilibrium structure; ΔE is the energy increase due to the presence of structural defects in the boundary; ν is Poisson's ratio; h is the distance between the defects; e is the base of the natural logarithm; and α is a parameter determining the energy of a dislocation core ($\alpha \simeq 1-2$).

The grain boundaries containing the grain boundary defects are under a structurally nonequilibrium state and move towards the equilibrium state when there is a recovery of their structure.

The presence of dislocation-type defects (TLD) and the products of their dissociation, viz., extrinsic GBD) has a significant effect on the properties of the grain boundaries. In this case, the grain boundaries usually have long-range elastic stress fields [2.70] which significantly affect the behaviour of the boundaries during plastic deformation and recrystallization (see below). In addition, the energy of vacancy formation decreases, because, in this case, the increase of the total grain boundary energy is elastic in nature. Thus, we obtained the following equation for the coefficient of grain boundary diffusion in the grain boundaries having a nonequilibrium structure [1.39]:

$$D_n = D_0 \exp\left(\alpha a^2 \Delta E/kT\right)\,, \tag{2.19}$$

where D_0 is the coefficient of grain boundary diffusion in the equilibrium grain boundaries; ΔE is the energy increase due to the dislocations present in the boundaries; α is a constant ($\alpha = 1$); a is the atomic radius; k is the Boltzmann constant; and T is the temperature, K.

The presence of the grain boundary defects can lead to accelerated (by 1–2 orders of magnitude) grain boundary diffusion [1.39].

Grain Boundaries During Superplastic Deformation

During superplastic deformation, the effects associated with the interaction between the grain boundaries and the lattice defects are of great significance. As was indicated in Sect. 2.1.2, during SP deformation, dislocation generation occurs mainly at the grain boundaries. In this case, development of the local stresses during the occurrence of grain boundary sliding leads to the generation of lattice dislocations at extremely low stresses that are characteristic of SP flow. The absence of visible dislocations in the grain interior after SP deformation is due to the facility of lattice dislocation absorption by the boundaries. The superplastic state occurs at temperatures exceeding the TLD relaxation temperature; therefore, the lattice dislocations are not stable in the grain boundaries during SP deformation and dissociate rapidly and form GBD. It is therefore not surprising that TLD are rarely observed in the specimens deformed in the superplastic state. It can only be observed using the procedure of "instantaneous" quenching (Fig. 2.7).

The importance of the phenomenon of absorption of the lattice dislocations during superplastic flow is confirmed by the experiments designed to study the effect of the state of the grain boundaries on the superplastic behaviour (Chap. 1). Thus, a rolled specimen of the Zn-0.4%Al alloy having grain size $d = 1\,\mu m$ completely lost its SP properties when subjected to aging (the SP properties were observed in the as-rolled alloy being deformed at ambient temperature, $\dot{\varepsilon} = 3 \times 10^{-4}\,s^{-1}$). The following experiment was conducted for determining the ability of the boundaries to absorb lattice dislocations. In order to introduce the lattice dislocations into the boundaries, the specimens of the alloy existing in the superplastic and aged conditions were deformed up to a strain of 3% at −50°C and were subsequently held at the temperature of SP deformation (+20°C). It was found that the trapped lattice dislocations remain stable in the grain boundaries in the aged alloy, but relax in the as-rolled alloy, i.e., aging diminishes the ability of grain boundaries to absorb dislocations. Such a difference in the behaviour of the lattice dislocations is apparently related to the occurrence of impurity segregation at the grain boundaries of the alloy during the aging processes. Structural studies showed that the aged alloy subjected to deformation under conditions corresponding to SP deformation of the as-rolled alloy contains a high dislocation density within the grains, and the extent of grain boundary sliding decreases abruptly. Evidently, these effects are associated with the hindered absorption of the lattice dislocations in the grain boundaries. Thus, the experiments performed provide convincing enough evidence that the realization of superplasticity requires that lattice dislocations should be easily absorbed in grain boundaries. The obtained data are also confirmed by the results of studies conducted on the model aluminium alloys containing different types of alloying elements (Sect. 1.1); these results established a correlation between the ability of the alloys to undergo superplastic deformation and the effectiveness of absorption of the dislocations at the grain boundaries.

The TLD relaxation process is significantly affected not only by the impurity content in the grain boundary, but also by the grain boundary spectrum related to the fraction of special boundaries as well as by the degree of their nonequilibrium due to the prior processing of the alloy [2.69, 70]. Apparently, the effect of the initial texture on the evolution of the dislocation structure and the superplastic properties of the Zn-22%Al alloy and other materials are due to these phenomena (Chap. 1). It is well-known that a close relationship exists between the type of crystallographic texture and the grain boundary spectrum in a polycrystal. The effect of the absorption of lattice dislocations on the development of superplasticity depends on the appearance of mobile GBD (which are formed during the TLD dissociation) and their influence on the development of grain boundary processes, viz., GBS, grain boundary migration and DC [2.54, 70, 71]. Let us consider these aspects in detail.

During plastic deformation of metals, sliding along the boundaries occurs concurrently with intragranular dislocation slip (IDS). There are several models in which GBS is treated as a result of different degrees of deformation of the adjacent grains [2.22], as a consequence of the movement of individual lattice dislocations [2.2] or as a dislocation layer in the vicinity of a grain boundary. In these cases, GBS requires an intensive near-boundary deformation. However, sliding usually occurs in association with the shear along the grain boundaries only (true GBS) and it can be observed, for example, under the conditions of dislocation creep and diffusion creep in the absence of dislocation movement within the grains. Thus, two types of sliding can be identified, viz., the "pure" GBS that is not related to intragranular slip and GBS developing simultaneously with IDS.

In this context, of special mention are the results of studies carried out on GBS of zinc bicrystals in which it was possible to observe and study these modes of sliding [2.46, 58, 71]. The scheme of specimen cutting is shown in Fig. 2.16. Since zinc has only one preferred slip plane, the experiments on the zinc bicrystals give favourable conditions for distinguishing the effects of the simultaneously occurring deformation processes at the grain boundary and within the grains. It was established that the "pure" GBS can be observed only at very low stresses (1–3 MPa) and that the parameter m_{GBS} is near unity, and the activation energy is near the activation energy for grain boundary diffusion. Furthermore, it was found that "pure" sliding occurs nonuniformly along the boundary which is a characteristic of a dislocation mechanism of the process. Probably, the observed "pure" GBS is due to the viscous movement of GBD

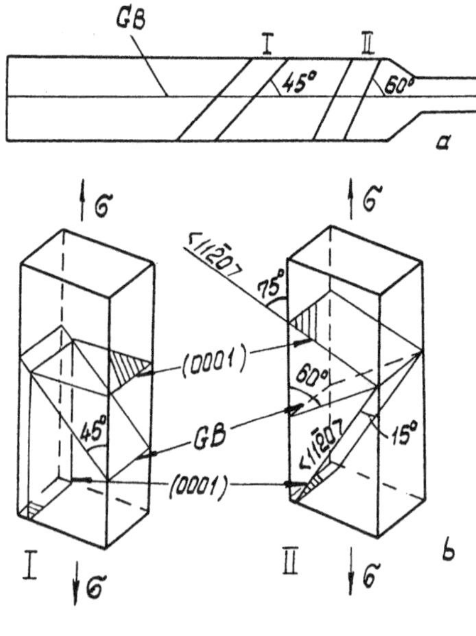

Fig. 2.16. Scheme of cutting (**a**) and the types of zinc bicrystals (**b**) used for studying "pure" GBS (I) and GBS related to intragranular slip (II); GB is the boundary between the grains of the bicrystals. From [2.58]

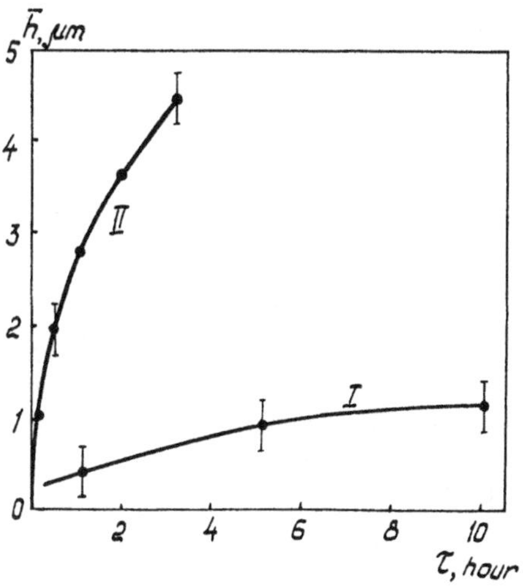

Fig. 2.17. Dependence of the magnitude of sliding $\bar{h}$ of I- and II-type bicrystals on the test duration under identical conditions. The shear stress is 1.4 MPa. From [2.58]

that are generated directly within the plane of the boundary. At the same time, the operation of this micromechanism is possible only at low stresses; upon increasing the shear stresses in the boundary up to a level exceeding 3 MPa, cessation of the process of GB sliding and a rapid generation of the lattice dislocations from the boundary (the accommodating IDS) were observed.

Furthermore, the zinc bicrystals were used for studying the regularities of sliding along the boundary (having the same structure) during its occurrence simultaneously with a slip within the crystals. It was found that in this case, at identical shear stress levels, the rate of sliding at the boundary increases (stimulation of sliding) as compared to that of "pure" GBS by 50–60 times (Fig. 2.17); and a decrease of the parameter m_{GBS} up to 0.4–05. takes place, because the rate of sliding and the magnitude of m_{GBS} depend on the extent of intragranular slip. It is interesting to note that in all cases considered, the magnitude of sliding was proportional to intragranular strain. Since the interaction between the dislocations and the grain boundaries occurred in the course of the experiment, and the test temperature exceeded the TLD relaxation temperature, it can be assumed that the mechanism of sliding is related to the movement of GBD formed during the dissociation of the lattice dislocations in the boundary. It is important to note that the decisive role of the extrinsic GBD in ensuring the sliding process was shown immediately at the special boundaries [2.72, 73] (Fig. 2.8), where GBD have large Burgers vectors and can be observed in an electron microscope.

Thus, there is convincing experimental evidence that GBS is accomplished via the movement of GBD. GBD generation can occur either directly in the boundary or by means of dissociation of trapped lattice dislocations. In the former case, it is "pure" GBS; in the latter, it is "stimulated" GBS. Thus, the higher rate of sliding observed when it operates simultaneously with IDS (as compared to "pure" GBS) can be attributed to the easier formation of GBD and the increased density of the mobile GBD in this case. In addition, the extrinsic GBD, being nonequilibrium defects of the grain boundaries, accelerate grain boundary diffusion, therefore, the mobility of GBD can increase when their density is high. However, we underline the fact that the unrelaxed lattice dislocations present in the boundaries (prior to dissociation) must hinder the development of sliding. This conclusion is in agreement with the data of [2.54, 71].

When studying the nature of superplasticity, one must consider the following questions which are of basic importance: What is the micromechanism of superplastic flow? How do the triple junctions affect the development of sliding? As mentioned above, most of the models of superplasticity do not consider the mechanism of GBS, because those models are based on the assumption that the strain rate is controlled by accommodation processes. The study of GB sliding in tricrystals is of special interest for the investigation of accommodation processes in triple junctions. Recently such a study [2.74] was conducted on zinc tricrystals with the $90° \langle 11\bar{2}0 \rangle$ boundary. The same boundary was investigated earlier in zinc bicrystals. It was established that when an active accommodation develops in a triple junction, the magnitude of sliding decreases insignificantly (less than twice) in tricrystals as compared to the magnitude of GBS in bicrystals.

Thus, the results of these new experiments indicate directly that accommodation processes in triple junctions do not necessarily control the rate of grain boundary sliding.

Let us analyze data referring to the regularities of GBS in bi- and tricrystals, and in superplastic materials. For this purpose, we compare the flow stresses observed during SP deformation and the stress level required for accomplishing different types of GBS. Let us compare the stress level required for ensuring identical sliding rates in the bi- and tricrystals of zinc and the Zn-0.4%Al alloy [2.5].

Figure 2.18 shows the initial portion of the stress-strain curve of the Zn-0.4%Al alloy. It can be seen that just as in other superplastic materials, the transition to the stage of stable deformation occurs after the occurrence of signif-

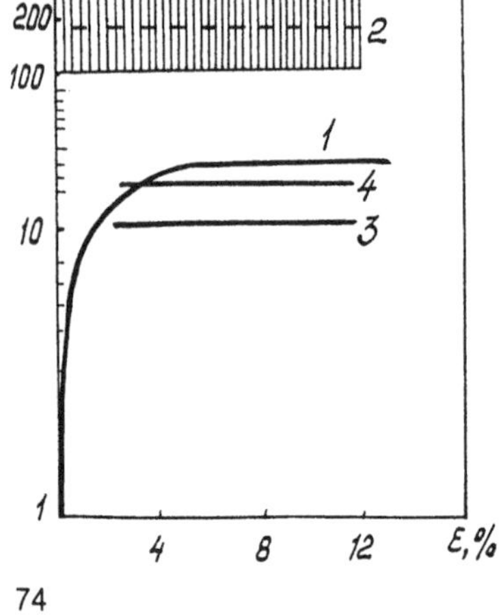

Fig. 2.18. The initial portion of the $\sigma - \varepsilon$ curve of the Zn-0.4%Al alloy during its deformation in region II (*1*) and the stress levels required for ensuring a sliding rate observed during SP deformation of the alloy in the cases of "pure" GBS (*2*), GBS induced by IDS (*3*), and GBS in zinc tricrystal (*4*)

icant strengthening during the initial stage. The stress level required for ensuring a sliding rate of $4 \times 10^{-4}\,\mu m/s$ that corresponds to the grain boundary sliding rate observed during superplastic deformation of the alloy [2.5] and in the zinc bicrystal during the development of "pure" GBS and the IDS-induced GBS (curves 2 and 3), and in zinc tricrystal (curve 4) are also indicated in Fig. 2.17. It can be seen that the flow stress in the case of "pure" grain boundary sliding is significantly higher than that of superplastic deformation; the stress level of superplastic flow is significantly closer to that required for the stimulated grain boundary sliding, and corresponds to that in the tricrystal. In addition, in the latter case, $m_{GBS} = 0.4$, i.e., it coincides with the magnitude of the parameter m observed under the conditions of superplastic flow. Therefore, the comparative data of the aforementioned study form important evidence for the occurrence of GBS induced by the intragranular slip, i.e., GBS due to the movement of the grain boundary dislocations during superplastic deformation. This GBS controls the flow stresses. It follows that a close relationship between the IDS and the GBS processes must exist at the microlevel, and this relationship can be observed experimentally (Sect. 2.1.4).

To induce the "stimulated" GBS, it is necessary to maintain a certain temperature above which the grain boundaries cease to be obstacles for the movement of the lattice dislocations but trap and actively absorb them. As a result of the entrance of lattice dislocations into the boundary, they relax and form GBD, their movement being an impetus for the development of GBS.

Thus, to provide for the development of GBS and to induce superplasticity, the temperature at which lattice dislocations in the boundary relax (this process is observed in an electron microscope as the spreading of TLD) must be lower than the temperature of SP deformation. This conclusion is confirmed by a number of experimental data mentioned above.

When examining the movement of the grain boundary dislocations along the surface of the boundary as a micromechanism of sliding, it must be remembered that it is only possible by means of glide and climb [2.2, 3]. Therefore, activation of grain boundary diffusion is required for facilitating a high mobility of GBD, which causes the relationship between GBS and the diffusional processes. However, in this case, other aspects of the problem are also important. The movement of GBD necessarily requires absorption and emission of vacancies. The local vacancy concentration depends on the density and the mobility of GBD. These factors are different at different boundaries; consequently, diffusion fluxes reducing the concentration gradient of vacancies are developed. The fluxes of diffusional mass transport are localized and are determined by the development of GBS, but these, in turn, facilitate the occurrence of sliding by creating conditions for the rapid movement of GBD. It is important to note that during the development of this mode of diffusion creep, its rate is determined by the rate of vacancy generation and absorption at the grain boundaries. This model agrees with the data of the macroscopic observations (Sect. 2.1).

The effect of the grain boundary defects on the process of migration and, consequently, on grain growth is another important aspect associated with the structural changes occurring in the boundaries. The effect of the nonequilibrium state of the structure on the migration rate of the grain boundaries is mainly stimulated by its effect on the mobility of the boundaries [2.58]. Evidently, the physical reason underlying this fact is that the presence of grain boundary defects has an effect on diffusion.

To summarize this Section we conclude that at present there are numerous data indicating that the interaction of the lattice defects with the grain boundaries has a significant effect on the structure and the properties of the grain boundaries, because of which there is a change in the behaviour and the role of the grain boundaries during the deformation processes. These results are important for understanding the nature of GBS during SP flow; GBS is probably accomplised by the cooperative movement of the grain boundary dislocations along the surface of the grain boundaries. However, this process is closely related to the movement of the lattice dislocations and the vacancies. The lattice dislocations create the necessary conditions for the formation of GBD whereas their movement ensures accommodation of the grains during palstic flow. Increasing the vacancy concentration provides the conditions for the movement of GBD at a fairly high rate, and the diffusional fluxes also relieve the local stress concentrations during the process of deformation. These facts must be considered when analyzing the physical model of superplastic flow.

2.2.3 A New Model of Structural Superplasticity

An extensive study of the experimental data shows that superplastic deformation is an integral process in which grain boundary sliding forms the predominant deformation mechanism that controls the flow stress, and the accommodating mechanisms of intragranular dislocation slip and diffusion creep are interrelated. Just as in the case of usual deformation, movement of the lattice defects (GBD, dislocations and vacancies) is essential for accomplishing superplastic deformation, but the specific features of this mode of deformation are in retaining the continuity of the material due to the absence of an accumulation of the lattice defects during deformation. This is possible when GBS acts not only as a deformation mechanism, but also as a recovery mechanism of the structure of the material, and the occurrence of IDS and DC takes place under the influence of localized stresses and leads to their relaxation.

In order to understand the nature of superplastic deformation and to develop its theory, it is necessary to consider the specific micromechanism of GBS, the atomistic aspects, and the related structural changes in the grain boundaries. Let us examine the methods of working out such a theory and the possibilities offered by it [2.75].

Description of the Model

First of all, let us identify all the facts which constitute the basis for the new approach.

1) Superplasticity of metallic materials is mainly related to the redistribution of the grains due to mutual sliding and the change of their neighbours. The shape changes of the grains have a cyclic nature, and the grains remain essentially equiaxial. This conclusion directly follows from data concerning specific features of the mechanisms of SP deformation and the evaluation of their contributions to the total deformation (Sect. 2.1.4).
2) According to the results of the investigations on the nature of GBS (Sect. 2.2.2), two types of grain boundary sliding are possible: the "pure" GBS that is apparently due to the movement of the grain boundary dislocations that are generated within the boundary and have Burgers vectors lying in its surface and the "stimulated" GBS related to the movement of GBD that are generated during the dissociation of the lattice dislocations in the boundary and whose Burgers vectors possess random directionality in the general case.
3) At the atomic level superplasticity can be treated as a result of the cooperative movement of the grain boundary dislocations, the lattice dislocations, and the vacancies, because their movement occurs with a close interrelationship due to the role of the grain boundaries as sources as well as the sinks of dislocations and vacancies, and the rates of various deformation processes show kinetic correspondence among themselves.

Treating superplasticity as a result of the movement of lattice defects, it can be assumed that the deformation process must start with the generation of grain boundary dislocations. In fact, vacancy generation and absorption in a boundary requires the movement of GBD, whereas the generation of lattice dislocations in the ultrafine-grained materials is difficult. The generation of GBD leads to the occurrence of "pure" grain boundary sliding, but this process can provide for an extremely small deformation since the boundary surfaces are not always flat and contain different obstacles and triple junctions which hinder the movement of grain boundary dislocations. Noting that the number of dislocations in the pile-ups $n \approx d\sigma/Gb$, where d is the length of the boundary, and inserting the typical values of the superplastic alloys $\sigma/G \approx 10^{-4}$ and $b/d = 10^{-6} - 10^{-5}$, we obtain $n \approx 10^{-2} - 10$; hence, the magnitude of the accumulated deformation $\varepsilon - bn/d \approx 10^{-4}$. For the subsequent occurrence of a deformation by the movement of GBD, it is necessary to "operate" some accommodation mechanism that would ensure resorption of the grain boundary dislocations in the pile-ups.

Apparently, under these conditions, the most effective process of accommodation is the generation of lattice dislocations which can be made substantially facile due to stress concentration at the head of the cluster of GBD. According to *Kurzydlowski* et al. [2.64], this local stress concentration can exceed the magnitude of the applied stresses by more than order of magnitude. (At the same time, at the beginning of deformation, diffusion cannot ensure the required accommodation of the grains during their sliding since the gliding grain boundary dislocations are neither the sources nor the sinks of vacancies [2.2]). The lattice dislocations generated at the grain boundaries of the superplastic materials pass through the grains and "reenter" the boundary where they get absorbed. During the development of the successive processes of generation, movement, and absorption of the lattice dislocations, the last one is the controlling process, and it determines the rate of recovery. However, as was emphasized earlier (Sect. 2.2.2), the formation of highly mobile GBD occurs during the process of absorption of the lattice dislocations which, in turn, leads to the development of the "stimulated" grain boundary sliding. This GBS plays a dual role (not only as a deformation mechanism, but also as a recovery mechanism), because it leads to the disappearance of grain boundary dislocations. Thus, during the initial stage of superplastic flow, the conditions are created for the gradual "operation" of GBS, induced IDS, and its increased role during superplastic deformation.

During the stage of stable deformation, a dynamic equilibrium is established between the rate of strengthening caused by the formation of pile-ups of the gliding grain boundary dislocations and the rate of recovery due to the annihilation of grain boundary dislocations formed during the absorption of the lattice dislocations. In this case, a kinetic correspondence occurs between the lattice defects entering the grain boundaries and those generated by them (Fig. 2.19). On the one hand, accumulation of GBD initiates the generation of the lattice dislocations, and, on the other, during the dissociation of the lattice dislocations trapped by the boundaries, formation of highly mobile GBD accomplishing grain boundary sliding takes place. At the same time the movement of these grain boundary dislocations is related to the generation or absorption of vacancies; therefore, local diffusional fluxes are developed due to the differences in the density and the mobility of the grain boundary dislocations at different grain boundaries, and diffusional accommodation occurs. The most favourable conditions for grain boundary sliding are created when the intragranular dislocation slip can occur along many slip systems.

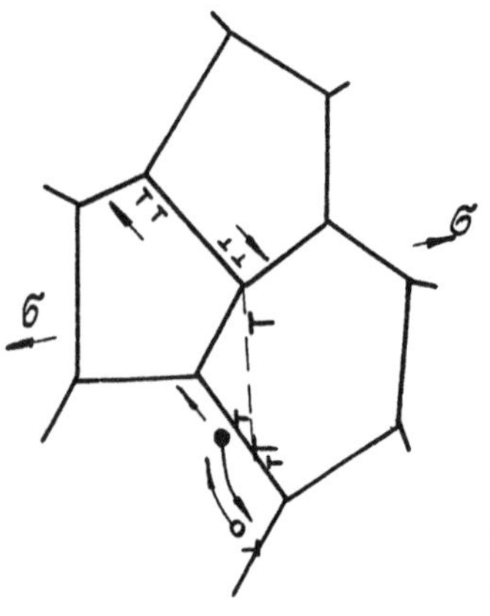

Fig. 2.19. Scheme of the development of the microprocesses of deformation during the stage of stable SP flow. From [2.75]

We note that the microscopic scheme under consideration agrees with the results of studies on the mechanisms of superplastic deformation (Sect. 2.1) according to which grain boundary sliding makes the main contribution to deformation, and intragranular dislocation slip and diffusional creep play essentially an accommodating role but have an effect on the kinetics of GB sliding.

Based on the developed concepts, the equations describing the basic regularities of superplasticity can be obtained.

It is known that the macroscopic deformation of a specimen can be described by the following equation:

$$d\sigma/dt = E^*\left(\dot{\varepsilon}_0 - \dot{\varepsilon}_{\mathrm{p}}\right) , \tag{2.20}$$

where $\dot{\varepsilon}_0$ is the strain rate set by the testing machine (cross-head speed); $\dot{\varepsilon}_{\mathrm{p}}$ is the rate of plastic deformation of the specimen; and E^* is the effective modulus of elasticity of the "specimen-machine" system.

According to the obtained experimental data (Sect. 2.1.4), the magnitude of the plastic strain of the specimens ε_{p} during superplastic deformation consists of three contributions of three mechanisms, viz., grain boundary sliding, intragranular slip, and diffusion creep, i.e.,

$$\varepsilon_{\mathrm{p}} = \varepsilon_{\mathrm{GBS}} + \varepsilon_{\mathrm{IDS}} + \varepsilon_{\mathrm{DC}}$$

and, consequently,

$$\dot{\varepsilon}_{\mathrm{p}} = \dot{\varepsilon}_{\mathrm{GBS}} + \dot{\varepsilon}_{\mathrm{IDS}} + \dot{\varepsilon}_{\mathrm{DC}} . \tag{2.21}$$

Under optimum conditions superplastic deformation (region II) and the contributions of IDS and DC are small; therefore, the rate of plastic deformation can be expressed as

$$\dot{\varepsilon}_{\mathrm{p}} = \dot{\varepsilon}_{\mathrm{GBS}} = K\bar{b}_{\mathrm{b}}\bar{\varrho}_{\mathrm{b}}\bar{v}_{\mathrm{b}} , \tag{2.22}$$

where $\bar{b}_{\mathrm{b}}$ is the average Burgers vector of the grain boundary dislocations; $\bar{v}_{\mathrm{b}}$ and $\bar{\varrho}_{\mathrm{b}}$ are the average rate of movement and volume density of GBD, respectively; and K is the averaging factor that considers the statistical nature of the geometric characteristics (grain sizes and orientations) of the polycrystalline materials.

Within the framework of the given model, the density of the grain boundary dislocations can be obtained using the kinetic equation

$$d\varrho_{\mathrm{b}}/dt = \varrho_{\mathrm{b}}^{+} - \varrho_{\mathrm{b}}/\tau_{\mathrm{b}} , \tag{2.23}$$

where $\varrho_{\mathrm{b}}^{+}(\sigma)$ is the rate of density increase of the generated GBD; τ_{b} is the average life period of GBD that is determined by their annihilation in the triple junctions; and τ_{b} is approximately equal to d/v_{b} (within an order of magnitude). The value of ϱ_{b}^{+} can be evaluated by looking at the relationship

$$\varrho_{\mathrm{b}}^{+} = \xi d(\delta n/\delta t) , \tag{2.24}$$

where ξ is the volume density of the sources of GBD that is approximately equal to $1/d^3$; $\delta n/\delta t$ is the generation rate of GBD; and δn is the number of dislocations generated within a time δt.

In order to compute $\delta n/\delta t$, a generalized force model of the dislocation source is used [2.76] which makes it possible to establish the necessary relationships without specifying the micromechanism of the generation of GBD. According to this model, the source generates dislocations until the effective stresses on the source $\sigma = \sigma - \sigma_{\mathrm{bk}}$ (where σ_{bk} represents the back stresses from the already emitted dislocations) exceed certain critical stress $\sigma_{\mathrm{i}} = \sigma - \sigma_{\mathrm{bk}} \geq \sigma_{\mathrm{i}}$. Since, within a time δt, the emitted dislocations move a distance $\delta L = v_{\mathrm{b}}\delta t$, the source will continuously emit dislocations until the back stresses due to δn dislocations (that are defined as $Gb_{\mathrm{b}}\delta n/\delta L = Gb_{\mathrm{b}}\delta n/v_{\mathrm{b}}\delta t$) are less than or equal to $\sigma - \sigma_{\mathrm{i}}$. Hence, the nucleation rate can be expressed as

$$\delta n/\delta t = (\sigma - \sigma_{\mathrm{i}})v_{\mathrm{b}}/Gb_{\mathrm{b}} \ . \tag{2.25}$$

Substituting (2.24, 25) into (2.23), we obtain the following relationship:

$$d\varrho_{\mathrm{b}}/dt + \varrho_{\mathrm{b}}/\tau_{\mathrm{b}} = (\sigma - \sigma_{\mathrm{i}})v_{\mathrm{b}}/Gbd^2 \ . \tag{2.26}$$

During the stable stage of SP deformation, σ = const, and (2.26) has the following solution:

$$\varrho_{\mathrm{b}}(t) = \frac{(\sigma - \sigma_{\mathrm{i}}}{Gb_{\mathrm{b}}d}\left(1 - \mathrm{e}^{-t/\tau_{\mathrm{b}}}\right) + \varrho_0\,\mathrm{e}^{-t/\tau_{\mathrm{b}}} \ , \tag{2.27}$$

where ϱ_0 is the initial density of grain boundary dislocations.

From (2.26) it follows that at $t = \tau_{\mathrm{b}}$, the density of GBD tends to attain the stationary value

$$\varrho_\infty = (\sigma - \sigma_{\mathrm{i}})/Gb_{\mathrm{b}}d \ . \tag{2.28}$$

In order to describe the macroscopic deformation of the specimen according to (2.20, 22), it is also necessary to present the expression for the rate of movement of GBD. According to experimental studies on bicrystals (Sect. 2.2.2), the movement of GBD is, as a rule, nonconservative and is determined by the diffusional processes; therefore, for v_{b}, one can use the expression describing the climb rate of dislocations having a length $\sim d$ and moving under the action of a stress $(\sigma - \sigma_{\mathrm{i}})$ [2.76]:

$$v_{\mathrm{b}} \approx D\Omega(\sigma - \sigma_{\mathrm{i}})/16dkT \ , \tag{2.29}$$

where D is the coefficient of grain boundary diffusion; and Ω is the vacancy volume.

Equations (2.22, 28, 29) lead to the following relationship for the stage of stable flow:

$$\dot{\varepsilon}_0 = (\sigma - \sigma_{\mathrm{i}})^2\Omega D/16Gd^2kT \tag{2.30}$$

or, in the conventional form, it can be written as

$$\dot{\varepsilon}_0 = \frac{A}{kT}\left(\frac{\sigma - \sigma_i}{G}\right)^2 \left(\frac{b_b}{d}\right)^2 D_0 \exp\left(\frac{Q}{RT}\right) , \quad (2.31)$$

where $A = \Omega G/16b_b^2$; Q is the activation energy for grain boundary diffusion; and, contrary to other models, b_b is the Burgers vector of grain boundary dislocations; and the stress required for superplastic flow at the given rate is given by

$$\sigma = \sigma_i + d\sqrt{16kTG\dot{\varepsilon}/\Omega D} . \quad (2.32)$$

These relationships show that $\dot{\varepsilon}_0 \sim d^2$ or $\sigma \sim d$ which definitely agrees with the experimental data for the optimum strain rates (region II) (Sect. 1.1).

The magnitude of the coefficient of strain rate sensitivity m can be calculated from (2.30)

$$m = \partial \ln \sigma / \partial \ln \dot{\varepsilon} = (\sigma - \sigma_i)/2\sigma . \quad (2.33)$$

It follows from (2.33) that at $\sigma \gg \sigma_i$, the coefficient m tends to 0.5 and at $\sigma \Rightarrow \sigma_i$ (which is a characteristic feature at low strain rates) $m \Rightarrow 0$. Thus, within the limits of the model under consideration, the coefficient of strain rate sensitivity varies in the range $0 \leq m \leq 0.5$ during the transition from region I to region II which agrees with the experimental data obtained on various materials (Sect. 1.1).

As was shown experimentally (Sect. 2.1.4), at high strain rates (region III) the contribution of intragranular dislocation slip to the total deformation increases significantly and intragranular slip begins to play an independent role. Therefore, according to (2.21), $\dot{\varepsilon}_{IDS}$ (with the coefficient $m \ll 0.3$) contributes significantly to the strain rate of the specimen which, evidently, leads to a decreased strain rate sensitivty of the flow stress of superplastic deformation under these conditions.

The usefulness of any model is assessed by its ability to explain experimental data on the variation of mechanical properties and microstructure during plastic deformation. However, the task of a physical theory is not only to explain the specific features of superplastic flow (including their microscopic causes), but also to predict and reveal new effects. Thus, it is important to note that the model being analyzed is based on the results of detailed studies on the characteristic structural changes occurring during SP deformation: the retention of equiaxial nature of grains, the absence of dislocations in the structure and the typical changes of texture. In addition, the model permits one to understand the underlying reasons and to provide for a quantitative description of the phenomena which have not found a satisfactory explanation up to now: accelerated diffusion during the process of SP deformation, the deformation-induced grain growth, the critical grain size ensuring the transition into the SP state, the effect of texture on SP deformation etc.

Structure of Grain Boundaries During Superplastic Deformation and the Reasons Underlying Accelerated Diffusional Processes

Development of grain boundary structural defects during SP deformation forms the main factor in determining the variation of their properties. According to the concepts developed on the process of SP flow, two processes occur simultaneously at the grain boundaries, viz., trapping of lattice dislocations by the grain boundaries and their relaxation associated with the dissociation into grain boundary dislocations. In this case, the surface density of the trapped lattice dislocations can be described by the equation obtained by *Valiev* et al. [1.39]:

$$\varrho = \alpha(\dot{\varepsilon}\tau/b) , \tag{2.34}$$

where α is a constant that considers the geometry of intragranular slip, $\alpha \approx 1$; τ is the TLD relaxation time; and b is the Burgers vector of TLD, $b \approx 2b_{\mathrm{LD}}$ [2.58].

During SP deformation, the duration τ does not exceed but a few seconds (considering the acceleration due to deformation). Substituting a value of $\tau \approx 1$ s into (2.34), we obtain a value of $\varrho \approx 10^5\,\mathrm{cm}^{-1}$ at $\dot{\varepsilon} \approx 8 \times 10^{-3}\mathrm{s}^{-1}$ which approximates the data obtained according to the "instantaneous" quenching method [1.23, 2.29]. An important fact is that the density of TLD does not depend on the degree of deformation but increases rapidly with increasing strain rate, and this agrees with the experimental results.

It is more difficult to evaluate the density of GBD because the mechanism of their annihilation in the grain boundaries is not completely clear. Based on a consideration of mutual annihilation of GBD, we obtained [1.39] a density value $\varrho_{\mathrm{b}} = 2 \times 10^5\,\mathrm{cm}^{-1}$ which also showed some increase with increasing strain rate. Unfortunately, it is not possible to compare these estimates with the experiments since, as was mentioned earlier, the resolution of an electron microscope is insufficient for observing GBD at the random boundaries. Nevertheless, the conducted evaluations are important for analyzing the properties of grain boundaries during SP deformation. In fact, the presence of defects (trapped lattice dislocations and products of their dissociation, i.e., grain boundary dislocations) in grain boundaries ensures a significant increase of the energy of the boundaries (Sect. 2.2.2). Using (2.18), we evaluated the magnitude of the energy increase ΔE, for example, in the magnesium alloy MA8 [1.39]. Substituting the characteristic values of the alloy $G = 3 \times 10^6\,\mathrm{N/cm}^2$, $\nu = 0.3$, $h = 1/\varrho$, and $h_{\mathrm{b}} = 1/\varrho_{\mathrm{b}}$ in (2.18), we obtain $\Delta E \approx 17 \times 10^{-6}\,\mathrm{J/cm}$, i.e., the values of ΔE reach approximately 30% of the initial boundary energy.

Such a change in the energy state of the boundaries must lead to an abrupt acceleration of the kinetic processes (including the diffusional processes) in them. According to our analysis [1.39] (2.19), the increase of the coefficient of grain boundary diffusion in a nonequilibrium boundary (as compared to that in an equilibrium boundary) is determined by the relationship $D_{\mathrm{b}}/D_{\mathrm{b}}^0 = \exp(\lambda d^2 \Delta E/kT)$. Hence, at $\Delta E = 17 \times 10^{-6}\,\mathrm{J/cm}^2$, $\lambda^2 d \approx 10^{-15}\,\mathrm{cm}^2$, $kT = 8 \times 10^{-21}$ J, the value of $D_{\mathrm{b}}/D_{\mathrm{b}}^0$ is approximately 10.

Since $\Delta E = f(\dot{\varepsilon})$, it can be expected that during superplastic deformation the diffusion coefficient would increase by 1–1.5 orders of magnitude, depending on the strain rate. The agreement between these results and the experimental data on the acceleration of diffusion during SP deformation [1.3, 2.52] permits one to assume that the changes occurring in the structure and the energy state of the grain boundaries due to their interaction with the lattice dislocations form the main reason underlying the changes occurring in the kinetics of the diffusional processes under the conditions of superplastic flow. In fact, [1.39] gives an analysis of other possible causes of accelerated diffusion (generation of vacancies by the steps of the moving screw grain boundary dislocations and vacancy generation during the climb of GBD showed that their role is insignificant).

Grain Growth and the Strain Rate Dependence of the Contribution of Grain Boundary Sliding

As was mentioned in Sect. 1.2, SP deformation is characterized by grain growth and is due to the deformation-induced grain boundary migration. The phenomenon can be explained on the basis of the present model. In this case, for analyzing the kinetics of grain growth, one can use the concepts concerning the effect of nonequilibrium structure of the boundaries on their migration characteristics (Sect. 2.2.2). According to these concepts, the migration rate $\partial d/\partial t$ is determined by the driving force of the process F and the mobility of the boundaries $M : \partial d/\partial t = MF$ in that nonequilibrium mainly affects the magnitude of mobility which depends on the kinetic parameters of grain boundary diffusion. In order to obtain a quantitative evaluation of grain growth, we shall take a simplified approach as described in [1.36, 2.5].

Thus, when other factors are held constant, the mobility of grain boundaries is determined by the vacancy concentration C_v, under the conditions of thermodynamic equilibrium:

$$M = KC_v \, , \tag{2.35}$$

where K is a constant depending on the type of grain boundaries. At the same time, the parameter $F \approx 1/d$; hence,

$$d^2 - d_0^2 = 2KC_v\tau \, , \tag{2.36}$$

where τ is the duration of deformation (of annealing).

Let us evaluate the parameters of the Zn-0.4%Al alloy that exhibits superplasticity at room temperature (Sect. 1.2). Under these conditions the equilibrium vacancy concentration $C_0 = 10^{-6}$ [1.3, 36]. From the condition of grain growth during annealing, the magnitude of K was determined to be $5.2\,\mu\text{m}^2/\text{s}$. Using (2.18), we could determine the values of C_v during superplastic deformation. They were found to approximate the experimental data [1.3] and amounted to 2×10^{-5}, 9×10^{-5}, and 1.4×10^{-4} at the strain rates 4×10^{-5}, 4×10^{-4}, and $4 \times 10^{-3}\,\text{s}^{-1}$, respectively. Based on these values and using (2.36), we evaluate the grain size of the alloy at different strain rates.

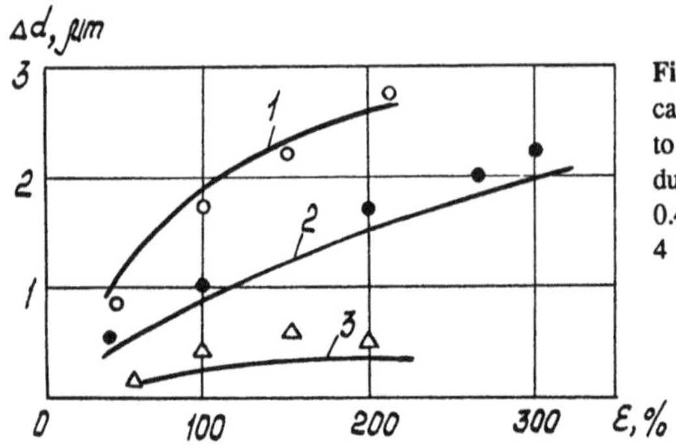

Fig. 2.20. Experimental values and the calculated plots of grain growth Δd due to deformation at different strain rates $\dot{\varepsilon}$ during SP tensile deformation of the Zn-0.4%Al alloy; (*1*) $\dot{\varepsilon} = 4 \times 10^{-5}\,s^{-1}$; (2) $4 \times 10^{-4}\,s^{-1}$; (*3*) $4 \times 10^{-3}\,s^{-1}$

Figure 2.20 shows the experimental values of Δd (Sect. 1.2) and the $\Delta d = f(\varepsilon)$ curves obtained from (2.36). It can be seen that the experimental and the calculated values at different strain levels agree satisfactorily with each other. Thus, it can be concluded that the grain growth occurring during SP deformation is a direct consequence of the changes taking place in the structure of the boundaries as a result of grain boundary absorption of the lattice dislocations, i.e., the basic concept of the present model.

There is yet another important structural effect which can be explained on the basis of the theory, viz., the experimental dependence of the contribution of grain boundary sliding towards the total deformation. The contribution of GBS is maximum at the optimum rates of SP deformation (region III) and decreases on increasing or decreasing the strain rate (region III and I, respectively) (Sect. 2.1). This effect has been reliably established experimentally and is a characteristic of different superplastic materials. Its explanation must be associated with the analysis of the effect of strain rate on the rate of GBS that is determined by the behaviour of the grain boundary dislocations. In fact, according to (2.22), the rate of GBS is proportional to the density of GBD and the rate of their movement which are determined by (2.28, 29) for region II. At present, the specific form of these relationships for regions I and II is not completely clear; consequently, it is not possible to give a quantitative description of the strain rate dependence of the contribution of GBS. However, for a qualitative evaluation of this aspect, it is extremely important to note that in region III the rate of movement of GBD is less than the value obtained from (2.31), because of the appearance of a significant density of unrelaxed lattice dislocations in the boundaries. At the same time, in region I, as a result of the increased role of the recovery processes in the grain boundaries (for example, because of rapid migration [2.60, 61]), the value of the density of GBD must be less than that given by (2.28). These effects lead to a decrease in the magnitude of the relative rate of GB sliding: $\dot{\varepsilon}_{GBS}/\dot{\varepsilon}_{tot}$ in regions I and III and, consequently, a decreased contribution of grain boundary sliding towards the total deformation.

Grain Size Corresponding to the Transition into the Superplastic Condition

Considering the behaviour of the grain boundaries during the process of SP deformation, it is possible to carry out a quantitative evaluation of the magnitude of the critical grain size at which the transition from the usual deformation into superplastic flow occurs. A satisfactory explanation of this aspect has not yet been given.

It is known that during the usual deformation of polycrystalline materials, the grain size dependence of the flow stress is conditioned by the transfer of strain from grain to grain and is determined by the Hall-Petch relationship

$$\sigma_{\mathrm{d}} = \sigma_0 + K d^{-1/2} , \tag{2.37}$$

where K and σ_0 are material constants.

During superplastic deformation, grain boundary sliding forms the main deformation mechanism; therefore, the conditions of strain transmission from grain to grain undergo changes and the Hall-Petch relationship becomes invalid in this case: $\sigma \sim d$ (2.32). The transition from the usual deformation to superplastic deformation is observed when the following equation is obeyed:

$$\sigma_{\mathrm{d}}(d_{\mathrm{cr}}) = \sigma(d_{\mathrm{cr}}) . \tag{2.38}$$

Using (2.32, 37), at $\sigma_{\mathrm{d}} \gg \sigma_0$, the following is obtained for determining d_{cr}:

$$d_{\mathrm{cr}} \approx \sqrt{K D \Omega / 16 \dot{\varepsilon} G k T} . \tag{2.39}$$

Equation (2.39) can be used for evaluating d_{cr}, for example, in the case of the Zn-0.4%Al alloy (Fig. 1.2), where $K = (2.1 \div 3.2) \times 10^{-4}\,\mathrm{N/cm^2}$ [1.3]. Considering the acceleration of diffusion, it is found that the coefficient of grain boundary diffusion during SP deformation $D = 2 \times 10^{-11} \div 2 \times 10^{-10}\,\mathrm{cm^2/s}$. Substituting the remaining values in (2.39), we note that d_{cr} would remain in the 8–18 μm range which is in satisfactory agreement with the experimental data of the Zn-0.4%Al alloy (Fig. 1.2). Such an agreement between the calculated and the experimental values forms yet another important argument in justifying the concepts treating the process of grain boundary sliding induced by intragranular slip as the controlling mechanism of superplastic deformation.

Inelasticity and the Initial Stage of Superplastic Deformation

The concepts being developed can be used not only for the purpose of a formal explanation of the strain rate dependence of the mechanical properties, i.e., the $\dot{\varepsilon} = f(\sigma)$ curve of superplasticity (2.31), but also for analyzing the true stress-strain curves which have not received enough attention in the literature.

As was mentioned in Sect. 1.1, two regions (stages) can be identified in the $\sigma - \varepsilon$ plots, viz., initial deformation and stable flow. According to the present approach, a study on the first stage is important for explaining the "switching on" processes responsible for superplastic flow, and thus of practical interest. As an

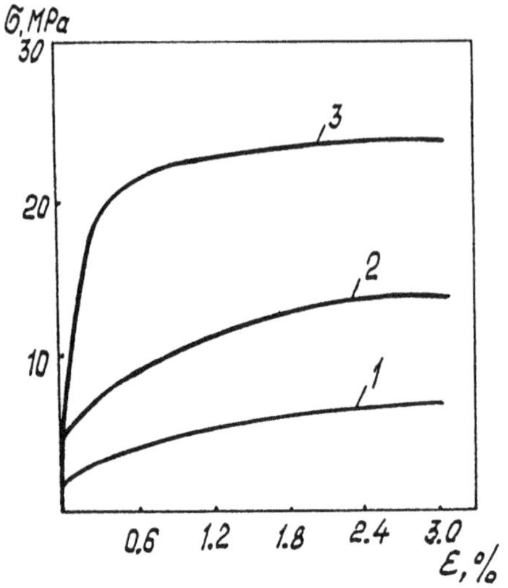

Fig. 2.21. Initial portions of the flow stress vs. strain ($\sigma - \varepsilon$) curves of the Sn-5%Bi alloy at t_{def} = 20°C and different (*1*) $\dot{\varepsilon}$ = $3 \times 10^{-5}\,\mathrm{s}^{-1}$; (2) $3 \times 10^{-4}\,\mathrm{s}^{-1}$; (*3*) $3 \times 10^{-3}\,\mathrm{s}^{-1}$. From [2.77]

example, the initial segments of the $\sigma - \varepsilon$ plots for Sn-5%Bi alloy are presented in Fig. 2.21 [2.77]. It can be seen from these curves that, during superplastic flow, the linear region characterizing elastic deformation is virtually absent and plastic flow of the alloy starts at very low stresses and leads to a relatively rapid increase of σ. The slope of the initial segment increases with increasing strain rate.

Similar effects were observed in other superplastic materials [2.78]. In addition, careful studies on the mechanical behaviour during the initial stage of superplastic deformation showed [2.77] not only strain hardening, but also a significant nonlinear elastic deformation in this case. This inelasticity is manifested in the formation of a linear portion in the $\sigma - \varepsilon$ curve after a short-term hold and subsequent deformation in that the modulus of nonlinear elasticity E_{c} can be evaluated from the slope of this portion. It was found that the magnitude of this modulus increases on increasing the strain rate and that its value approxiamtes that of the grain boundary modulus. Thus, in the Sn-5%Bi alloy, the values of E_{c} are equal to 3.1×10^3, 3.8×10^3, and 4.5×10^3 MPa at the strain rates 3×10^{-5}, 3×10^{-4}, and $3 \times 10^{-3}\,\mathrm{s}^{-1}$, respectively. The observed values of E_{c} are considerably less than the modulus of elasticity of a single crystal of tin (equal to 1.9×10^4 MPa), but approximate the value of the grain boundary modulus which is less by an order of magnitude. These data confirm the conclusion [2.78] that the elastic behaviour observed during SP deformation is probably determined by the elastic effects occurring at the grain boundaries. This conclusion agrees with the concepts being developed in the present model according to which superplastic flow starts with the generation and the movement of grain boundary dislocations. In addition, the model makes it possible to explain the appearance of the linear segment in the $\sigma - \varepsilon$ curve during a sudden cessation and subsequent deformation of the specimen and, also, the strain rate dependence of the modulus E_{c}.

As was mentioned, superplastic deformation occurs at a certain stationary grain boundary dislocation density corresponding to the kinetic equilibrium between GBD formed during the interaction of the dislocations with the grain boundaries and their recovery due to the development of grain boundary sliding induced by intragranular dislocation slip. The elastic properties of the boundaries increase in the presence of grain boundary dislocations in them [2.58]. During abrupt unloading of the specimen, an excess density of GBD is fixed in the structure which hinders the development of "pure" GB sliding during subsequent loading and, in turn, leads to the formation of the linear segment on the $\sigma - \varepsilon$ curve.

The density of grain boundary dislocations increases with increasing strain rate (2.34). This fact can be used to explain the subsequent increase of the modulus E_c with increasing strain rate.

According to the present model, the duration of the region of transition into the stage of stable flow is determined by the time taken by the parameters ϱ_b and V_b to attain their steady-state values which, as follows from (2.27, 29), are related to time τ_b and the initial density of the grain boundary dislocations ϱ_0. At $\sigma \gg \sigma - \sigma_i$, one can evaluate the nature of deformation in the transient region.

Let us rewrite (2.20) in the following form:

$$dQ/d\varepsilon_t = E^*\left(1 - \dot{\varepsilon}_p/\dot{\varepsilon}_t\right) . \tag{2.40}$$

Using (2.22, 27, 29), we obtain

$$\dot{\varepsilon}_p = \dot{\varepsilon}_t\left(1 - e^{-t/\tau_b}\right) = \dot{\varepsilon}_t\left(1 - e^{-\varepsilon_t/\varepsilon_i}\right) , \tag{2.41}$$

where $\varepsilon_i = \dot{\varepsilon}_t \cdot \tau_b$ is the magnitude of the initial stage.

Substituting (2.41) into (2.40), we obtain the equation of the deformation curve $\sigma(\varepsilon)$ in the transitional region

$$d\sigma/d\varepsilon_t = E^* e^{-\varepsilon_t/\varepsilon_i} . \tag{2.42}$$

The solution of (2.42) is as follows:

$$\sigma(\varepsilon) = \sigma_f\left(1 - e^{-\varepsilon_t/\varepsilon_i}\right) , \tag{2.43}$$

where ε_t is the strain of the specimen; and σ_f is the flow stress during the stable stage.

Let us compare the obtained results with the experimental data from the magnitude of the first stage. Taking the typical values of superplastic deformation, viz., $d \leq 10\,\mu m$, $V_b \leq 10^{-6}\,cm/s$, and $\dot{\varepsilon}_t = 10^{-4} \div 10^{-3}\,s^{-1}$, it can be seen that $\varepsilon_i = \dot{\varepsilon}_t\tau_b \approx 10^{-1} \div 10^{-2}$, i.e., 1–10%, that agrees with the experimental data within an order of magnitude (Sect. 1.1).

Effect of the Conditions of Deformation on Superplastic Deformation

The present theory makes it possible to explain a number of specific features of the stress-strain curve, i.e., deformation when varying the conditions of deformation, in particular, the decrease of (superplastic) flow stress after short-term deformation at increased strain rates and the effect of pre-straining on superplastic deformation.

According to the described model, the parameters of SP deformation can be varied by changing the stationary (steady-state) density of grain boundary dislocations which, in turn, depends on the deformation conditions. For example, after short-term deformation at a high strain rate $\dot{\varepsilon}_1(\dot{\varepsilon}_2 > \dot{\varepsilon}_1)$, an increased density of GBD must exist at the grain boundaries which ensures a higher rate of grain boundary sliding (2.22). Under the conditions of the given strain rate $\dot{\varepsilon}_1$, the change in the rate of sliding leads to a decrease in the flow stress. This assumption was recently verified experimentally in a number of materials [2.79, 80]. Figure 2.22 shows the scheme of the conducted experiments. It was established that after short-time deformation at an increased strain rate $\dot{\varepsilon}_2$, there is a decrease in the flow stress during subsequent deformation at a strain rate $\dot{\varepsilon}_1$ (compare points C and E). The magnitude of this effect can be characterized by two parameters, viz., the σ_E/σ_C ratio (Fig. 2.22 for notations) and the recovery time τ_r which was defined as $(\varepsilon_D - \varepsilon_E)/\dot{\varepsilon}_1$. It was found that the stress reduction is characteristic only for alloys existing in the superplastic condition, and the magnitude of this effect depends on the deformation conditions, the range of the varied strain rates, and the degree of deformation. In the Sn-5%Bi alloy we observed nearly equal values of $\sigma_E/\sigma_C \approx 1.18$ when the strain rate is changed by an order of magnitude in regions I and II; but the effect decreased in region III. The magnitude of the effect increases with increasing range of the varied strain rates (by 2–3 order of magnitude), where the strain rate $\dot{\varepsilon}_2$ corresponds to region III and the strain rate $\dot{\varepsilon}_1$ corresponds to region II or I.

The structural studies conducted on the MA8 alloy showed that the grain size does not change after short-term deformation at a high strain rate. At the same time, one observes a sharp increase in the density of the lattice dislocations which are present mainly at the grain boundaries. Since these dislocations form

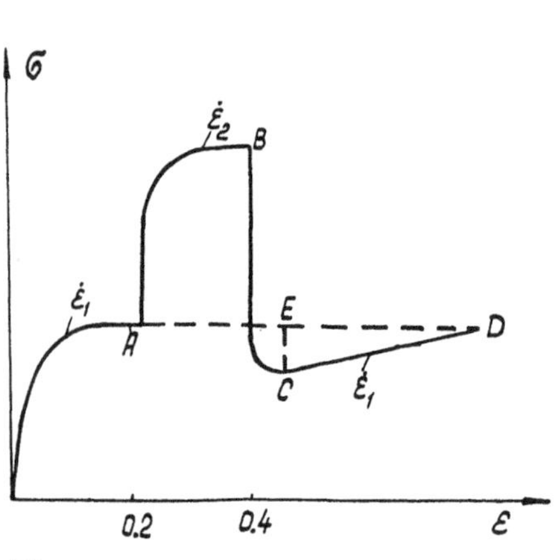

Fig. 2.22. Scheme of the flow stress vs. strain ($\sigma - \varepsilon$) curve of SP alloys during an abrupt change of the strain rate. From [2.80]

(due to their dissociation) the direct cause for the appearance of highly mobile grain boundary dislocations, the obtained results can be considered as a direct confirmation of the concepts concerning the decisive role of the grain boundary defects in realizing grain boundary sliding during superplastic deformation.

Additional introduction of GBD into the structure of the grain boundaries can be accomplished not only by deforming at high strain rates, but also by pre-straining. In fact, if the specimens are subjected to cold working, accumulation of lattice dislocations occurs in the structure. They will be present at the boundaries as well as within the grains. When heating up to the temperature of superplastic deformation and during the process of superplastic flow, the lattice dislocations trapped by the grain boundaries will be absorbed (by dissociating into grain boundary dislocations). Thus, an increase in the density of mobile grain boundary dislocations must lead to accelerated grain boundary sliding and, consequently, to a decreased flow stress. In order to verify these assumptions, we conducted the following experiment: specimens of the MA8 alloy were subjected to 3–5% pre-straining at room temperature. Figure 2.23 shows the stress-strain curves under

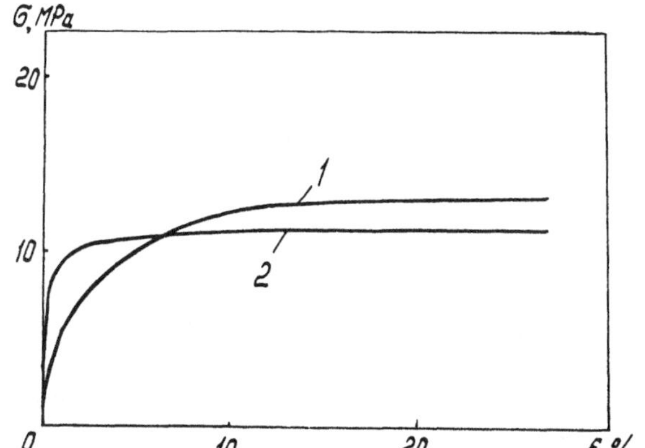

Fig. 2.23. Initial portions of the flow stress vs. strain ($\sigma - \varepsilon$) curves of the MA8 alloy during SP deformation in the original state obtained by recrystallization annealing (*1*) and after 5% cold pre-straining (2). From [2.47]

the conditions of SP deformation of the alloy in different states, i.e., without pre-straining (1) and with additional 5% cold deformation (2). A comparison of these relationships indicates the development of two phenomena associated with the effect of pre-straining: a more rapid hardening during the initial stage and a decrease in the stress of superplastic flow during the stage of stable deformation. It is important that in this case, the grain size after SP deformation of the alloy is virtually identical in both states. Electron microscopy showed that pre-straining led to the formation of dislocations in the structure, but in the alloy having an ultrafine-grained structure, they were located mainly at the grain boudaries. On heating up to the temperature of SP deformation, the lattice dislocations trapped by the grain boundaries showed relaxation which is apparently due to their dissociation into GBD (Sect. 2.2.2). In addition, the presence of GBD not only makes grain boundary sliding easy, but also leads to an increased nonlinear elasticity. Evidently, this fact explains the opposite effect of cold work on the magnitude of the stresses during the initial and the stable stages of superplastic deformation.

Effect of Texture

As was shown in Sect. 1.1, crystallographic texture has an unusual effect on the superplastic behaviour. The location of the optimum range of SP deformation and the flow stress depend on the type of texture, and a significant anisotropy of ductility is observed. The following explanation of these results can be given in the framework of the concepts developed herein.

Texture affects both the development of IDS and the grain boundary structure. In both cases texture influences the development of GBS. If the orientation is favourable for slip, generation of lattice dislocations will be easily generated facilitating the development of GBS (which is the main mechanism of superplastic flow); the flow stress must decrease in this case. In the opposite case, when the orientation is unfavourable for slip, the generation of dislocations is hindered and the flow stress must increase. In order to verify theoretical conclusions, a special experiment [2.79] was carried out. In a magnesium alloy MA15 (Mg-3.1%Zn-1.6%Cd-0.45%La-0.6%Zr) with a grain size of 9 μm, a fibrous texture with a (0001) basal plane perpendicular to the tensile axis was developed by means of a preliminary treatment. At t = 450°C in the longitudinal and transverse directions, the alloy exhibits flow stresses $\sigma_{\|} = 16.5$ MPa and $\sigma_{\perp} = 10.7$ MPa, respectively, i.e., the coefficient of the flow stress anisotropy is $R_\sigma = 1.54$. The relative elongations are $\delta_{\|} = 480\%$ in the longitudinal direction and $\delta_{\perp} = 280\%$ in the transverse direction. The optimum strain rate shifts slightly if at all towards higher strain rates in the transversely cut samples as compared with the longitudinally cut samples. Thus, the effect of texture on SP flow is explicit in the alloy. These experimental data can be explained only if we assume that flow stress during SP deformation is controlled by the development of "stimulated" GBS. In fact, an orientation unfavourable for IDS impedes the generation of lattice dislocations and, consequently, hinders the movement of GBD. This situation must lead to an increase in the value of σ_i according to (2.32) and to an increase in the flow stress. However, the level of flow stress can be decreased if we somehow succeed in introducing lattice dislocations into grain boundaries. Specifically, this can be achieved by a short-term deformation at a high strain rate (in region III). The results of experimental verification are shown in Fig. 2.24. It can be seen that the change of strain rate led to a decrease in the flow stress of the longitudinal samples down to that of the transverse samples, i.e., it was directly shown that pre-straining at a high strain rate (generation of an excess density lattice dislocation) causes a sharp decrease in the flow stress anisotropy. This experiment can be considered not only as evidence of the validity of the reasoning offered to explain the effect of texture on superplastic flow which follows from the theory, but also as a confirmation that "stimulated" GBS is a

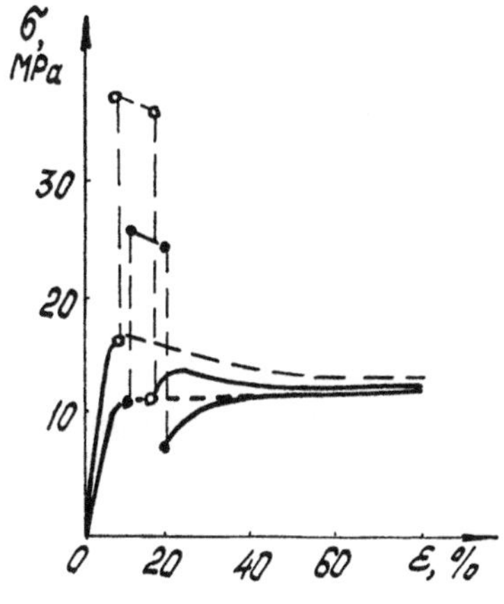

Fig. 2.24. Stress-strain curves of the magnesium-based MA15 alloy at a stepped change in the strain rate at t = 450°C. Adopted from [2.79]; o – longitudinal samples; • – transversal samples; (—) loading and behaviour after switching the strain rate; (– – –) switching of the strain rate; (- - -) trend of curves after deformation without switching the strain rate

deformation mechanism controlling strain rate during SP flow (a basic principle underlying the model).

However, in a general case, crystallographic texture not only affects the development of intragranular dislocation slip, but also affects the structure of the grain boundaries. In the presence of a specific texture in the alloy it is possible to increase the role of the special grain boundaries. In the general case this situation must be taken into account when analyzing the effect of texture on SP deformation.

Thus, the theory of superplastic deformation developed by taking into account the interaction of lattice dislocations with grain boundaries during deformation makes it possible not only to describe a number of most important regularities in the variation of the mechanical properties and the structural changes that are characteristic of this phenomenon but also to predict new effects.

3. Microcrystalline Materials and Microstructure Refining

3.1 Structure and Properties of Microcrystalline Materials

SP deformation is a result of the macroscopic influence of grain boundaries on the properties of polycrystalline materials. Hence, it can be assumed that the refining of microstructure is accompanied by both quantitative and, probably, qualitative changes in the structure of these materials.

Indeed, as shown in previous chapters, the structure of grain boundaries considerably differs from that of the grain interior. In coarse-grained materials the fraction of atoms in the boundary is extremely small as compared to the total amount of atoms, and their influence on the properties is insignificant. However, the situation radically changes with the refinement of microstructure and acquisition of grains of a certain size. As the grain boundary area increases, the fraction of the boundary atoms continuously occupy more and more space in the microstructure. In fact, in such materials, boundaries can regarded as a phase with different structure and properties, i.e., a grain boundary phase [3.1].

Experimentally, the change in the state of atoms near grain boundaries is already explicit in materials with submicrocrystalline structures. In iron with a grain size of 0.22 μm, Mössbauer spectroscopy revealed the superposition of two decomposed magnetic subspectra with a considerable difference in the hyperfine structure parameters (Fig. 3.1, Table 3.1) [3.2]. Figure 3.1 shows that the difference in the intensities and hyperfine magnetic fields and a uniform shift of

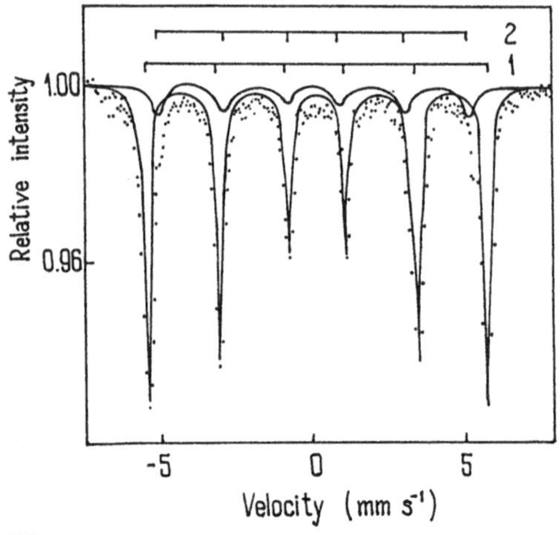

Fig. 3.1. Mössbauer spectrum of submicron-grained iron at room temperature: the experimental spectrum is indicated by *points*. *Continuous lines* represent two subspectra obtained as a result of decomposition

Table 3.1. Parameters of the hyperfine structure of subspectra 1 and 2 in Fig. 3.1

Parameter	Subspectrum 1	Subspectrum 2
Relative integral intensity of subspectra	0.89 ± 0.02	0.11 ± 0.02
Hyperfine magnetic field H_{ef} [kOe]	329 ± 5	299 ± 5
Isomeric shift with respect to the reference α – Fe, δ [mm/s]	0.00 ± 0.03	-0.25 ± 0.03
Band width at half maximum [mm/s]	0.33 ± 0.05	0.54 ± 0.05

subspectra can be easily discerned. Analysis indicates that the parameters of the more intensive subspectrum coincide with those observed in coarse-grained iron. The second subspectrum is associated with a specific state of atoms near the boundary. From Table 3.1 it follows that the fraction of such atoms is $11 \pm 2\%$. Assuming that the atomic density of both the phases is identical, the thickness of the near-boundary layer is 8.4 ± 1.5 nm.

Taking into account the latter results, *Valiev* and *Mulyukov* calculated by means of computer simulation the amount of grain boundary phase with different boundary thicknesses (Fig. 3.2). The following thicknesses were used: 1 nm, corresponding to the crystallographic thickness of a boundary in a conventional material, and 10 nm, a value which is very similar to that obtained in the above described experiment.

It can be seen that even with grain sizes of a few micrometers and decimals of a micrometer, the fraction of grain boundary phase can constitute several percent and when the grain size is a few hundreds of a micrometer, it is tens of percent. At a grain size of 6nm the fraction of boundary phase becomes 50%.

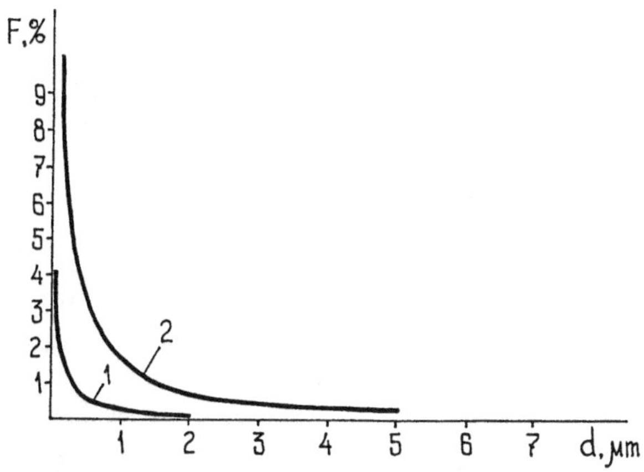

Fig. 3.2. Dependence of the quantity of grain boundary phase (F) on the size of grains with different thickness of the boundary: Curve (*1*) 1 nm; (*2*) 10 nm

In such a nanocrystalline material, the different structure of the near-boundary areas is reflected by an abrupt change in the phonon spectrum and in a number of physical and mechanical properties [3.1].

Thus, starting with a grain size of about 1 μm or less, the physical state of materials substantially changes because of the effect of grain boundaries on the position of atoms in the boundary area. Such a material consists of two phases: crystalline and grain boundary. With the microstructure refinement, the fraction of the grain boundary area abruptly grows, and it begins to exert a certain influence on the properties of polycrystals.

Unfortunately, at present, there are no systematic data on the complex of mechanical properties of materials with small grain sizes. It would be interesting to study the effect of grain refinement on both the mechanical properties at room temperature and their variation with the temperature and rate of deformation. Table 3.2 presents data on the mechanical properties of a number of materials when submicrocrystalline and nanocrystalline structures are created in them [3.3–7]. Though these data hardly lead to any clear conclusions, they point to the most important effects accompanying the microstructure refinement. Table 3.2 shows that the reduction of grain size abruptly enhances the material strength. In copper, nickel alloy, and steels with submicron grains, the strength is twice as much, and

Table 3.2. Effect of grain size on the mechanical properties of materials

Materials	Grain size [μm]	t[°C]	σ_t [MPa]	$\sigma_{0.2}$ [MPa]	δ [%]	References
Cu	100/0.2	20	200/410	140/400	64/71	a
Ti	150/0.4	20	350/510	–	30/33	b
	150/0.4	550	180/90	–	30/190	b
RSR Rene80	100/0.3	20	375/850	–	–	[3.3]
Al-Mg-Li-Sc-Zr	10/0.2	20	450/600	330/550	5/6	[3.4]
Mg	20/0.2	20			18/80	[3.5]
Fe-0.1%C-0.51%W -0.17%Zr	15/0.3	20	–	245/705	35/20	[3.6]
Fe-1.8%C	crystal/ nanocrystal	20	5600/67200			[3.7]
Fe-18%Cr-10%Ni -0.5%Ti-0.12%C	40/0.2	20	650/1180	250/970	50/22	c
Fe-25%Cr-0.2%Ti -0.13%C	50/0.2	20	485/665	300/600	26/17	d
Ti-38.5%Al	30/0.4	700	365/730	–	3.3/130	e

[a] R.Z. Valiev, R.R. Mulyukov, and N.A. Akhmadeev
[b] A. Salishchev and O.R. Valiakhmetov
[c] K.G. Farkhutdinov and R.G. Zaripova
[d] A. Salishchev and V.D. Afanasyev
[e] A. Salishchev, R.M. Imayev, and V.M. Imayev

[(a–e): Institute for Metals Superplasticity Problems, USSR Academy of Sciences, UFA USSR] in the nanocrystalline moustache of the Fe-1.8%C alloy it increases twelve times. A considerable increase in strength is also observed in the aluminium alloy and titanium.

It is interesting that, contrary to substructural hardening, during grain boundary hardening ductility in certain materials remains unchanged or even increases in spite of an abrupt increase of strength. However, data on room temperature ductility are rather contradictory. When the microstructure is refined, the ductility of cooper, titanium, aluminium alloy, and magnesium increases slightly; while in stells, it becomes lower (Table 3.2). This is probably due to the effect of the second phase particles on the ductility of steels. In austenitic steel it is the effect produced by the grain size on the subsequent transformation during deformation. Nevertheless, there are data [3.8] on the increasing lamellation at 70–180°C even in TiO_2 ceramics and in CaF_2 salt, when nanocrystalline structures are created.

Different ductilities of the fine and coarse grained states are well-manifested when the test temperature rises. Thus, for example, at 550°C, in titanium, when the grain size decreases from 150 to 0.4 μm, the ductility becomes six times as high. The effect of the microstructure refinement on ductility is especially explicit in intermetallics and ceramics. In the TiAl intermetallic compound at 700°C ductility increases 30 times when the strength doubles (Table 3.2).

Generally, the latter fact is associated with the influence of the microstructure refinement on the temperature of ductile-brittle transition (Chaps. 9, 10). The temperature of ductile-brittle transition becomes lower when the grains are refined [3.9]. However, the effect considerably increases when microcrystalline or still finer structures are created in materials.

Thus, in a steel containing 0.45%C, the refining of the microstructure from 6 to 0.8 μm decreases the temperature of ductile-brittle transition by 60°C. In this case, the impact strength at −40°C practically doubles, increasing from 35 to 60 kJ/cm^2. In the TiAl intermetallic compound the refinement of microstructure from 30 to 0.4 μm decreases the temperature of ductile-brittle transition by 200°C. Under the temperature-rate conditions of SP deformation, a transition to SP flow is observed in both intermetallics and ceramics (Chaps. 9, 10).

The microstructure refinement exerts considerable influence on the internal friction and, hence, the damping properties of materials. Thus, in copper at room temperature, the refining of the grain size from 1 to 0.2 μm makes the value of the logarithmic decrement of damping increase from 30 to 120 units (Fig. 3.3) (R.R. *Mulyukov* and N.A. *Akhmadeev*; private communication). Thus, the refinement of microstructure strongly affects the complex of mechanical properties and, under certain testing conditions, induces an abrupt increase of strength, ductility, and viscosity. What causes a change in the properties?

The increase of strength with the diminishing grain size is usually accounted for by the grain boundary hardening of material based on the Hall-Petch relationship. Indeed, when the grain size is of an order of a few microns or larger, the variation of the yield stress agrees with the variation of the grain size [3.3, 6, 10]. However, when the grain size is smaller, the data on the dependence $\sigma_{0.2} - d^{-1/2}$ are contradictory. Many papers report [3.3, 6, 10] a deviation from this depen-

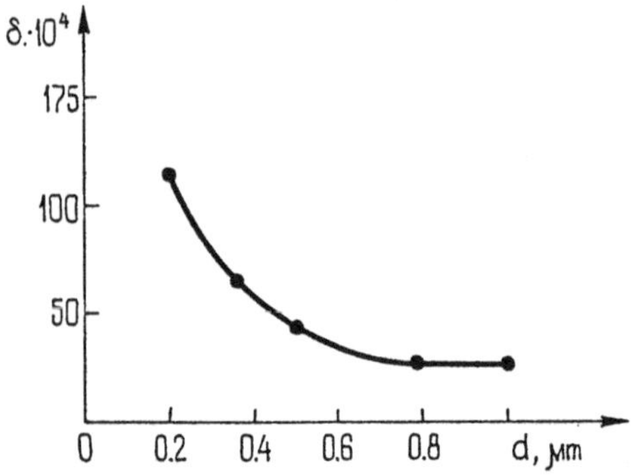

Fig. 3.3. Dependence of the logarithmic decrement damping factor δ of copper on the grain size at 30 Hz

dence, occurring when the grain size is a few decimals of a micrometer. For example, Fig. 3.4 shows that starting with a certain grain size, there is practically no growth of the yield stress in ferritic steel (G.A. *Salishchev* and V.D. *Afanasyev*; private communication). In [3.11] the decrease of strength with the grain refining in the nanocrystallites of copper and palladium is also noted. At the same time, in [3.3, 12], continuous hardening was observed during the grain refining.

At present, there is no satisfactory explanation of the nature of the Hall-Petch relationship. There are at least three main factors contributing to grain boundary hardening. The first one is related to the necessity of increasing the actual stresses to enable the transference of deformation from grain to grain because of the hopping change of orientation while crossing the grain boundary. The second factor, usually associated with the material hardening with the diminishing grain

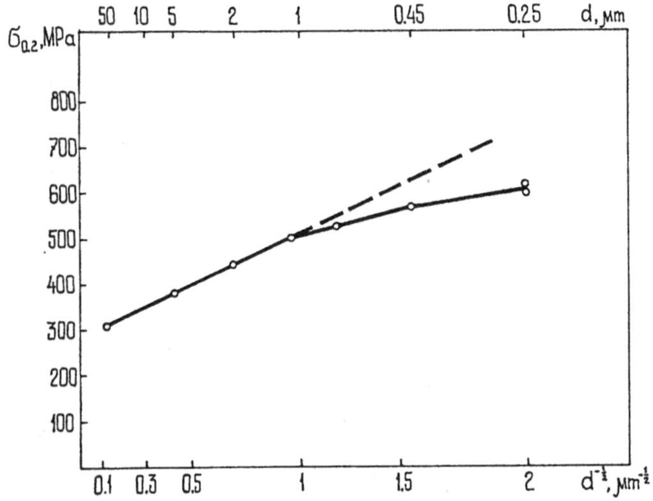

Fig. 3.4. Dependence of yield stress of ferritic Fe-25%Cr-0.2%Ti-0.13%C steel on the grain size at room temperature. The *dashed line* indicates the calculated values of σ, while the *solid line* corresponds to experimentally obtained results

size, is the formation of dislocation pile-ups near the grain boundaries. However, in [3.13], no relation between the yield stress and pile-ups was reported. The observation of the initial stage of plastic flow indicated that lattice dislocations are first trapped by the boundary; then they are accumulated in it; and only then are dislocation pile-ups formed. Due to trapped lattice dislocations, the grain boundaries are surrounded with the fields of long-range stresses which, in their turn, slow down dislocations in the near-boundary area. The third factor inducing the growth of the flow stress with the diminishing grain size is the difficulty of the nucleation of dislocations in the grain interior. This is due to the fact that at small grain sizes the minimum size of the source of dislocations of the Frank-Reed type is equal to the grain size. In this case, it is possible to assess the increase of stresses when the grain sizes are of dozens and units of nanometers. A well-known expression for stresses required for the operation of the Frank-Reed source of dislocations is

$$\sigma = Gb/r \, , \tag{3.1}$$

where r is the radius of the source. If $r \sim d$ then σ is approximately $(0.1 \pm 0.01)G$, i.e., in such structures the values of stresses required for the nucleation of the sources of dislocations are very similar to the theoretically predicted values of the shear stress. This is probably an explanation of the high strength of the nanocrystalline alloy Fe-1.8%C (Table 3.2).

As the temperature of deformation changes, the physical situation also changes, and instead of being a factor of hardening, grain boundaries become a factor of softening. The grain size at which the flow stress decreases probably depends on the homologous temperature and rate of deformation.

In earlier attempts to explain this fact different temperature dependence of the strength of grain boundaries and grain interior were given. It was supposed that there was a certain so-called equicohesive temperature at which the strength of grain boundaries and grain interior was identical. This strength is higher than that of the grains below the equicohesive temperature, and the boundaries become a hardening factor. Vice versa, above the equicohesive temperature, the strength of boundaries is lower than that of the grain interior, and the boundaries become a softening factor[1].

At present, these suppositions acquire a physical meaning. The properties of polycrystalline materials are determined by the interaction between dislocations and grain boundaries. If there is no relaxation of trapped lattice dislocations in grain boundaries, the long-range stress fields resulting from the extrinsic lattice dislocations are retained and in this case, the increasing area of grain boundaries enhances the strength of the polycrystalline material. Vice versa, if the relaxation of trapped lattice dislocations is rapid enough, the flow stress decreases with the diminishing grain size.

[1] In [3.11] the decrease of flow stress with diminishing grain size is explained by the Coble diffusion creep. Similar argumentation was once used to explain SP flow but received no experimental confirmation.

Above a certain temperature the relaxation of lattice dislocations occurs with a rate commensurable with the rate of the sample deformation. The concurrent formation of grain boundary dislocations stimulates the development of grain boundary, which, in its turn, results in an abrupt decrease of flow stress and enhances ductility. In Chap. 2 the critical grain size required for transition to SP flow is calculated. It should also be noted that the critical grain size is characteristic not only of the Zn-0.4%Al alloy but of other materials as well.

The grain size affects not only the structure-sensitive mechanical properties, but also such fundamental constants as the elastic modulus and the Curie and Debye temperatures. All these characteristics are related to the electron structure which is evidently different in the grain interior and grain boundaries. This is why the structure refinement can affect these parameters. Indeed, the Curie temperature in nickel is known to be 358 K, but when the grain size is 70 nm, it decreases to 322 K [3.14]. The Debye temperature in microcrystalline iron is 467 K and in nanocrystalline iron 345 K [3.15].

In copper, at room temperature, the normal elastic modulus E is 145 GPa with the grain size being 100 μm (R.Z. *Valiev*, R.R. *Mulyukov*, and N.A. *Akhmadeev*; private communication). When the grain size is 0.2 μm, the elastic modulus is 130 GPa. The change of the shear modulus is still more substantial during the transition to a nanocrystalline state. In [3.15] it was reported that in the palladium crystallites, E = 123 GPa and in the nanocystals, E = 88 GPa.

Such a change in the elastic modulus can be associated with the fact that in the nanocrystalline material its minimum will be determined by the value of the grain boundary elastic modulus. Since in such a material the amount of grain boundary phase approaches 50% of the material volume, the average value of the elastic modulus decreases.

These data yield the following practical and important result: the refinement of structure can be used as a hardening factor until it causes the reduction of the material shear modulus, i.e., there is a certain optimal level of the microstructure refinement inducing the improvement of mechanical properties.

Hence, the microstructure refining strongly affects the mechanical and physical properties of materials. The most substantial change of properties is observed when the grains are smaller than 1 μm, which is due to the growth of the grain boundary phase with a specific structure. However, this aspect requires a systematic investigation.

Based on the results discussed, a classification of materials according to the geometric dimensions of their crystals can be proposed. *H. Gleiter* referred to materials with a grain size of about 0.01 μm as nanocrystalline. Materials with grain sizes of decimals of a micrometer, being different in their properties to nanocrystals and microcrystals, can be regarded as a separate group. Thus, it would be quite reasonable to offer the following classification based on the sizes of grains which is used in this book:

1) microcrystalline – 1–10 μm;
2) submicrocrystalline – about 0.1 μm;
3) nanocrystalline – 0.01 μm.

Let us consider the methods for obtaining such structural states in materials. Methods of microstructure refinement can be divided into two groups: casting and powder metallurgy and methods based on heat treatment (phase transformations, thermocycling, and recrystallization).

3.2 Metallurgical Methods of Refinement

Crystallization. During crystallization, the microstructure of ingots can be refined using various techniques, such as controlling the conditions of solidification, exposing an alloy under crystallization to ultrasonic waves and alloying with special modifying elements. Using these techniques and combinations thereof, a remarkable level of microstructure refining can be attained in practically all commercial alloys.

The overcooling of liquid metal provides for a considerable growth of the number of the crystallization nuclei. This method of microstructure refinement is successfully applied to produce ingots. The advantages of this method are illustrated by experiments involving quenching from liquid state at high rates of cooling. In [3.16] melted copper crystallized at a fast rate acquired a microcrystalline structure with a grain size of 0.8 to 3.4 μm. Using plates with high thermal conductivity, it is possible to reach cooling rates up to 10^6 degrees and grain sizes as small as tenths of a micrometer [3.17]. However, the dimensions of thusly obtained samples are rather small, the ingot thickness being only several tens of micrometers.

When large quantities of liquid metal are cooled, it is difficult to ensure a high level of melt overcooling. At low levels of overcooling, the grain size in the ingots is known to be virtually stable. In [3.17] it is reported that the grain size of pure nickel changed very little until the overcooling temperature reached about 175°C, in which case the grain size abruptly decreased virtually by an order of magnitude, i.e., from 2000 from 200 μm. An analogous change was observed in iron-based ingots. When the overcooling temperature is about 160°C, the grain size reduces to 100 μm. Even such a level of overcooling is difficult to obtain uniformly throughout the commercial ingot cross sections.

The microstructure refinement in large-dimension ingots is remarkable when ultrasonic treatment of crystallizing liquid metal is applied [3.17, 18]. In this case the grain size of one-phase alloys decreases twice or thrice and in a number of multiphase alloys it is from 5 to 10 times as low while the ingot homogeneity increases. For example, in the alloys Al-4%Cu-0.5%Zr and Al-0.5%Zr, the grain size reduced from 3000 μm to 100 μm [3.18]. The effect of the ultrasonic treatment is the most efficient when the melt is subjected to it during casting.

The ultrasonic treatment, though ensuring the actual refinement of grains in the ingots, fials to yield a microcrystalline structure. In addition, ultrasonic irradiation requires high power, and the continuation of the ultrasonic effect after solidification may cause a cracking of the ingot.

The melt modification is one of the most conventional methods of obtaining refined cast structures. A typical example is the addition of titanium and boron or their aluminium compounds to the melt. Contacting with the melt, these additives form intermetallic compounds $TiAl_3$ and AlB_2 that stimulate grain refinement [3.19]. In modified ingots, grains were two to three times as small, and the average size od dendritic cells reduced to 100–200 μm. The considerable refinement of the grain size was accompanied here with a complete elimination of the diversity of grain sizes and with the formation of a comparatively fine and equiaxial microstructure throughout the cross section.

To prevent the growth of the needle-shaped crystals of silicon in the Al-Si system alloys, the melt was impregnated with sulphur and phosphorus additives. Being adsorbed in the interphase boundaries and reducing the surface energy, these elements provided the formation of a structure with grain sizes of a micrometer [3.20].

These data indicate that by varying the conditions of crystallization, it is possible to refine considerably the ingot microstructures, but, as a rule, it is impossible to obtain microcrystalline structures in massive billets of commercial alloys. Nevertheless, the microstructure refinement during crystallization is important for the creation of the microcrystalline structure during the subsequent solid phase treatment. This is confirmed in [3.18].

There are a number of other methods for producing microcrystalline materials similar to recrystallization. These are, specifically, methods of galvanic layer-by-layer deposition [3.21] and condensation in vacuum [3.22]. Both methods offer better opportunities than casting to control the size of generated grains; for example, in copper, the grain size can be refined down to 0.05 μm [3.22]. However, the samples obtained are no more than ten, sometimes a hundred, micrometers thick.

Powder Metallurgy. The difficulty in obtaining a homogeneous fine-grained microstructure via casting gave an impetus to the development of powder metallurgy. At present, a number of powder metallurgy techniques are used, providing an extremely wide range of grain sizes, from coarse to nanocrystalline, attainable in a variety of materials. In spite of such a broad range of possible applications, as in the case of casting, there are limitations to obtaining a specific grain size depending on the billet volume. In industry, massive blanks are usually produced by means of powder metallurgy, which includes pouring and spraying the melt on the rotating cylinder or on the refrigerator walls, followed with the compression of the resulting granules, their sintering, and pressing into a semifinished product. This method is usually applied to process iron, nickel, aluminium and titanium based alloys. The size of the powder particles is from tens to hundreds of micrometers. The grain size of the compacted body is determined by high temperatures of sintering which substantially exceed the temperatures of recrystallization and usually amounts to dozens and hundreds of micrometers [3.23]. This is why the refinement of microstructure is usually accomplished during the pressing of the blank via subsequent heat treatment. Hence, the final grain size will be determined by the intensity of plastic deformation (Sect. 3.3). While being prepared, the powders can oxidize which may prevent SP, as, for example, in the sintered aluminium alloys (Chap. 1).

Recently, there has been great interest in obtaining nanocrystalline materials, by a method of condensation in gas [3.1]. This involves the evaporation of the material atoms in gas (helium). The atoms of the material lose their kinetic energy by collisional cooling with the bath gas and condense on the cold rod. Since the particles produced are very fine, all the subsequent operations are performed in high vacuum (10^{-7} mm of the mercury column) to prevent their contact with air. The particles are scratched off the rod, compacted, and pressed into blanks. At this stage, it should be noted that, due to the large particle surface area and interparticle friction, intensive heating is possible which can result in grain coarsening. However, by observing certain regimes of mechanical loading, this can be avoided [3.24]. At present, materials produced in this way have grain sizes as fine as 10 to 20 nm, irrespective of the nature of materials. However, the thickness of the objects never exceeds hundreds of micrometers. Hence, the finest possible microstructure can be obtained using this method.

The formation of the nanocrystalline and submicrocrystalline structure and its retention under further treatment largely depend on the presence of disperse intermetallic, oxide, or carbide particles impeding the grain growth. The formation of such particles is often hindered because of their high melting points and their coarsening during the melt cooling. In this connection, the method of mechanical alloying [3.25] which combines the techniques of powder metallurgy and "quasi-alloying" in the solid phase is of special interest. As reported in [3.26], powders of aluminium alloy containing 4.4%Cu and 2%Mg were produced. These powders were mixed with fine oxide and carbide particles. Thereafter, the mixture was processed in the ball-mill, and the powders were not only subjected to more refinement, but, as the particles came in contact under pressure, the oxide and carbide particles interefered in the cold worked aluminium powders. The resulting fine powder was sintered. In view of the restraining effect produced by the disperse particles on the grain growth, a microstructure was formed with the matrix grain size of 0.5 μm and with the size of oxide and carbide particles of 10 to 25 nm, the spacing between the latter being from 30 to 45 nm.

Another possible variant of this method consists of placing the mixture of powders in a special container and subjecting it to plastic deformation, according to *Bridgeman* [3.27]. As a result, a massive blank can be obtained immediately after deformation.

Nano- and microcrystalline states can be created by other means. Reports of samples prepared by crystallization from an amorphous phase are given in [3.15, 28]. When an amorphous alloy $Fe_{73.5}Cu_1Nb_3Si_{13.5}B_9$ is crystallized, a structure with a grain size of 10 nm is formed. Other techniques applied are rapid spraying, molecular-ray epitaxy, among others [3.15]. But the application of these techniques is limited because of the difficulty of large blanks. In many cases, to eliminate inevitable cavitation in materials produced by powder metallurgy, the temperatures of sintering and pressing must be increased, which results in the grain growth and coarsening of the microstructure.

3.3 Heat Treatment

A microcrystalline structure can be obtained by a heat treatment utilizing the features of phase transformations, spheroidization, recrystallization, and transformation of the lamellar microstructure into equiaxial. The assets of each of these methods for optimizing the microstructure refinement are considered below.

Phase Transformations. The utilization of phase transformations to refine the microstructure of materials is the easiest method of heat treatment. For example, a microduplex structure can be formed in the Zn-22%Al alloy by quenching from the one-phase region. According to [3.29], this is due to the monotectoid dissociation of the high-temperature phase into a fine-disperse mixture of the α- and β-phases which, when heated during the transformation, fragments and spheroidizes. In this case, the alloy grain size is refined to a submicrocrystalline value of 0.5 µm. Another method is the phase transcrystallization (austenization) of carbon steels [3.18]. Depending on the carbon content, pearlite steels are heated above the temperatures A_1 and A_3. A microcrystalline structure results from these two processes, i.e., the formation of the austenite grains during the phase transcrystallization and the primary recrystallization developing due to cold working. Such treatment of the steels yields a microstructure with a grain size of a few micrometers [3.30].

When the post-eutectoid steels are heated above the temperature A_1, and when a microcrystalline structure is formed, spheroidization is also observed. As a result, the steels acquire the microstructure of granulated pearlite which is similar to the microduplex structure [3.18].

Spheroidization. The lamellar structure can be transformed into microduplex only in materials with identical or very similar volume fractions of phases by means of spheroidization annealing [3.31]. However, this process is very slow. It is accelerated by preliminary cold working. Under heating, low- and high-angle boundaries are formed in the lamellas of the phases. They become sites for the accelerated development of diffusion, and less time is required for spheroidization. The authors of [3.32] consider that these conditions are sufficient for the formation of a microduplex structure in the titanium and zinc-aluminium alloys and in brasses. However, there is experimental evidence which shows that it is impossible to produce a microduplex structure in titanium alloys by means of annealing. In [3.32], a two-phase titanium alloy Ti-6.6%Al-3.5%Mo-1.7%Zr-0.27%Si (VT9) was used to study the annealing of the lamellar structure of 950°C (which is 50°C lower than the temperature of the polymorphous transformation). The alloy was in two states: annealed from the β-area and 15% cold-strained after annealing.

Figure 3.5 shows the results of the assessment of the right-side microstructure parameters: the nonequiaxial level, length, thickness, and volume fraction of the α-phase particles depending on the duration of annealing. It can be seen that

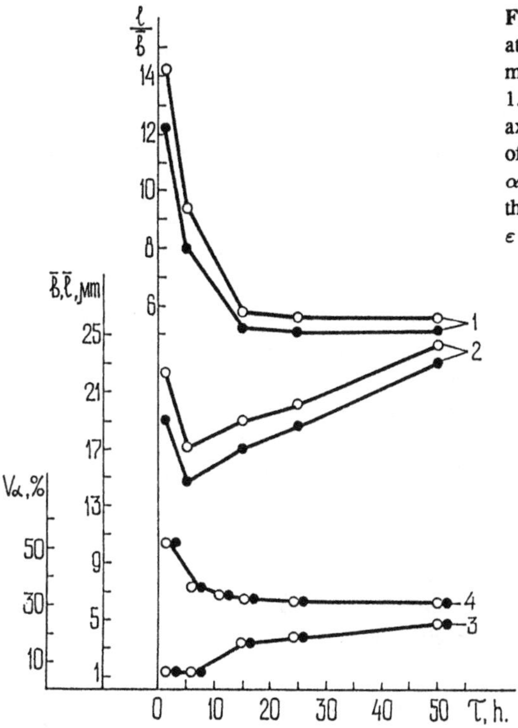

Fig. 3.5. Influence of the time of exposure at 950°C on the parameters of the microstructure of the Ti-6.6%Al-3.5%Mo-1.7%Zr-0.27%Si. Curve (*1*) the ratio of the axes l/b; (2) the length l; (*3*) the thickness of α-plates b; (*4*) the quantity of initial α-phase V_α; ○ – the annealed state; • – the post annealing state after deforming to $\varepsilon = 15\%$

in both states, under annealing in the $(\alpha + \beta)$ region, the length of the lamellas is considerably shortened and then, as the time of annealing is increased, they acquired their initial size again. In this case, the nonequiaxial level of the α-phase lamellas substantially decreased, which is due to the remarkable growth of their thicknesses. In addition, it turned out that the preliminary cold working of the alloy, done to create low-angle boundaries in the α-phase lamellas, caused no changes in the examined microstructure parameters as compared to the as-annealed state. It is interesting that the refinement of lamellas observed at the initial stage of annealing took place simultaneously with the change of the alloy phase composition due to the $\alpha - \beta$ polymorphous transformation.

Spheroidization annealing can thus be considered to be applicable for the refinement of lamellar structures only in some individual alloys. The authors of [3.33] considered this situation and undertook a thermodynamic analysis of factors affecting the high stability of lamellar structures. It was found to be due to the high anisotropy of interphase energy, i.e., the difference in the interphase energies of the lamella edge boundaries and the parallel plane habitus. This was observed in some systems, such as titanium alloys and zincals. Hence, to transform lamellar precipitates into equiaxial, a regime capable of enhancing the energy of the low-energy interphase boundaries and creating high-angle intergranular boundaries is required. Such conditions are possible during thermocycling and recrystallization.

Thermocycling. In steels, thermocycling is accomplished by multiple cycles of heating and cooling at or near the temperature of phase transformation. The microstructure is refined by the processes of phase transcrystallization and recrystallization or spheroidization and recrystallization. There are two factors causing the recrystallization: phase cold working due to heat stresses and stresses arising from the difference between the heat expansion coefficients of different phases [3.34]. This method can be used to obtain a microcrystalline structure in iron and carbon and alloyed steels. In armco-iron, a microcrystalline structure with a grain size of 1 to 5 μm was created by multiple cycles of heating and cooling near the phase transformation point [3.35]. After five cycles in the temperature range of $A_3 + 50°C–A_1 + 70°C$, a microstructure with a grain size of 7 to 8 μm was formed in the 0.4%C + 1%Cr + 1%Ni alloy [3.18].

The microstructure refinement is affected by the rate of heating and cooling. In [3.36], the microstructure of a steel containing 0.42%C and 1.8%Mn was refined by multiple cycles, rapid austenization, and quenching to a grain size of 1 to 2 μm. In carbon steels, the microstructure refining is observed after one-time quenching and high tempering. However, in this case, the grain of the 5%Ni + Cr + Mo + V steel is only 17 μm [3.37]. The efficiency of refinement grows with the increasing number of the heating and cooling cycles. In this steel, a homogeneous structure with a grain size of 3 to 5 μm is formed only after 5 cycles.

After 5–10 cycles in the A_1 temperature range, granulated pearlite with the ferrite grain size of 3 to 6 μm can be obtained in steel [3.38].

In recently years, efficiency of thermocycling for the refining of the microstructures of titanium and even aluminium alloys has been demonstrated [3.32, 34]. It was established that in titanium alloys it considerably enhances the dispersion of the α-phase lamellar precipitates, but fails to induce the recrystallization of the matrix β-phase [3.32]. Similar changes were observed in the Al-Si system alloys. Thermocycling considerably increased the dispersion of the silicon phase, but caused no refinement of the matrix microstructure [3.34].

In a general case, the microstructure refinement under the phase transcrystallization is efficient if the latter is accompanied by the developing recrystallization. In this case, it can be supposed that the formation of the microduplex structure in the Zn-22%Al alloy is due not only to the monotectoid dissociation proper, but to the cold working during the transformation and subsequent recrystallization of the fine-lamellar mixture of phases. In this regard, it is important to analyze the possibilities of creating a microcrystalline structure during the recrystallization of materials.

Static Recrystallization. The final size of grains formed under static recrystallization depends on the rate of nucleation and nuclei growth. Many researchers [3.39, 40] believe that the minimum grain size depends on the size of subgrains formed under plastic deformation. The size of subgrains in its turn depends on the stacking fault energy, temperature, and rate and level of prestraining and varies from several micrometers to decimals of a micrometer [3.10, 41]. However, the minimum grain size obtained under static recrystallization is from 0.2 to 0.3 μm [3.42]. The main obstacle in creating a submicrocrystalline structure is the difficulty of preventing the growth of the nuclei. To overcome this obstacle, either the time of heating at the temperature of recrysstallization must be abruptly shortened or the precipitation of the second-phase particles must be induced. This can be accomplished by means of rapid heating and preliminary regulated heterogenization of alloys.

In [3.18], pure nickel, one-phase nichrome, and austenitic steel 0.12%C-18%Cr-10%Ni-1%Ti were used to study rapid heating as a method of creating a microcrystalline structure. As-annealed rods made of these materials were drawn with 50% reduction and subjected to recrystallization annealing in a special plant which enabled the heating of the samples from 250–25000 K/s. The effect of the rate and temperature and of the time of exposure before quenching on the formation of microstructure during recrystallization were studied. For the sake of comparison, furnace annealing was conducted, the rate of heating being 0.5 K/s. The result was that under rapid heating, the microcrystalline structure can be formed in pure nickel. After heating with a rate of 400 K/s up to 630°C and essentially instantaneous cooling, the size of the nickel grains never exceeded 4 μm. Under furnace annealing, a microstructure with a grain size of less than 10 μm is never formed. When nichrome and steel are subjected to rapid heating, a structure with a grain size of approximately 2 μm can be formed, which is considerably smaller than in nickel.

Under slow heating (V = 0.5 K/s) of nichrome and steel, a microstructure with a grain size of 4 to 6 μm is formed by the end of primary recrystallization, but there is a considerable diversity of grain sizes and some of the grains are elongated.

The final grain size is substantially affected by both the rate and time of heating before quenching in water (τ), because under rapid heating the time of exposure of metal to high temperature during the heating proper is very short and is comparable to the period of time from heating to cooling. In 0.12%C-18%Cr-10%Ni-1%Ti steel, when the heating rate is within 35–500 K/s and the period of time between the end of heating and quneching in water is τ = 0.03 s, the refinement of grains is insignificant, i.e., from 2.5 to 2 μm. The dependence of the grain size on the time τ is more pronounced. Thus, the same rate of heating of V = 500 K/s, the change of τ from 0.4 to 0.03 s results in a change of the grain size of steel from 4 to 2 μm. Hence, both the rate of heating and the time of exposure at the temperature of recrystallization are important for the formation of the microcrystalline structure.

When rapid heating was applied in commercial aluminium alloys, the grain size was refined to 7–9 μm (Chap. 5). For comparison, under furnace annealing the grain size was $14l$, μm at the end of recrystallization.

Thus, rapid heating is an efficient means of microstructure refinement. However, in this case, it is probably impossible to create a submicrocrystalline structure because of rapid grain growth.

The presence of the second-phase particles facilitate the formation of a stable microcrystalline structure. At the same time, it should be noted that the role of these particles in recrystallization depends on their volume and size and on the spacing between them. The microstructure refinement under recrystallization is considered the most efficient in materials with noncoherent coarse (> 1 μm) precipitates of second phases [3.43].

In a general case, the critical size of the newly formed grains D_{cr} under the given size and volume fraction of particles can be defined as [3.43]

$$D_{cr} = 4r/3f \tag{3.2}$$

where f is the volume fraction and r is the particle radius. The analysis of this expression offered in [3.44] indicated that the transition from the impeding of recrystallization by small particles to its acceleration by coarse precipitates was accomplished when the parameter f/r decreased to about $0.25\,\mu m^{-1}$. In this case, according to (3.2), it is possible to obtain a grain size of 5 μm, the interparticle spacing being defined as

$$L = 0.0554r(4\pi/3f)^{1/3} \,. \tag{3.3}$$

It should be noted that, depending on the size of particles, the microstructure refinement is caused by the development of either continuous or discontinous recrystallization. In the case of disperse particles (0.1–0.2 μm or less), discontinuous recrystallization is in progress which is sometimes referred to as recrystallization in situ [3.43]. This process is accompanied by subgrain coarsening via the coalescence of their and formation of high-angle boundaries. Discontinuous recrystallization is due to the nucleation of new grains with their subsequent growth. It is clear that it can be observed only in alloys with the coarse precipitates of second phases. It should be noted that continuous recrystallization, e.g., in the Al-Cu-Zr "Supral" system alloys, is rather slow, because the subgrains are stabilized by the zirconium aluminides. However, during plastic deformation, it substantially accelerates and at 100% strain, a completely recrystallized microstructure with a grain size of 6 μm is observed in the alloy [3.45].

The situation is similar in the Al-Mg-Sc system alloy where the disperse particles of Al_3Sc (5–100 nm) produce a strong stabilizing effect on the microstructure (*Yu.B. Timoshenko* and *R.R. Sahitova*; private communication). The cold straining and annealing of the alloy at 350°C induce the formation of a partially recrystallized microstructure with the size of grains and subgrains about 0.3 μm. After SP deformation at 350°C, the microstructure becomes homogeneous and the grain size is 1 μm. Due to the original submicrocrystalline structure, the alloy displayed SP features at this temperature which is unusually low for aluminium alloys (Chap. 5).

In real materials there are both coarse and fine precipitates. In this case, recrystallization is nonuniform and develops predominantly near the coarse precipitates. Hence, the final grain size will be determined by the distribution of grains in the alloy. The recrystallization of the cold-rolled Al-0.75%Al_2O_3-0.2%Fe-0.17%Si alloy with uniformly distributed disperse (0.1 μm) and coarse (0.2–4 μm) particles was discussed in [3.46]. It was established that recrystallization nuclei were formed predominantly near the boundaries of original grains. Coarse particles of $FeAl_3$ were active nucleation centers only if they were located near such boundaries. The results obtained indicated that both types of recrystallization were in progress in the alloy (continuous and discontinuous) since the subgrain coalescence induced formation of high-angle boundaries which took place in areas with fine substructured particles. Recrystallization yielded a microstructure with a grain size of 5 μm.

In precipitation-hardening alloys, the required heterogenization of microstructure is attained via the dissociation of the oversaturated solid solution. The grain size and volume fraction can be controlled by selecting the temperature of dissociation. This temperature must be lower than the solvus temperature, but higher than the temperatures at which the excessive phase precipitates from a less disperse phase, i.e., near the "nose" of the S-shaped diagram of the dissociation of the oversaturated solid solution.

In [3.47], a method of producing semifinished parts with a microcrystalline structure out of an alloy Al-5.8%Zn-2.2%Mg-1.6%Cu, using the above described technique, was described. The method involved the heterogenization of the quenched-annealed alloy with the objective of producing a large amount of intermetallic particles of required dispersion; cold and warm deformation creating local cold working around the intermetallics; and rapid heating above the solvus temperature. Such treatment ensures the formation of a microcrystalline structure with a grain size of a few micrometers. The volumes of phases precipitated during the heterogenization are controlled by prestraining [3.48].

If the volume fractions of the phases are similar, then the task is to refine the grains of each of the phases and to distribute them uniformly within the alloy volume, i.e., to create a microduplex structure. The authors [3.49] studied the formation of the microduplex structure in the Fe-9.8%Ni-0.04%C steel. It is impossible to obtain a microstructure by annealing martensite in the $\alpha + \gamma$ area at 620°C. However, the annealing of preliminary cold worked martensite abruptly increases the number of the γ-phase nuclei and results in the formation of the microduplex. A similar treatment was used to produce a microduplex with a grain size up to 2 μm in the Fe-25%Cr-5.7%Ti-0.1%Al-0.02%C alloy [3.50].

The martensite decay can be combined with plastic deformation, which enhances the dispersion of the microstructure. Such treatment was applied to carbon steels with 0.4%C subjected to 650% deformation after annealing. The resulting grain size was from 0.5 μm. A similar method was suggested in [3.31] to process titanium alloys. In the preliminary martensite-annealed alloy Ti-5%Al-3%Mo-1%V deformed at 750°C, the grain size was 1–2 μm.

Thus, static recrystallization is rather promising as a method of creating sub- and microcrystalline structures in materials.

Dynamic Recrystallization. This method of microstructure refinement is interesting for two reasons. First, in the course of dynamic recrystallization a microcrystalline structure can be formed; hence, further hot straining can induce a transition

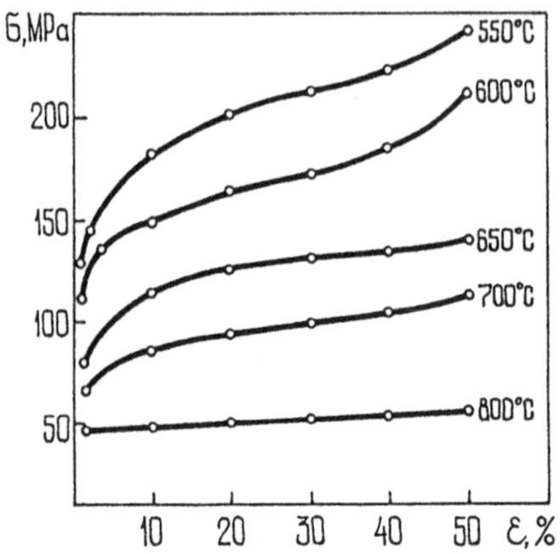

Fig. 3.6. Influence of the test temperature on the flow stress and deformation of commercial-grade iron at $\dot{\varepsilon} = 1.0 \times 10^{-3}\,\mathrm{s}^{-1}$

to SP flow [3.51]. Second, contrary to static recrystallization, the formation of new grains under dynamic recrystallization is accomplished in addition to the continuous nucleation of dislocations and their interaction with subboundaries and boundaries of original and recrystallized grains, which makes it possible to control the microstructure quantitative parameters in a broad range by varying the temperature-rate conditions of deformation. Consider the peculiarities of the effect of temperature, rate and level of deformation on the mechanical behaviour and formation of microcrystalline structure under dynamic recrystallization. This aspect was investigated on the commercially pure iron with an original grain size of 40 μm. (Results obtained in cooperation with *R.M. Galeyev*.)

Figure 3.6 shows the stress-strain curves of iron at different temperatures of compression test and $\dot{\varepsilon} = 1.0 \times 10^{-3}\,\mathrm{s}^{-1}$. In the temperature range of 550–700°C, the flow stress increases with the increasing strain level, the iron hardening being the most intensive when the temperature of deformation decreases. A similar change in the character of the stress-strain curves is observed when the strain rate increases. At 600 and 800°C and $\dot{\varepsilon} = 1.0 \times 10^{-3}\,\mathrm{s}^{-1}$, the coefficient m is 0.1 and 0.2, respectively.

Dynamic recrystallization was observed in iron at all the considered temperatures of deformation, the size and volume fraction of the recrystallized grains increasing with the rising temperature and decreasing strain rate (Table 3.3). These parameters are strongly affected by the strain level, i.e., the larger the strain level, the finer the grains and the larger the recrystallized volume. If at

Table 3.3. The average size of recrystallized grains d_r and recrystallized volume V_r in armco-iron after deformation at different temperatures ($\varepsilon = 75\%$)

ε [s^{-1}]	1.0×10^{-3}					2.2×10^{-4}
t [°C]	550	600	650	700	800	800
d_r [μm]	1.2	2.1	3.1	4.8	10.4	15.6
V_r [%]	12.0	20.0	–	44.5	71.0	80.0

800°C, $\dot{\varepsilon} = 1.0 \times 10^{-3}\,s^{-1}$, and $\varepsilon = 20\%$, the average size of the recrystallized grains and their fraction are 15 μm and 16%, respectively and at $\varepsilon = 50\%$ the respective parameters are 12 μm and 45% reaching 10.4 μm and 71% at 75%.

The evolution of the iron microstructure at different temperatures of deformation was studied by means of electron microscopy. The initial stage of deformation is characterized by a high density of dislocations, subgrains, and dislocation cells. Under 20% strain, the first recrystallized grains appear in the grain boundaries. As the strain level increases, the character of microstructure is preserved, although the number of recrystallized grains slightly increases. Some of the recrystallized grains contain dislocation pile-ups.

As the temperature of deformation increases up to 800°C, the density of dislocations decreases. At $\varepsilon = 20\%$, the formation of twisted grain boundaries and of subgrains was observed in the microstructure. The recrystallized grains appear the original boundaries. The size of the recrystallized grains varies from 5 to 20 μm. A substructure is formed in the recrystallized grains. The grain boundaries contain lattice dislocations trapped by them. As the strain level increases, the fraction of recrystallized grains considerably increases but causes no substantial changes in the dislocation structure.

The temperature of deformation affects the parameters of the iron substructure. At 600°C, the subgrain size is 1.2 μm, and at 800°C, it is 3 μm. The analysis indicates that at lower temperatures the size of subgrains approaches that of grains, while at higher temperatures their sizes are considerably different, which is probably associated with the acceleration of grain growth with the rising temperature.

The mechanical behaviour of iron under hot straining is determined by the developing hardening and softening. At lower temperatures of deformation, in spite of the effects of dynamic recovery and recrystallization, work hardening prevails; hence, the flow stress increases with the growing strain level. At high temperatures (800°C), the flow stress and strain are in a dynamic equilibrium, which results in the increases length of the steady stage flow region on the stress-strain curve. In the latter case, the curve is similar to that observed during SP deformation.

Let us consider the possibilities of transition from dynamic recrystallization to SP flow during hot straining. In iron, such transition is possible if the strain rate decreases. Thus, if a deformation at 800°C started at a strain rate of $1.0 \times 10^{-3}\,s^{-1}$ and continued after $\varepsilon = 50\%$ at a lower strain rate of $4 \times 10^{-4}\,s^{-1}$, the m value increases up to 0.31. There are, probably, two factors impeding the continuous transition from dynamic recrystallization to SP flow in iron. First, the size and recrystallized volume of the new grains are insufficient for inducing SP flow at a rate specified by the machine. Second, the microcrystalline structure of commercial iron is unstable and it quickly ceases to be superplastic because of the grain growth. Intergranular slip and dynamic recrystallization are again in progress in the coarsened grains.

It is clear that if the rates of the reproduction of the microcrystalline structure (dynamic recrystallization) and SP flow coincide, the development of SP in pure and one-phase materials is quite possible. In iron, the rates of these processes do not coincide and the continuous transition from dynamic recrystallization to SP flow is not observed. However, in a β-alloy VT30, which is virtually one-phase, SP is displayed immediately in the course of dynamic recrystallization (Chap. 6). In this case, during hot straining, the sequence of structural changes is as follows: grain refinement, their growth, and their refinement again, the average microcrystalline size of grains being the same at any given moment. It can be supposed that the rates of SP and dynamic recrystallization coincide because grains are generated via the mechanism of the local migration of the high-angle boundary. In this case SP and dynamic recrystallization are controlled by processes that are in progress in the grain boundaries, which, probably, ensures the similarity of their strain rates.

The stabilization of the microstructure by the second-phase particles slows down the growth of grains and enables an immediate transition from dynamic recrystallization to SP flow. Thus, in the MA14 alloy, this transition is observed during the high-temperature plastic flow without any changes in the strain rate (Chap. 4).

Consider the possibilities of employing dynamic recrystallization to create micro- and submicrocrystalline structures. Experiments with iron indicated that at low temperatures and high strain rates, a grain size commensurable with the minimum size of subgrains at a given temperature can be obtained during dynamic recrystallization. In the experiments the minimum grain size was about 1 μm. During static recrystallization in pure metals, even under rapid heating, such grain sizes are practically unattainable. However, there is a possibility of creating a submicrocrystalline granular structure. This requires conditions for intensified plastic deformation and formation of a completely recrystallized structure. Thus, for example, in [3.52], nickel and copper were subjected to deformation to $e = 5 - 7$, according to Bridgeman. The grain size in it was 0.1 μm, which is similar to the smallest size of subgrains attainable under plastic deformation.

A specific feature of dynamic recrystallization is that it is very difficult to obtain a fully recrystallized structure. It is important to identify structural factors affecting the recrystallized volume.

States with grain sizes of 85 and 24 were obtained in the magnesium alloy MA14 by R.O. *Kaibyshev*. The effect of the material grain size before dynamic recrystallization on the volume fraction of recrystallized grains was studied. After 50% strain at 250 and 350°C at a strain rate of $2.8 \times 10^{-3}\,s^{-1}$, the recrystallized volumes were 47 and 25% and 80 and 42% in the fine and coarse grained alloys, respectively.

The temperature increase to 450°C stops the recrystallization process and induces the formation of a microcrystalline structure in fine grained samples after 40% strain, while in order to stop recrystallization in coarse grained alloys, a considerably larger strain level is required ($\varepsilon = 80\%$). It should be noted that in this material new grains appear exclusively near the original boundaries and their number per unit length of a high-angle boundary is almost the same in both the MA14 states. Thus, the increasing rate of dynamic recrystallization with the diminishing original grain size is associated with the growing number of the nucleation sites.

Let us now consider the effect of alloying and disperse second-phase particles on dynamic recrystallization.

Alloying elements reducing the stacking fault energy abruptly accelerates recrystallization during deformation [3.53]. If elements introduced into the solid solution produce little effect on the stacking fault energy, they can both accelerate and slow down dynamic recrystallization [3.54]. At the same time, the size of the recrystallized grains always decreases with the growing alloying level [3.55].

Fine precipitates of second phases ($< 0.3\,\mu m$) usually curb dynamic recrystallization impeding the nuclei formation mobility of grain boundaries [3.56]. In contrast, coarse inclusions ($> 0.6\,\mu m$) accelerate dynamic recrystallization, because the formation of nuclei is facilitated near the strongly misorientated interphase boundaries [3.57]. The influence of the grain size on dynamic recrystallization is rather complex. In MA14, at room temperature, as detected by R.O. *Kaibyshev*, the particles of excessive phases, with their grain size and spacing between them being commensurable with the size of recrystallized grains, accelerate dynamic recrystallization; and vice versa, when the temperature of deformation increases up to 250°C, they impede it. The size of recrystallized grains always decreases when the second phase particles appear, and their fraction increases in the material structure due to the Zener force impeding the grain boundary migration.

The Transformation of Lamellar Microstructure into Equiaxial. Under the hot deformation of microduplex alloys, dynamic recrystallization can be accomplished in both phases. Hence, the final grain size of the phases will largely depend on the conditions of recrystallization in each of them. Let us consider the development of this process in a titanium alloy with the lamellar structure and phases [3.32]. As shown above, no microduplex structure can be created by spheroidization annealing.

Figure 3.7 shows the stress-strain curves of the alloy calculated for the strain levels of 10 and 50%. It can be seen that in a certain strain rate interval σ_{50} increases more rapidly than σ_{10}. The coefficient m_{10} is independent of the strain rate and is 0.25. The m value depends on the strain rate and attains its maximum of 0.52 at $\dot{\varepsilon} = 2 \times 10^{-3}\,s^{-1}$. This is an indication that in the VT9 alloy with an originally lamellar microstructure, a transition to SP flow is observed after a certain stage of hot deformation.

The alloy microstructure undergoes remarkable changes already at the initial stages of its hot deformation at $\dot{\varepsilon} = 5.5 \times 10^{-4}\,s^{-1}$. When the strain level is 30%, the interphase boundaries become sinuous with grooves formed in them, and the α-lamellas bulging (Fig. 3.8a). Under strains below 50%, at all the studied strain rates, the α-lamellas turn in the direction of plastic flow, which results in the formation of a metallographic texture in the alloy. As the strain level increases, equiaxial grains of the phases appear in the microstructure (Fig. 3.8b). After 85% strain, the microstructure is radically different, i.e., in the course of deformation,

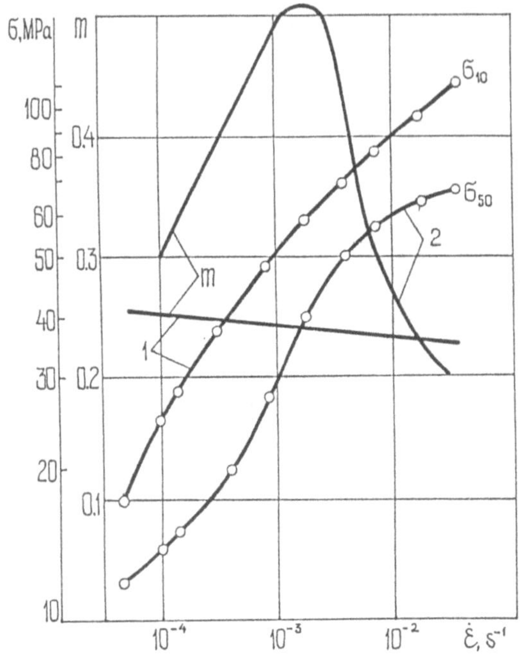

Fig. 3.7. Dependence of flow stress σ and the coefficient of strain rate sensitivity m on the deformation rate for the Ti-6.6%Al-3.5%Mo-1.7%Zr-0.27%Si at 950°C and levels of deformation of (*1*) 10%; (2 50%

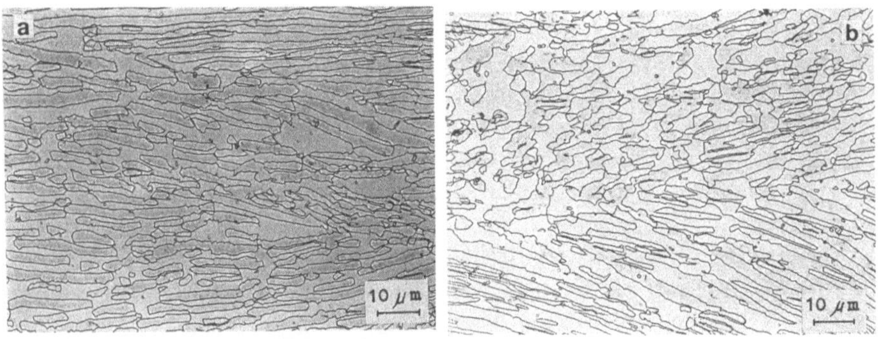

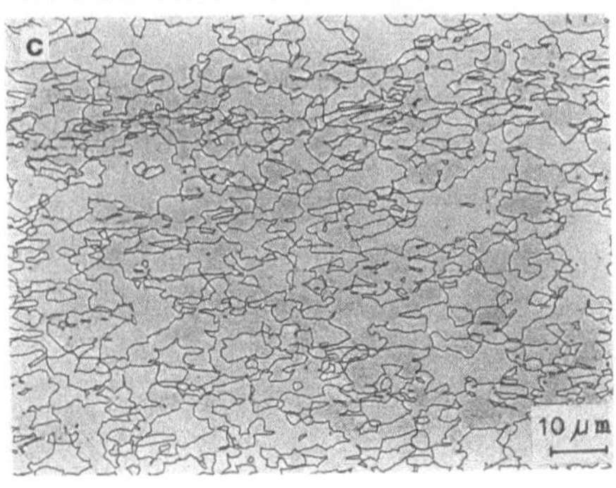

Fig. 3.8. Microstructure of the deformed Ti-6.6%Al-3.5%Mo-1.7%Zr-0.27%Si; **(a)** ε = 30%; **(b)** ε = 50%; **(c)** ε = 85%

instead of lamellar duplex, it becomes microduplex (Fig. 3.8c). In this case, at the optimal rate of SP deformation, the α- and β-phase grains are arranged in chains, with interphase boundaries being oriented at an angle of approximately 45°C with the flow direction. Such an arrangement of the α- and β-phase grains is typical of the SP flow of two-phase alloys (Chap. 2). Under higher strain rates, the metallographic texture is retained even after 85% strain.

The statistical analysis of the microstructure indicated that at 85% strain rates of deformation similar to those of SP are matched with the shortest and the least nonequiaxial lamellas (Fig. 3.9a,b) and with the largest fraction of the equiaxial α-phase particles (Table 3.4). It should also be noted that the fractions of the fine α-phase particles, 10 µm or less long, are 68, 84, 88, and 48% at strain rates of 5.5×10^{-2}, 5.5×10^{-3}, 5.5×10^{-4}, and $5.5 \times 10^{-5}\,s^{-1}$, respectively.

The metallographic analysis indicated that the hot deformation of the alloy is accompanied by the $\alpha - \beta$ phase transformation, the rate of the latter being higher than under conventional exposure at test temperatures. Thus, an alloy with a lamellar microstructure, held an hour at 950°C, contains 52% of the α-phase, while after 85% strain at a rate of $5.5 \times 10^{-3}\,s^{-1}$ (the time of deformation is less than 3 min), the α-phase content decreased to 41%.

The analysis of the alloy thin foils indicated that a dislocation structure is formed in the packs of α-lamellas at the initial stage of hot straining, which is accompanied with the fragmentation of the α-lamellas. With the growing strain level, the misorientation angle between two adjacent fragments increases and the misorientation of interphase boundaries grows from 1 to 10–15°.

The analysis of the texture formation in an alloy strained at a rate of $5.5 \times 10^{-3}\,s^{-1}$ to different levels of deformation indicated a remarkable enhancement of texture maxima in the α-phase under 10% strain and their rapid spreading with the rising strain level [3.32].

The transformation of the lamellar microstructure is most complete under the SP temperature-rate conditions. The deviation from the optimal conditions, in particular the growth of the strain rate, impedes the transformation of the originally lamellar microstructure into equiaxial.

The complexity of the microstructural variations is due to the operation of different mechanisms of deformation at different stages of plastic flow. At the initial stage of the deformation with a lamellar microstructure, the main role is

Table 3.4. The variation of the fraction of the equiaxial α-phase particles depending on level and rate of the VT9 alloy deformation at 950°C

ε [%]	Fraction of equiaxial α-phase particles [%]			
	$\dot{\varepsilon} = 5.5 \times 10^{-2}\,s^{-1}$	$\dot{\varepsilon} = 5.5 \times 10^{-3}\,s^{-1}$	$\dot{\varepsilon} = 5.5 \times 10^{-4}\,s^{-1}$	$\dot{\varepsilon} = 5.5 \times 10^{-5}\,s^{-1}$
10	4 ± 1	4 ± 1	4 ± 1	4 ± 1
30	5 ± 1	7 ± 1	8 ± 1	13 ± 1
50	10 ± 1	11 ± 1	16 ± 1	18 ± 1
70	15 ± 1	22 ± 1	44 ± 1	43 ± 1
85	20 ± 1	36 ± 1	52 ± 1	44 ± 1

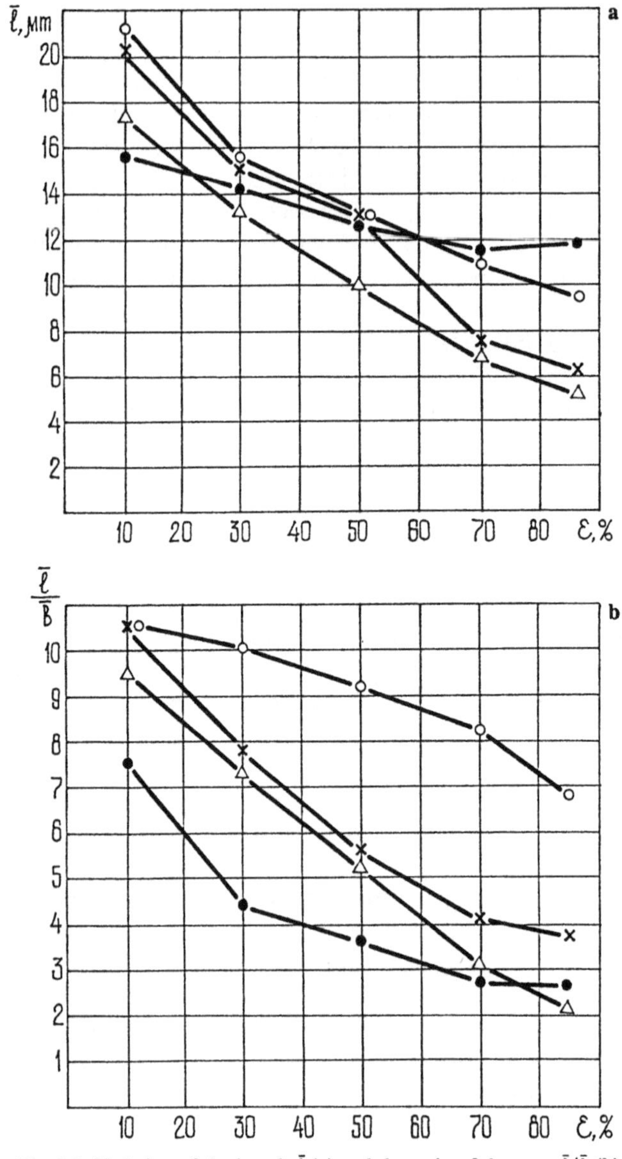

Fig. 3.9. Variation of the length $\bar{l}$ (a) and the ratio of the axes $\bar{l}/\bar{b}$ (b) of the plate of the α-phase of the Ti-6.6%Al-3.5%Mo-1.7%Zr-0.27%Si alloy after deformation at 950°C at different strain rates: • – $\dot{\varepsilon} = 5.5 \times 10^{-2}\,s^{-1}$; x – $\dot{\varepsilon} = 5.5 \times 10^{-3}\,s^{-1}$; $\triangle$ – $\dot{\varepsilon} = 5.5 \times 10^{-4}\,s^{-1}$; o – $\dot{\varepsilon} = 5.5 \times 10^{-5}\,s^{-1}$

played by dislocation slip. This is proved by the formation of a sharp crystallographic and metallographic texture and substructure under low strains. In this case, the alloy hardening results in the growth of flow stress; hence, the strain rate sensitivity coefficient m is 0.25 or less. A further growth of the strain level induces dynamic recrystallization. Dislocations interact with interphase boundaries, and their energy and misorientation increase enabling the development of grain boundary sliding as indicated by the spreading of crystallographic texture. The enhancement of diffusion due to a change in the state of interphase boundaries [3.58] causes their bulging and increases their mobility.

When fragmentation is in progress in the α-phase, the development of grain boundary sliding in new grains makes the lamella fragments rotate and shift. The changes result in the alloy softening. At this stage of deformation flow stresses decrease and the coefficient m reaches its maximum value. The rapid development of grain boundary sliding with the growing strain level eliminates the metallographic structure and results in the uniform distribution of the phases. This is why the fraction of the equiaxial grains of the phases is the largest under 85% strain at strain rates similar to the optimal rates of SP deformation.

A specific aspect of spheroidization in a titanium alloy is the development and acceleration of the $\alpha - \beta$ transformation during deformation. Due to the enhancement of diffusion processes under straining, a redistribution of the alloying element is accomplished; hence, the phase transformation is accelerated. In [3.59], this phenomenon was studied using the Ti-5%Al-4.7%Mo-4.5%V alloy, which yielded a state with an original phase nonequilibrium, where the polymorphous $\alpha \rightarrow \beta$ transformation is accomplished during deformation. No phase transformation is observed in the equilibrium state of the hot strained alloy. Note that, in the former state, the fraction of equiaxial grains is much larger than in the latter.

These data indicate that during the deformation of titanium alloy, the polymorphous $\alpha \rightarrow \beta$ transformation is caused by the original phase nonequilibrium and accelerates the transformation of the lamellar structure into equiaxial. In its turn, the intensive redistribution of alloying elements during the hot deformation is probably induced by diffusion creep.

Thus, in titanium alloys, the transformation of the lamellar microstructure into equiaxial turns out to be a complex process induced by dynamic recrystallization, phase transformation and grain boundary sliding. The combination of these processes affects spheroidization and, thus, it is more enhanced under the optimal temperature-rate conditions of SP deformation. The above described change of the microstructure is characteristic of other alloys as well. Similar processes take place, for example, during the hot deformation of the MA21 alloy (Chap. 4), which results in the transformation of the duplex microstructure into microduplex.

The difference in the deformation parameters of phases can substantially affect spheroidization under hot straining. An aluminium alloy with 12% silicon was studied (S.S. *Nechayev*, private communication). In the cast state it contained lamella-shaped silicon precipitates. At the temperature of deformation of 480°C the difference in the homologous temperatures of the alloy phases was $\sim 0.5\,T_m$, i.e., this temperature is "cold" for silicon and "hot" for aluminium. Silicon displayed traces of SP deformation. However, during deformation, the particles of this phase were destroyed. The sites of ruptures were filled with the matrix material. At the same time, the aluminium phase was a site where dislocations were accumulated and recrystallization developed. After 75% strain at $\dot{\varepsilon} = 5 \times 10^{-4}\,s^{-1}$, a microduplex type structure is formed, with the matrix grain size and the size of the silicon particles being 7–8 and 3–5 μm, respectively.

The temperature-rate mode of the deformation of alloys with duplex microstructures substantially affects the final grain sizes of the phases. As in alloys with matrix structures, there can be a grain size similar to the size of subgrains. For example, after the Ti-6.7%Al-3%Mo two-phase alloy has been deformed to a level of 75% or more at 600°C and $\dot{\varepsilon} = 7.4 \times 10^{-4}\,s^{-1}$, a substructure is formed with the phase grain sizes of 0.1 μm [3.60].

Thus, the refinement of material microstructure via dynamic recrystallization is a universal method and can be applied to alloys with both matrix and duplex microstructures. Varying the temperature-rate conditions and strain level, micro- and submicrocrystalline structures can be created in metallic materials.

Dynamic recrystallization enables the transition from conventional hot deformation to SP flow. In a general case, such a transition requires the rates of the microstructure refinement under dynamic recrystallization and of SP flow to be quite similar. This aspect requires further systematic investigation.

4. Superplastic Deformation of Magnesium Alloys

Magnesium alloys are distinguished by their greater stiffness and specific strength as compared to other metallic materials. Due to these characteristics, magnesium and its alloys can withstand higher impact loads than aluminium alloys. Consequently, magnesium alloys can be efficiently used in instruments and components which are subjected to such loads. The high chemical resistance of magnesium and its alloys to fuels, mineral oils, and alkalis and their characteristic physical properties such as low neutron absorption capacity and good thermal conductivity make them rather promising for applications in various instrumental components.

However, the limited use of magnesium and its alloys in engineering and relatively small volume of deformable magnesium alloys (wrought products) in industrial applications is due to their relatively low ductility: (their relative elongation does not exceed 30–70% even under the conditions of hot deformation) and high degree of anisotropy in mechanical properties [4.1,2]. The coefficient of anisotropy of the MA15 alloy, taken as the ratio of the strength properties in the longitudinal and the transverse directions, is as high as 4 and 2.3 with respect to the yield stress and the ultimate tensile strength, respectively [4.3,4].

The crystal structure is the main reason for the low ductility and anisotropy. Magnesium has an hcp structure with an axial ratio $c/a = 1.624$ that is near the theoretical value. In view of this, its deformation at the room temperatures takes place almost entirely due to slip along the basal (0001) planes in the $\langle 11\bar{2}0\rangle$ direction [4.3]. The existence of only three slip systems adversely affects the conditions of simultaneous deformation of grains in polycrystalline magnesium and results in its low ductility. The limited number of slip systems and the presence of texture in magnesium alloys are the main reasons for the high degree of anisotropy in their mechanical properties.

Numerous attempts have been made to improve the properties of deformable magnesium alloys by the additional alloying with different elements, but no significant improvement was achieved. At best, alloying makes it possible to improve the strength properties, but low ductility is retained. As a rule, an improvement of the strength characteristics is accompanied by an intensified anisotropy of other properties [4.4] due to the oriented precipitation of the second-phase particles.

A significant improvement in ductility, specific strength, and stiffness could be attained in magnesium-lithium alloys [4.4,5]. The main advantage of alloying magnesium with lithium is that a bcc β-phase is formed which is capable of abruptly increasing the room temperature ductility of the alloys. The ductility is increased not only by the formation of the β-phase, but also by the decreased c/a ratio in the hcp lattice. In the Mg-4.5 at.%Li alloy, the axial ratio c/a of the α-phase decreases to 1.61. As a result, a nonbasal slip along the prism and pyramidal planes becomes possible in the α-phase [4.6].

However, the mechanical properties of Mg-Li commerical alloys deteriorate with time at room temperature [4.3]. Their application may become possible only when simple and efficient methods of stabilizing their mechanical properties are found.

One of the radical methods of improving the mechanical properties is their processing in a superplastic state. In this case, two problems can be solved simultaneously: (1) the ductility can be increased, allowing production of components of intricate shapes and (2) specific features of SP deformation can be utilized to obtain desired structures and properties.

Superplasticity in magnesium alloys of test compositions was first reported in [4.7–11]. The authors described superplasticity in the binary eutectic alloys of magnesium with nickel, copper, and aluminium and in commercial magnesium alloys MA8 and Zk60 with an initially fine-grained structure.

Systematic data concerning the SP effect in commercial magnesium alloys were first presented in [4.5]. However, important issues such as the preparation of the magnesium alloy structures and the effect of a number of factors on SP deformation are still vague. The latest data concerning ductility and strength in a number of magnesium alloys under the conditions of conventional and superplastic deformation are presented in the following sections.

4.1 Effect of Chemical Composition, Structure, and Temperature-Rate Conditions of Deformation on Ductility

Magnesium alloys of different chemical and phase compositions (Table 4.1) were used for the investigations. The commercial grade Mg90 is the closest to pure magnesium. The MA8, MA14, and MA15 alloys have matrix type structures and are distinguished by the level of alloying. Thus, MA8, containing a small number of dispersed second-phase particles, is considered to be low-alloyed and MA14 and MA15, with a large amount of various elements both in the solid solution of magnesium and in the excess phases, are high-alloyed. The MA21 alloy possesses a two-phase microstructure with similar volume fractions in the phases.

It can be easily seen that such a selection for investigation embraces virtually the whole variety of magnesium alloys. It can be expected that the results obtained in studying superplasticity in these materials will be of universal significance for all the other magnesium alloys.

Let us examine the effect of the temperature-rate conditions on the mechanical properties of Mg90, MA8, MA15, and MA21. To obtain comparable results, all the alloys were pretreated to create a microstructure with a grain size of about

Table 4.1. Chemical and phase compositions of magnesium alloys

Material	Components [wt.%]							Phase
	Mg	Al	Cd	Li	Mn	Zn	Others	composition
Commercial magnesium Mg90	99.9	–	–	–	–	–	0.1	α-Mg
MA8	97.95	–	–	–	1.8	–	0.25Ce	α-solid solution
MA14	93.35	–	–	–	–	6.0	0.65Zr	β-Mn;Mg,Ce α-solid solution Mg_2Zn_3, MgZn,Zr_3Zn_2
MA15	93.7	–	1.6	–	–	3.1	Zr 0.65 La 0.95	α-solid solution $(MgZn)_3La$, Zr_3Zn_2, ZrH_2
MA21	80.41	5.2	4.7	8.1	0.21	1.38	–	α-solid solution β-solid solution AlLi

Table 4.2. Mechanical properties of the magnesium alloys at room temperature

Mechanical properties	Mg90	MA8	MA15	MA21
σ_u [MPa]	200	250	300	210
$\sigma_{0.2}$ [MPa]	90	170	260	150
δ [%]	11.5	18.0	16.0	26.0

15 µm. The mechanical tests indicate that alloying with such elements as Mn, Ce(MA8), Zn, Zr, La, and Cd(MA15) considerably enhances the material strength at room temperature (Table 4.2).

The alloying of magnesium with lithium significantly softens the alloy even in the presence of a large number of relatively strong strengthening agents such as aluminium, zinc, cadmium, and manganese. The strength of MA21 is greater than that of pure magnesium by only 15–30 MPa and is considerably less than the strength of the MA8 and MA15 alloys. Alloying has a positive effect on the ductility of magnesium. The relative elongation of the MA8 and MA15 is 1.5 times greater than that of pure magnesium.

Thus, chemical and phase compositions have a fairly strong effect on the room temperature properties. The increase of the deformation temperature leads

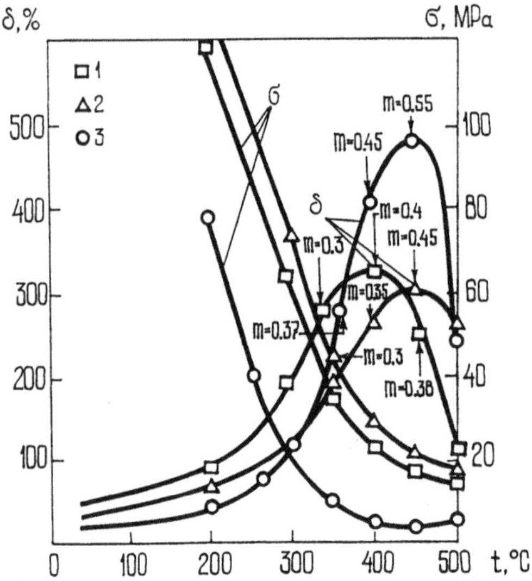

Fig. 4.1. Temperature dependence of the flow stress σ, relative elongation δ and the strain rate sensitivity coefficient m of magnesium alloys ($\dot{\varepsilon} = 4.1 \times 10^{-2}\,s^{-1}$): Curve (*1*) MA8; (*2*) MA15; (*3*) MA21

to different changes of the relative elongations and flow stresses of magnesium and its alloys.

The microstructure of commercially pure magnesium is unstable at elevated temperatures. Starting at 200°C, rapid grain growth is observed during holding, before the start and in the course of deformation. This is the reason why at high temperatures the Mg90 properties only slightly depend on the original microstructure. The relative elongation of magnesium is 80% or less throughout the temperature range, with the strain rate sensitivity coefficient being 0.12 or less, and the material straining to failure is accompanied with rapid strengthening. The results obtained indicate that originaly fine-grained pure magnesium does not display even minor symptoms of superplasticity.

The situation is different in magnesium alloys (Fig. 4.1). Their ductility depends nonmonotonically on the temperature of deformation. While at 20–250°C, the relative elongation to rupture of magnesium alloys gradually increases from 10 to 70%, an abnormally high ductility is observed in the temperature range from 350–450°C. In the MA8, MA15, and MA21 alloys, the relative elongations are 320, 300, and 450%, respectively. At temperatures above 450°C, ductility slightly decreases. At 500°C, the relative elongation is 100, 250, and 230%, respectively. A strong strain rate dependence of flow stress is observed in the range of $2 \times 10^{-3} - 2 \times 10^{-2}\,s^{-1}$. The coefficient m amounts to 0.4, 0.45, and 0.55 in alloys MA8, MA15, and MA21, respectively.

These data indicate that when fine-grained magnesium alloys are tensile strained at 400–450°C at strain rates ranging from 2×10^{-3} to $2 \times 10^{-2}\,s^{-1}$, they display all the features of SP flow. These features are common for all the SP materials, viz., large elongations at 350–450°C and strain rates from 10^{-4} to $10^{-2}\,s^{-1}$; specific form of the stress-strain curve; high strain rate sensitivity; and strong rate dependence of elongation and coefficient m.

At the same time, superplasticity with relatively coarse grained magnesium-based alloys is not typical of other materials.

Let us examine the effect of chemical composition on the properties of magnesium alloys under the conditions of SP flow. The alloys MA8 and MA15, whose room temperature mechanical properties are significantly different, have similar values of flow stress and relative elongation under SP flow conditions (Fig. 4.1, 2). Based on this fact, it can be assumed that, under the conditions of SP, the properties of the investigated magnesium alloys basically depend on the microstructure and that the effect of chemical composition is manifested indirectly through the microstructure stability.

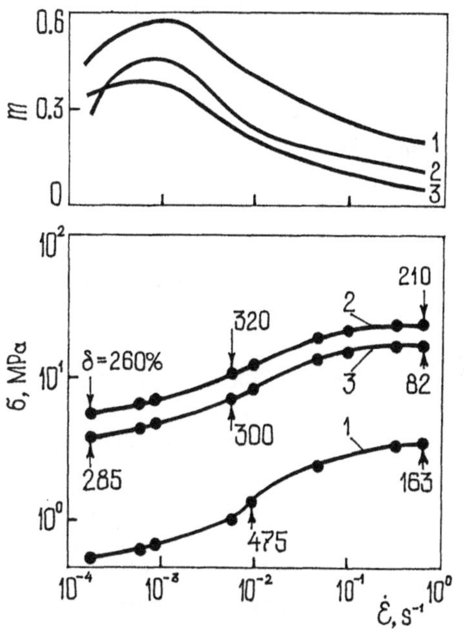

Fig. 4.2. Strain rate dependence of the flow stress, σ, relative elongation δ and coefficient m on the strain rate in magnesium alloys: Curve (*1*) MA21 at 450°C; (2) MA15 at 450°C; (*3*) MA8 at 400°C; (*4*) MA8 at 450°C

The chemical composition of the alloys has a significant effect on the location of the optimum temperature range of superplasticity. In fact, in the MA8 alloy in which stabilization of the microstructure as a result of alloying with cerium and manganese is effective up to 400°C and the maximum elongation is observed in the 380–400°C range; on the other hand, in the complex MA15 alloy containing a large quantity of the excess phases having zinc, zirconium, and lanthanum whose stabilizing effect is greater, the maximum elongation is attained at 450°C.

As for the two-phase MA21 alloy, in the temperature range in which the SP effect appears (350–450°C), its flow stress values are less than the corresponding values of the MA8 and MA15 alloys. As one would expect, the relative elongation and the magnitude of the coefficient of strain rate sensitivity of the two-phase MA21 alloy are greater than the respective parameters of the MA8 and MA15 alloys. The alloy is characterized by a stronger strain rate dependence of flow stress, and the range of the increased strain rate sensitivity is shifted towards the region of higher strain rates. The maximum values of m and δ (0.55 and 475%, respectively) are attained at $\dot{\varepsilon} = 8.3 \times 10^{-3}\,s^{-1}$ (Fig. 4.2). Such behaviour of the two-phase MA21 alloy can be attributed to the presence of a more favourable (from the standpoint of developing superplastic deformation) microstructure having the $\alpha{:}\beta$ phase ratio near 1:1, which make the alloy microstructure stable at high temperatures.

There are numerous reasons for the nonmonotonous character of the temperature dependence of the magnesium alloy ductility. The significant increase of ductility at a temperature exceeding 250°C is related to the development of multiple slip on basal planes [4.12].

The transition to SP flow at 350–450°C is associated with the growth of the contribution of grain boundary sliding to the alloy strain (Chap. 2), which results in a considerable increase in the strain rate sensitivity coefficient and relative elongation and in a decrease in the flow stress. Thus, in commercially pure magnesium, it is not possible to realize the SP effect because of the instability of the fine-grained microstructure. Insignificantly small alloying additions (MA8) make it possible to obtain all the symptoms of SP flow in the originally fine-grained alloys.

The increasing level of magnesium alloying leads to the stabilization of microstructure during heating and straining in the high temperatures range and facilitates the conversion of magnesium alloys to a superplastic state. The broad temperature range in which high alloys display superplastic features is of great practical importance for such easily inflammable alloys such as MA21, providing for SP deformation at lower temperatures.

It can be seen in Fig. 4.1 that at 350°C, which is 100°C lower than the optimal temperature, the MA21 alloy exhibits symptoms of SP flow, i.e., the flow stresses are strongly dependent on the strain rate, the strain rate sensitivity coefficient is rather high (m = 0.35), and there are large relative elongations (δ = 270%). However, the maximum m and δ values shift towards a lower strain rate of $\dot{\varepsilon} = 1.7 \times 10^{-3}\,s^{-1}$. In addition, this temperature is safe from the point of view of inflammability [4.13].

A systematic study of the effect of the original microstructure on the superplasticity of magnesium alloys was performed on the MA15 alloy. Different treatments were used to obtain states with grain sizes of 4.5, 9, 20, and 80 μm and identical crystallographic texture. Mechanical tests at room temperature in the direction of extrusion (Table 4.3) indicate that grain refinement improves both the strength and ductility of the MA15 alloy.

As the temperature of deformation increases, interesting peculiarities can be observed in the behaviour of alloys with different grain sizes. In the temperature range from 325 to 500°C an alloy with a grain size of 4.5 μm has the lowest flow stress compared to other states (Fig. 4.3). At the same time, in the same temperature range, the largest flow stress is characteristic of an alloy with a grain size of 9 μm.

Table 4.3. Mechanical properties of the MA15 alloy with different initial grain sizes ($t = 20°C$)

d_0 [μm]	σ_u [MPa]	$\sigma_{0.2}$ [MPa]	δ [%]
4.5	305	268	16.2
9	278	233	15.3
20	264	218	12.4
80	223	188	9.7

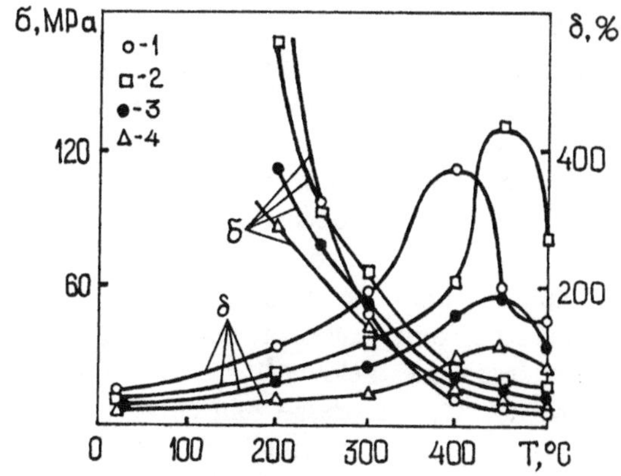

Fig. 4.3. Temperature dependence of the flow stress σ and relative elongation δ of the MA15 alloy at $\dot{\varepsilon} = 2.1 \times 10^{-3}\,s^{-1}$: Curve (*1*) 4.5 μm; (*2*) 9 μm; (*3*) 20 μm; (*4*) 80 μm

Grain size considerably affects the material ductility. Below 400°C, the relative elongation increases with the decreasing grain size, the temperature dependence of the relative elongation being more pronounced in MA15 with a grain size of 4.5 μm. However, with a further rise in the deformation temperature, the ductility of the alloy with a grain size of 4.5 μm abruptly drops, and at 450°C it is remarkably lower than in the alloy with a grain size of 9 mm.

The strain rate dependencies of flow stress, relative elongation, and the coefficient m of the different states of the MA15 alloy characterize the material for SP deformation (Fig. 4.4). In a coarse grained alloy ($d = 80$ μm), the value of m never exceeds 0.16 even at 450°C, i.e., no superplasticity is observed. The refinement of grains to 20 μm results in the strain rate sensitivity of flow stress and relative elongation in the range of strain rates from 5.1×10^{-4} to $2.1 \times 10^{-3}\,s^{-1}$, where m has its maximum value of 0.36 and the relative elongation reaches 260%.

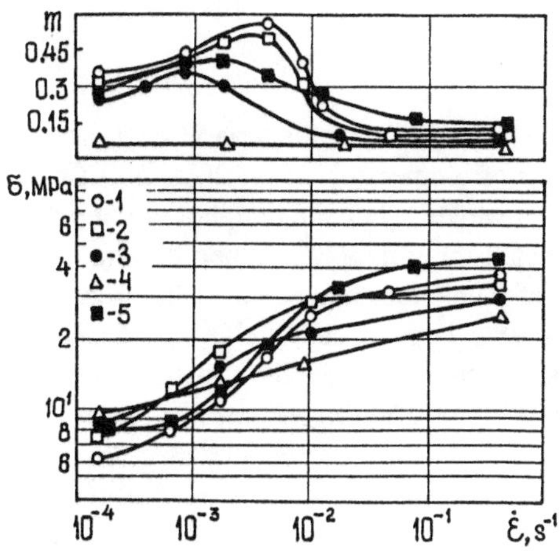

Fig. 4.4. Strain rate dependence of the flow stress σ and coefficient m of MA15: Curve (*1*) 4.5 μm, 450°C; (*2*) 9 μm, 450°C; (*3*) 20 μm, 450°C; (*4*) 80 μm, 450°C; (*5*) 4.5 μm, 400°C

When the grain size is 9 μm, the strain rate dependence of properties, characteristic of superplasticity, abruptly increases. The optimal strain rate of SP deformation shifts towards higher rates and is $5.1 \times 10^{-3}\,s^{-1}$. At this strain rate, the relative elongation is 475% and $m = 0.48$.

However, further refinement of grains to 4.5 μm reduces the alloy ductility at 450°C, its strain rate dependence being slight. At optimal strain rates the relative elongation never exceeds 200%, and at low and high strain rates it is 95–120%. At the same time, the strain rate sensitivity of flow stress also increases. When compared to other states of the alloy, the value of m increases at small strains and at strain rates below $10^{-2}\,s^{-1}$ and turns out to be much larger than 0.3.

At 450°C, the stress-strain curves (Fig. 4.5) of the MA15 alloy, with the original grain sizes varying from 9 to 80 μm, differ from the analogous curves of the alloy with a grain size of 4.5 μm. If in the former case the material plastic flow is accompanied with softening, in the latter case, after a strain of 40%, strengthening begins. When the temperature is decreased by 50°C, the character of the stress-strain curves of the MA15 alloy with an original grain size of 4.5 μm changes. No hardening is observed during deformation. The steady flow stage follows after a small peak of flow stresses.

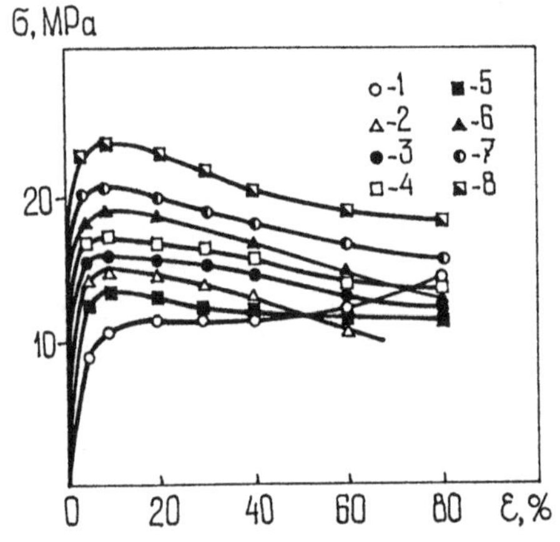

Fig. 4.5. True $\sigma - \varepsilon$ curves of the MA15 alloy ($\dot{\varepsilon} = 2.1 \times 10^{-3}\,s^{-1}$): Curve (*1*) 4.5 μm; (2) 9 μm; (*3*) 20 μm; (*4*) 80 μm; (*5*) 4.5 μm; (*6*) 9 μm; (*7*) 20 μm; (*8*) 80 μm

The results obtained indicate that the reduction of grain size affects the mechanical properties of MA15 in the range of temperature from 20 to 400°C, the same being observed in other materials. However, when the temperature of deformation is increased by 50°C, an unexpected thing happens. When the grain size becomes smaller than the optimal size of 9 μm, the material ductility decreases and its flow stress increases in the areas with large strain levels.

The results of the mechanical tests can be explained by structural changes occurring during deformation. During the hot straining of the coarse-grained states of the MA15 alloy (d_0 = 20 and 80 μm), dynamic recrystallization develops, which causes the reduction of the mean grain size and creates a structure with a variety of grain sizes. Alloys with d_0 = 9 μm at 450°C and with d_0 = 4.5 μm at 400°C, subjected to SP deformation, have a homogeneous microstructure. The original grains grow and acquire a slightly oblong shape. Their average longitudinal sizes increase to 14 and 8.2 μm, and the transverse sizes become 11 and 6.9 μm, respectively. The situation is different when the original grain size of the material strained at 450°C is 4.5 μm. In this case, some grains grow to 100 μm which is abnormal; after 100% straining, the average grain size becomes 30–40 μm, and the structure created is excessively heterogeneous.

At 450°C, the highest SP parameters are exhibited by the alloy with a grain size of 9 μm, while at 400°C there is a grain size of 4.5 μm. At 450°C and small strains, the latter has the highest strain rate sensitivity. At the same time, as the strain level increases, the growth of individual grains results in a considerable increase in the flow stress and reduces the ductility resource. At this temperature, the material with a grain size of 9 μm is preferable. When the grain size increases to 20 μm, SP symptoms can be observed only in the area of high temperatures and low strain rates. Materials with coarser grains display no SP symptoms under any conditions of deformation.

Hence, the reduction of the original grain size exerts different effects on the mechanical properties of the magnesium alloy, depending on temperature. At room temperature, the refining of grains causes a smooth increase in the flow stress and ductility. At high temperatures, changes in the original microstructure result in a nonmonotonous variation of the flow stress and ductility.

The MA15 alloy is specific in that there is a certain optimal grain size which provides for the highest properties at high temperatures. This is not typical of superplasticity in general for it is known that the smaller the grain size, the higher the SP parameters (Chap. 1). This peculiarity is due to the change of the microstructure stability, which in turn depends on the amount of second phases in the alloy. When the grain size is refined to 4 mm, the second phases are partially dissolved. In the alloy with $d = 9$ μm, the fraction of second phases is 5.5–6%, and in that with $d = 4.5$ μm it is only 3.8–4.0%.

On the other hand, the reduction of the grain size is accompanied by the growth of the grain boundary area, which requires a larger amount of the second phase disperse particles to maintain the structure stability [4.14]. It is the decrease of the structure stability that induces an abrupt grain growth during deformation and makes the material non-superplastic. This is why, at 450°C, the SP parameters of the alloy with a grain size of 9 μm are better than those with a grain size of 4.5 μm.

In actual practice, methods of metal forming are based on complex schemes of deformation. In this regard, the effect of the loading scheme on the character of SP flow must be assessed. Hence, in addition to tensile tests, there is much interest in the modes of SP deformation under conditions resembling those of die forging. The MA15 and MA21 alloys were used to study the possibility of superplasticity and the basic modes of SP deformation under compression.

The material ability for superplastic flow was assessed by the stress-strain curves, by the flow stress values and coefficient m. The results were that in the MA15 and MA21 alloys with a grain size of 15 μm, at 450 and 350°C, respectively, there is practically no difference in the variations of the true stress-strain curves during the tensile straining and compression. In both cases, the steady flow stage starts after 10–15% strain.

The strain rate dependence of the coefficient m during the tensile straining and compression of the materials is similar to that obtained experimentally. It should be noted that during the compression, the maximum m shifts towards higher strain rates as compared to tensile straining. During the tensile straining of the MA15 alloy, the maximum coefficient $m = 0.45$ is observed at $\dot{\varepsilon} = 4.1 \times 10^{-3}\,s^{-1}$ ($t = 450°C$) and in MA21 at $\dot{\varepsilon} = 1.7 \times 10^{-3}\,s^{-1}$ ($t = 350°C$). During the alloy upsetting, the maximum values of m are 0.48 and 0.33 at $\dot{\varepsilon} = 1.6 \times 10^{-2}\,s^{-1}$ and $\dot{\varepsilon} = 8.3 \times 10^{-3}\,s^{-1}$ in the alloys MA15 and MA21, respectively.

Therefore, during compression, as well as during tensile straining, the alloys display symptoms of SP flow, i.e., a specific character of the variation of the true stress-strain curve and a high strain rate sensitivity of the flow stress. The fact that the optimal rate of SP deformation during compression is higher than during tensile straining is of great practical importance for it allows forming at higher strain rates.

4.2 Refining of the Magnesium Alloy Structure

In magnesium, there are no phase transformations in the solid state, which prevents the formation of a microcrystalline structure by means of a heat treatment. The relatively small volume variations of magnesium and its alloys with temperature make thermocycling inefficient. In spite of the latest achievements in the granule metallurgy, the smallest possible grain size obtained by means of metallurgical methods is 15 μm. This is why the main ways of preparing the microcrystalline structure in magnesium alloys are based on recrystallization.

Cold straining followed with recrystallization annealing leads to considerable refining of the structures of low magnesium-based alloys. However, after such a treatment, the smallest grain size will be 10 μm [4.4]. In addition, while in sheets it is possible to create a fairly homogeneous structure using large prestrains, in bulk samples, recrystallization annealing results in the formation of a structure with a variety of grain sizes. It is very difficult to increase the level of prestraining in the bulk samples to eliminate the diversity of grain sizes, because at room temperature the ductility of magnesium alloys is low and the flow stresses are high.

A method of grain refining based on the dynamic recrystallization of metals and alloys has none of the above mentioned drawbacks [4.14–17]. The authors [4.12] have shown that during the slow hot straining of the low magnesium-based alloy, fine recrystallized grains appear in the boundaries of original grains. However, these papers provide no data on the optimal temperature-rate conditions of the fine grained structure formation and ways of converting the material to a superplastic state.

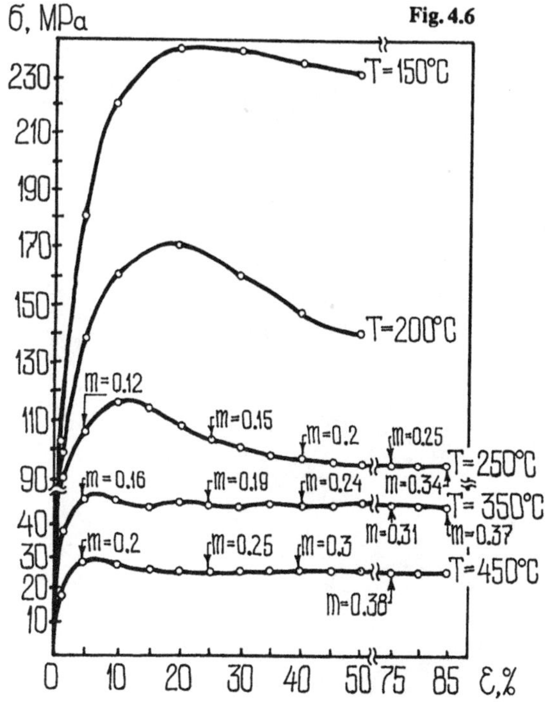

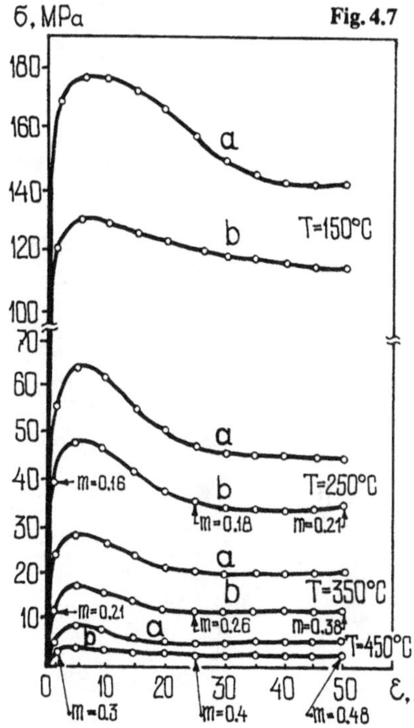

Fig. 4.6. True $\sigma - \varepsilon$ curves of the MA14 alloy: ($\dot{\varepsilon} = 2.8 \times 10^{-3}\,s^{-1}$)

Fig. 4.7. True $\sigma - \varepsilon$ curves of the MA21 alloy: (*a*) $\dot{\varepsilon} = 4.2 \times 10^{-3}\,s^{-1}$; (*b*) $\dot{\varepsilon} = 4.2 \times 10^{-4}\,s^{-1}$

The as-cast MA14 and MA21 alloys were used for a systematic study of the formation of fine grained structure during deformation [4.18–20]. Their original grain sizes were 85 and 50–60 μm, respectively. The samples were upset at temperatures ranging from 150 to 500°C with instantaneous quenching immediately after stopping the testing machine. At room temperature the deformation was performed on the Bridgeman anvil.

The stress-strain curves of both alloys (Fig. 4.6, 7) are typical of dynamic recrystallization [4.16, 17]. It can be seen that when the deformation temperature is increased, the flow stress decreases, especially remarkable in MA21. At the same time, the maximum of the flow stress and the corresponding strain level also decrease. The effect of temperature on the strain rate sensitivity coefficient is also pronounced. The coefficient m increases with the rising temperature. In the original state, at 450°C, the value of m of the magnesium-lithium alloy becomes 0.3, which is an indication of the "natural" superplasticity of this material at this temperature.

The dynamics of the variation of the coefficient m during plastic flow is determined by the temperature of deformation. In MA14, m exceeds 0.3 at 450°C after 40% strain, and at 350 and 250°C the strain levels must be 70 and 80%, respectively. In the magnesium-lithium alloy, the values of m after 40% strain

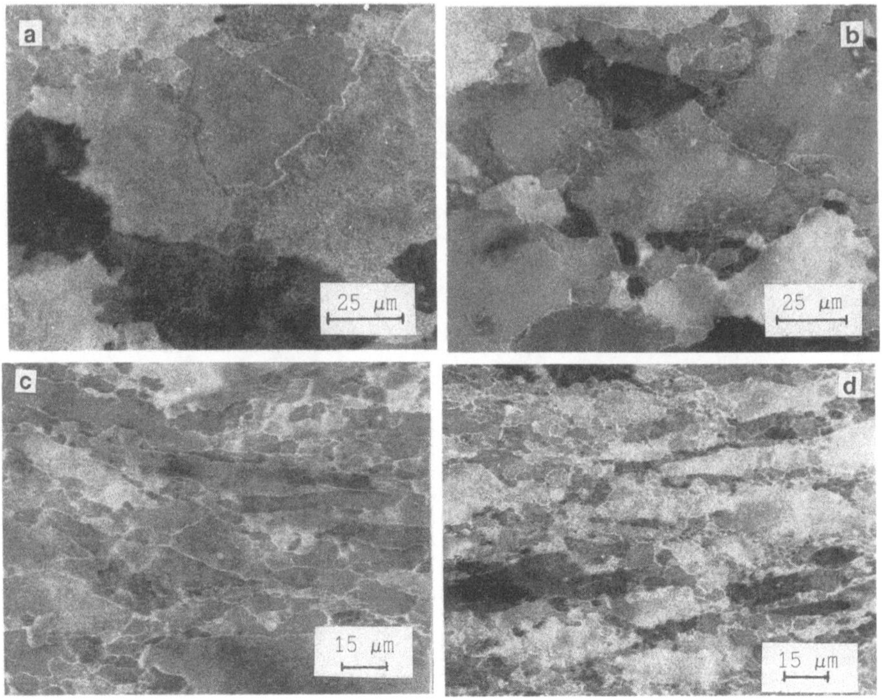

Fig. 4.8a–d. Microstructure of the MA14 alloy after hot deformation (T = 450°C, $\dot{\varepsilon} = 2.8 \times 10^{-3}\,s^{-1}$): (**a**) ε = 15%; (**b**) ε = 25%; (**c**) ε = 50%; (**d**) ε = 75%

attain 0.38 and 0.48 at 350 and 450°C, respectively. Plastic flow at a lower temperature (250°C) does not convert the material to an SP state.

Microstructural changes are different in the two magnesium alloys. In the MA14 alloy, in the temperature range from 150 to 500°C, the recrystallized grains are formed predominantly near the original high-angle boundaries. The formation of new grains in the place where twins intersect occurs at 150–200°C, but their number is insignificant when compared to the total amount of recrystallized grains. Below is a detailed consideration of the structure evolution at 450°C.

Changes in the microstructure become pronounced at 15% strain (Fig. 4.8a). The grain boundaries become wavy, local migration of their segments takes place, and they become the sites of the nucleation of new grains with high-angle misorientation. Grains with orientations similar those of the original grains, appear near some of the triple junctions. It should be noted that this is characteristic only of this temperature. After 25% strain, the shape of the original grains changes (Fig. 4.8b); they become elongated in the direction of the metal flow. They are surrounded with chains of new grains, the triple junction areas being recrystallized more rapidly. At 50% strain, the fraction of recrystallized grains increases in the alloy (Fig. 4.8c). At the same time, fine, equiaxial recrystallized grains are formed. Further deformation causes no qualitative change in the character of microstructure (Fig. 4.8d). The only thing worth mentioning is the increas-

Table 4.4. Temperature and strain dependence of the size and specific volume of recrystallized grains in the MA14 alloy at $\dot{\varepsilon} = 2.8 \times 10^{-3}\,s^{-1}$. The numerator designates the mean grain size of recrystallized grains [μm]; the denominator designates their volume fraction [%]

t [°C]	Size and specific volume of recrystallized grains							
	True strain (e)							
	0.05	0.16	0.29	0.5	1	1.38	4	7
20	–	–		$\frac{0.15}{1}$	$\frac{0.14}{4}$	–	$\frac{0.14}{42}$	$\frac{0.14}{94}$
250	–	$\frac{3}{4.7}$	$\frac{2.6}{8.5}$	$\frac{2.6}{20}$	–	$\frac{2.5}{41}$	–	–
350	$\frac{11.7}{1.8}$	$\frac{12.5}{9.5}$	$\frac{9}{19}$	$\frac{9.1}{42}$	–	$\frac{7.9}{93}$	–	–
450	$\frac{19.4}{3}$	$\frac{19}{4}$	$\frac{18.2}{32}$	$\frac{18}{42}$	–	$\frac{16}{96}$	–	–

ing similarity of the orientations of the groups of recrystallized grains with the growing strain level and the appearance of new subgrains observed in polarized light.

Similar structural changes are typical for all temperatures of deformation except room temperature, in which case the formation of new grains is observed throughout the material volume.

The quantitative parameters of the microstructure formed during plastic deformation are presented in Table 4.4, which shows that the specific volume and average size of the recrystallized grains depend on the temperature of deformation and increase as it rises. The growth of the strain level results in a continuous growth of the volume of the recrystallized grains and in a certain reduction of their size. A drop in temperature, especially below 300°C, slows the recrystallization down. The formation of a completely recrystallized structure at low temperatures is possible only after large strains ($\varepsilon = 4$–7).

The analysis of the deformation relief and fine structure of the MA14 alloy [4.18, 19] indicated that temperature determines the mechanism of deformation and the type of substructure formed at the early stages of plastic flow. At high temperatures, the development of multiple slip in the vicinity of original grain boundaries leads to the formation of low-angle boundaries (Fig. 4.9a). With the growing strain level, the density of dislocations in them increases. As a result, the subboundary misorientation angle grows, and the low-angle boundaries are transformed into high-angle.

During deformation at room temperature, dense mutually misoriented flat dislocation pile-ups appear in the original grain interior. These pile-ups become sites on which recrystallized grains, sized 0.1–0.2 μm, are formed under higher strains, which is probably due to the hopping rearrangement of the dislocation pile-ups into high-angle boundaries (Fig. 4.9b, c). At moderate temperatures, a substructure of an intermediate type is formed.

Comparing the above described results of mechanical tests and structural studies, it can be seen that, depending on the temperature of deformation, the

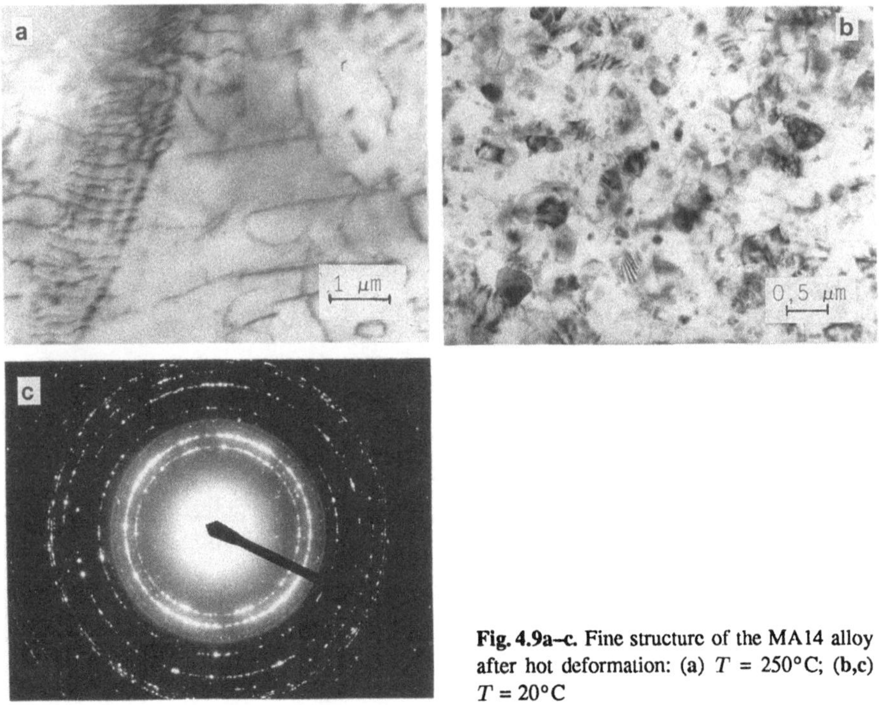

Fig. 4.9a–c. Fine structure of the MA14 alloy after hot deformation: (a) T = 250°C; (b,c) T = 20°C

MA14 alloy starts to exhibit SP symptoms with different specific volumes of recrystallized grains. The lower the temperature of deformation, the larger must be the volume fraction of recrystallized grains required for transforming the alloy to the superplastic state. The highest superplasticity is manifested by a material with a completely recrystallized structure.

We now consider the influence of the microstructure of high magnesium alloys formed under different temperature-rate conditions of deformation on the SP parameters. To do this, completely recrystallized structures with the grain sizes of 16 and 2.5 µm were obtained in MA14 by means of plastic deformation at large strain levels at 450 and 250°C, respectively. The mechanical tests of the two states at these temperatures indicated that the material displayed SP features (Fig. 4.10). The main distinctions are observed in the levels of the flow stress which considerably increase with the decreasing temperature of deformation, in spite of the simultaneous reduction of the grain size. At the same time, the values of the strain rate sensitivity coefficient m are similar, and in the interval of optimal strain rates they are substantially larger than 0.3.

Structural changes in MA21 are of a different character. At high and moderate temperatures (250–350°C), recrystallized grains are formed throughout the β-phase volume and in a zone free of the Al-Li lamellar precipitations in the vicinity of the α/β interphase boundaries in the α-phase. In the latter, recrystallization begins only when the temperature of deformation exceeds 350°C and the thin Al-Li-based lamellas of the phase dissociate.

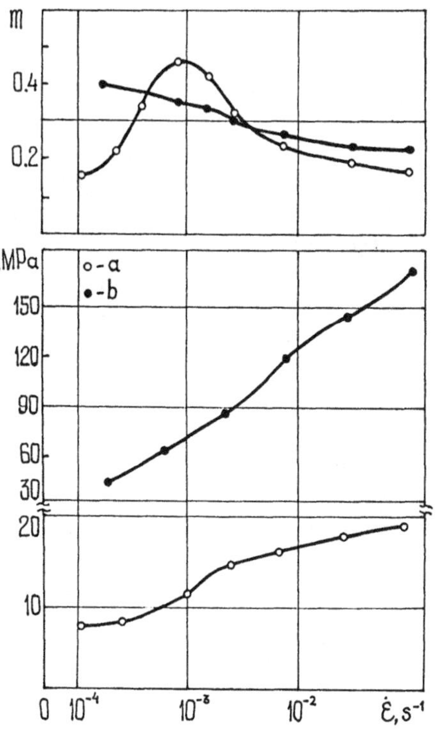

Fig. 4.10. Temperature dependence of the flow stress σ and coefficient of strain rate sensitivity m of the MA14 alloy at (o) T = 450°C; (•) T = 250°C

A detailed study of the microstructure evolution in the magnesium component at 450°C and at a strain rate of $4.2 \times 10^{-3}\ s^{-1}$ indicates that, at the initial stage of plastic flow, a heterogeneous substructure is formed in the α-phase (Fig. 4.11a) both near the interphase boundaries and within the α-grains. As the strain level increases to 25%, high-angle grain boundaries are formed in the vicinity of interphase boundaries and across the α-lamellas. The recrystallized grains are from 2 to 4 times larger than the subgrains.

Further deformation results in the growth of the interphase boundary area in the α-phase and in the precipitation of the α-phase in them, a process which is nonuniform within the α-phase. There is experimental evidence for the shift of the α-grains against one another. As a result, after 85% strain and more, a microduplex structure is formed (Fig. 4.11b).

In the lithium component, the symptoms of recrystallization are displayed starting at 250°C. At the initial stage of deformation (ε = 10%), a substructure is created throughout the β-phase (Fig. 4.11c). As the strain level increases, grains with high-angle boundaries are formed at the site of the subboundary structure (Fig. 4.11d). Their size coincides with the size of subgrains detected at early stages of plastic flow. As the testing temperature rises, the sequence of the microstructural changes in the β-phase is retained; however, the size of the recrystallized grains increases and new grains appear at lower strains.

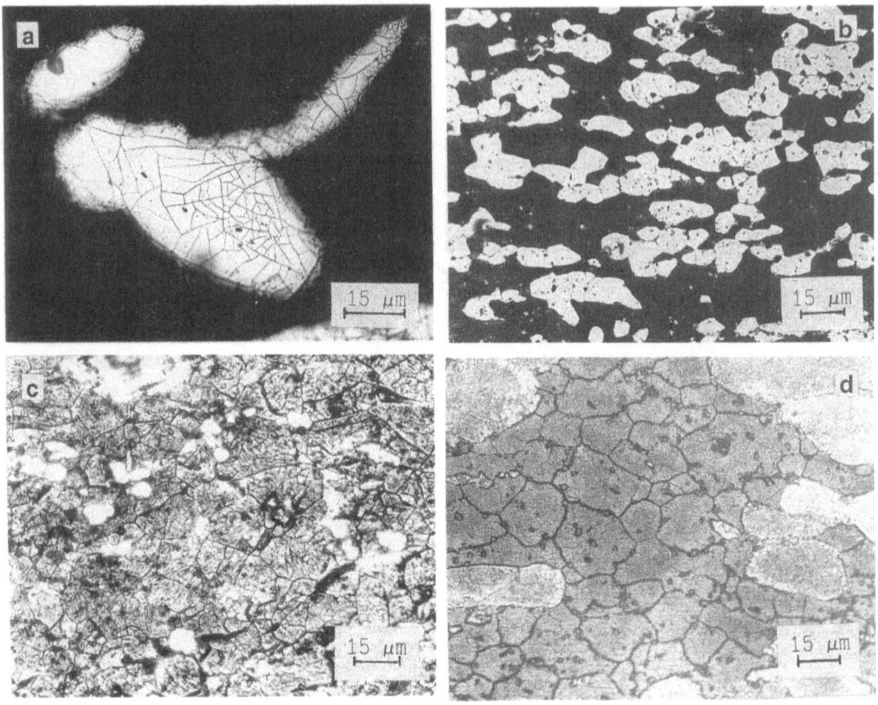

Fig. 4.11a–d. Microstructure of the MA21 alloy after plastic deformation ($\dot{\varepsilon} = 4.2 \times 10^{-4}\,s^{-1}$); (a) $T = 450°C$, $\varepsilon = 50\%$; (b) $T = 450°C$, $\varepsilon = 85\%$; (c) $T = 350°C$, $\varepsilon = 10\%$; (d) $T = 350°C$, $\varepsilon = 75\%$

The regularities of the transition of materials to an SP state established by the analysis of the MA14 alloy behaviour during plastic deformation are valid for the magnesium-lithium alloy with a duplex structure and yield rather interesting results. At temperatures of 350°C or higher, MA21 displays all the SP symptoms after recrystallization has been accomplished only in the β-phase, while at lower temperatures, a complete recrystallization of the lithium phase, though inducing a slight increase in the coefficient m value, fails to transfer the MA21 alloy to an SP state. The highest values of the coefficient m and the lowest flow stresses are observed in MA21 after a microduplex structure has been formed in it.

Contrary to MA14, in MA21 new grains are formed relatively uniformly throughout both the α- and β-phases. Such a difference in the character of dynamic recrystallization is associated with the difference in the actual mechanisms of deformation operating in the two alloys.

In MA14, the multiple slip is localized near the original boundaries where dynamic recrystallization is in progress [4.18]. In MA21, the multiple slip uniformly develops in both phases [4.6, 20], which results in the formation of recrystallized grains uniformly distributed throughout the material volume. The coexistence of dynamic recrystallization in the α-phase with the dynamic precipitation of the β-phase in the α/α boundaries observed in this magnesium-lithium alloy and the shear of the α-grains against one another under the effect of grain boundary sliding provide for the formation of the microduplex structure.

Thus, during plastic deformation dynamic recrystallization develops in magnesium alloys with both matrix and duplex structures. The size of recrystallized grains depends on the temperature-rate conditions of deformation and can vary in a broad range. The specific volume of the recrystallized grains is determined by the strain level. In magnesium alloys, a microcrystalline structure can be obtained by varying the temperature-rate conditions and strain level.

The smooth growth of the coefficient m during hot deformation is associated with the continuously increasing recrystallized volume fraction in the microstructure of magnesium alloys. In these materials it is possible to combine the preparation of structure and SP deformation in one technological process.

4.3 Effect of Superplastic Deformation on the Properties of the Alloys

From the practical standpoint, it is necessary to evaluate the effect of a strain under the conditions of SP flow on the service characteristics of the magnesium alloys, which are known to exhibit low strength and poor ductility; hence, the level of properties attained during processing is of practical importance not only for evaluating the efficiency of the latter, but also in determining the applicability of these alloys for manufacturing products.

The temperature regimes of superplastic deformation of industrial magnesium alloys (Sect. 4.1) are near the temperatures of their heat treatment [4.4, 5]; consequently, there exists the possibility of using superplastic deformation not only as a method of forming products from these alloys, but also for improving the set mechanical properties when used in combination with heat treatment [4.21]. In this case, it is important to establish whether superplastic deformation affects the variation of the service properties, whether this effect is related to the microstructural refinement of the alloys, and, finally, whether superplastic deformation has an effect on the subsequent phase transformation in the precipitation hardened alloys.

In order to solve these problems, we carried out a study on the MA15 and MA21 alloys. The choice of these alloys is not accidental. The MA15 alloy cannot be hardened thermally and serves as an example for evaluating the effect of superplastic deformation on the multitude of properties especially on their anisotropy in the "pure form" in the absence of phase transformations during the process of cooling.

The MA21 alloy can be hardened thermally, and its structure and properties undergo significant changes during heat treatment. In view of this, there exists the possibility of not only establishing the complex effect of superplastic deformation and subsequent heat treatment on the mechanical properties, but also evaluating the effect of SP deformation on the nature of phase transformations. The MA21

Table 4.5. The room temperature mechanical properties of the MA15 alloy after different types of treatments

Treatment	In the longitudinal direction			At an angle of 45° relative to the axis of pressure forging			In the transverse direction		
	σ_u [MPa]	$\sigma_{0.2}$ [MPa]	δ [%]	σ_u [MPa]	$\sigma_{0.2}$ [MPa]	δ [%]	σ_u [MPa]	$\sigma_{0.2}$ [MPa]	δ [%]
Hot pressure forging (initial state)	305	263	17.5	242	153	16	240	130	12
Annealing at 450°C, 1 h	298	265	20	238	140	18	230	121	15
SP deformation ($\varepsilon = 50\%$) at 450°C, $\dot{\varepsilon} = 4.1 \times 10^{-3}\,s^{-1}$	256	151	25	254	149	25	251	141	25

was also interesting because there had been no investigations concerning the lack of stability of its mechanical properties.

Let us examine the effect of SP deformation on the mechanical properties of the MA15 alloy. In this case, it is important to compare the room temperature mechanical properties of the alloy in its original condition (hot pressure-forged rods), after deforming in the regime of SP flow, and after annealing at the temperature of SP deformation (450°C). The test results presented in Table 4.5 show that in the original hot pressure-forged condition, the MA15 alloy exhibits extremely high anisotropy of mechanical properties in that the strength and ductility characteristics are relatively high in the longitudinal direction and are significantly higher than those in the transverse direction. The most significant difference is observed in the flow stress level. It is interesting to note that a decrease in the total level of the strength properties occurs when the angle between the direction in the specimen is cut out and the direction of pressure forging increases from 0 to 45°C; a subsequent increase of the angle from 45° to 90° affects σ_u and $\sigma_{0.2}$ insignificantly. Maximum and minimum values of ductility are observed in the direction of pressure forging and at an angle of 90°, respectively, and the relative elongation differs by approximately 1.5 times. Annealing after forging slightly decreases their strength and increases their ductility, but does not virtually affect their anisotropy. After superplastic deformation, the properties of the MA15 alloy become isotropic. In this case, there is a reduction in the level of strength in the longitudinal direction and an increase (by 10–20 MPa) of the properties in the transverse direction as compared to the corresponding properties of the original hot-forged bars.

Prior superplastic deformation has the most significant effect on the ductility of the alloy. After SP deformation, relative elongation increases by 1.5–2 times. The large reserve of ductility creates favourable conditions for strengthening the alloy by work hardening. It is possible to increase strength at the expense of a slight decrease of the reserve of ductility. While essentially small deformations (5–15%) are used during work hardening of the magnesium alloys, the isotropy of the mechanical properties obtained after working in the SP state is retained.

The presented data show that SP deformation has a favourable effect on the mechanical properties of the MA15 alloy, ensures isotropy of the properties, and increases its room temperature ductility.

Let us now describe the reasons for the positive effect of superplasticity on the mechanical properties of the magnesium alloys.

The increased ductility and the significant decrease of the anisotropy of mechanical properties cannot be attributed simply to the effect of annealing alone. This fact emerges from a comparison of the properties of the pressure-forged and annealed alloys. Metallographic analysis showed that the microstructure of the MA15 alloy in the original (hot-forged) condition is characterized by a certain degree of nonuniformity. In addition to the fine-grained recrystallized structure (grain size of about 15 μm), one encounters narrow deformation bands (with nonrecrystallized structure) that are dispositioned parallel to the axis of forging. In the experimental material the volume fraction of the deformation bands is not high (15–20%). Electron microscopic studies established that these bands contain numerous dislocation pile-ups and dislocation tangles, i.e., a dislocation structure in the bands that resembles the structure of cold worked materials. After annealing at 450°C, the structure becomes more uniform and a recrystallized fine-grained structure is developed within the bands. On the whole, the microstructure does not undergo significant changes during annealing and resembles the original structure of the material.

The microstructure of the SP deformed alloy was studied on the specimens obtained by upsetting the original hot-forged bars, maintaining a deformation of 50%. It was found that the microstructure of the alloy subjected to SP deformation does not differ significantly from that of the alloy subjected to annealing. During the deformation process, the grain size changes insignificantly (the grains coarsen by 1–2 μm on the average), and the equiaxial nature of the grains is retained. However, the structural nonuniformity of the original bar disappeared. In addition, the dislocation pile-ups and the sub-structure are absent after superplastic flow, and only individual dislocations are found in the grains of the alloy. Based on the significant differences observed in the properties of the two alloys, it can be assumed that the property variation of the alloy is associated with the change in the crystallographic texture.

A comparison of the crystallographic texture (Fig. 4.12) shows that the initial deformation texture does not significantly change during the annealing process. One observes only an insignificant blurring of the pole density in the transverse direction and some weakening of the texture maxima at the periphery of the pole figure. Due to this and the negligible microstructural changes occurring during annealing, the alloy retains its original level of the mechanical properties.

In addition, during the process of SP flow in the MA15 alloy, significant changes in the crystallographic texture are observed. Due to intense grain bound-

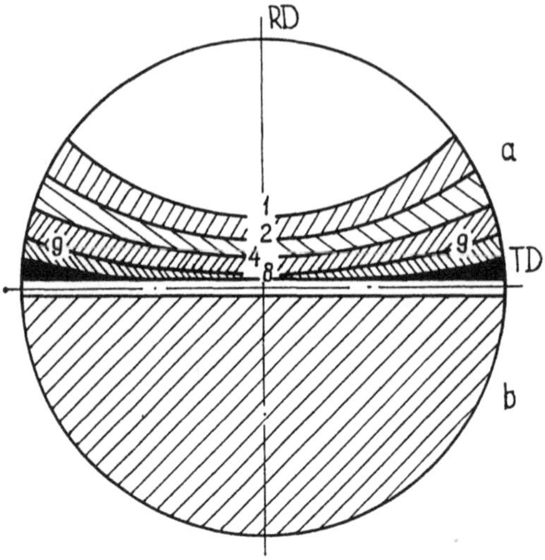

Fig. 4.12. Pole figures (0001) of the MA15 alloy after annealing at 450°C (*a*) and SP deformation at 450°C and $\dot{\varepsilon} = 4.1 \times 10^{-3}\,s^{-1}$ (*b*)

ary sliding, the original texture peaks are severely blurred, and after superplastically straining the alloy by at least 50%, it shows the formation of texture with uniform distribution of crystallographic grain orientation in the space. The sharp decrease in the anisotropy of the mechanical properties and the increased ductility of the MA15 alloy can be attributed to these facts.

Thus, superplastic deformation of the magnesium alloys makes avoids the shortcomings of other strengthening treatments, because SP deformation not only retains the fine-grained microstructure, but also leads to a virtually textureless condition and, thereby, ensures a high degree of isotropy of the mechanical properties. Therefore, SP deformation can be recommended as an effective method for eliminating the anisotropy of mechanical properties of magnesium alloys. In a given sample, the variation of properties is due to specific changes occurring in the microstructure during SP deformation. It would be interesting to establish the effect of superplasticity on subsequent phase transformations of the alloys.

Let us now examine the effect of SP deformation on the structure of the properties of the precipitation hardened MA21 magnesium-lithium alloy. It is known that these alloys have virtually no industrial application without a special strengthening treatment, because low constructional strength, and, particularly, an instability of mechanical properties as a function of time [4.3, 5]. In the forgings obtained under industrial conditions, σ_u = 210–230 MPa and $\sigma_{0.2}$ = 120–140 MPa, which is significantly less than the properties of the alloy after strengthening treatments [4.5]. Using a special thermal or thermomechanical strengthening treatment is not effective for improving the strength characteristics of the alloys since the level of the properties obtained in this case are unstable.

It should be noted that in the literature only a few suppositions concerning the factors determining the variation of mechanical properties under prolonged holding or service of magnesium-lithium alloys can be found. The absence of reliable structural data does not permit one to clearly establish the reasons for the variation of mechanical properties with time. Thus, special investigations are required.

The mechanical properties of the MA21 alloy were studied in the following states: (i) in the hot pressure-forged (original) state; (ii) after SP deformation at 350°C with subsequent water quenching; and (iii) after strengthening heat treatment (heating up to 350°C, holding at this temperature, and quenching and natural aging). Since quenching temperature coincides with the temperature of SP deformation, it was possible to distinguish the effect of SP deformation only on the properties of the alloy.

The obtained experimental data (Tables 4.6, 7) indicate that SP deformation significantly enhances the mechanical properties of the MA21 alloy at room temperature. Quenching from 350°C, followed by aging also has a favourable effect on the strength. However, this effect is less than that of SP deformation, whose effect is most pronounced in the stability of mechanical properties. It is known that MA21 tends to soften after long-term service or holding exhibited sufficient stability of its mechanical properties after SP deformation. The properties obtained after holding at room temperature for three years were at the same level as those found immediately after SP deformation. The properties of the alloy after heat treatment (quenching and aging) are unstable at ambient temperature and within two-three months, they become quite similar to the properties of the original hot pressure-forged bar (Table 4.7).

Let us now examine the possible reasons for the favourable effect of SP deformation on the mechanical properties of magnesium-lithium alloy MA21. These properties are enhanced by a combination of two effects, viz., strengthening caused by quenching of the alloy from the tempereature of SP deformation and the improvement of the properties due to the specific structural changes occurring during the SP flow. In order to establish the role and the nature of thermal strengthening, let us analyze the phase constitution of the alloy after different types of treatments.

According to published data on the Mg-Li-Al ternany diagram, the MA21 alloy consists of the magnesium-rich α-solid solution, the lithium-rich β-solid solution, and a certain amount of the excess AlLi phase and the manganese-based

Table 4.6. Mechanical properties of the MA21 alloy after different types of treatments

Treatment	σ_u [MPa]	$\sigma_{0.2}$ [MPa]	σ_{-1} [MPa]	δ [%]	ϕ [%]
Hot pressure-forging (initial state)	231	167	–	18	26.0
Heat treatment, quenching from 350°C and natural aging	268	201	–	17	23
Superplastic deformation with subsequent water quenching and natural aging	286	220	111	16	26.0

Table 4.7. Effect of prolonged holding at room temperature on the stability of mechanical properties of the MA21 alloy

Treatment	7 days			2 months			6 months			12 months			24 months			36 months		
	σ_u [MPa]	$\sigma_{0.2}$ [MPa]	δ [%]	σ_u [MPa]	$\sigma_{0.2}$ [MPa]	δ [%]	σ_u [MPa]	$\sigma_{0.2}$ [MPa]	δ [%]	σ_u [MPa]	$\sigma_{0.2}$ [MPa]	δ [%]	σ_u [MPa]	$\sigma_{0.2}$ [MPa]	δ [%]	σ_u [MPa]	$\sigma_{0.2}$ [MPa]	δ [%]
Heat treatment, quenching from 350°C and natural aging	268	201	16	252	187	18	242	172	18.5	240	170	19.5	238	171	19	240	172	18
Deformation under SP conditions with subsequent quenching and natural aging	286	220	16	284	212	18	277	213	17	285	214	17	280	217	16	290	217	16

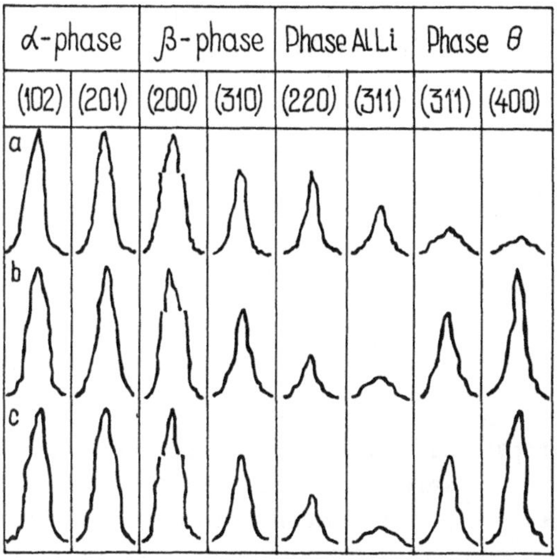

Fig. 4.13. Influence of different treatments on the phase composition of the MA21 alloy. The intensities of X-ray reflection lines are given for different phases of the alloy after three types of treatment: *a, b,* and *c*. (*a*) initial hot-extruded state; (*b*) heat hardening; (*c*) SP deformation followed by quenching

phase. The additions of zinc and cadmium do not change the phase constitution of the alloy [4.22]. The presence of the equilibrium AlLi phase in the material does not lead to strengthening during heat treatment [4.22]. Thus, the mechanism of the thermal strengthening of the MA21 alloy remains unclear.

In order to understand the reasons underlying the change in the mechanical properties, we carried out a detailed study of the phase composition of the alloy. According to X-ray data (Fig. 4.13), the diffraction patterns of the alloy in its original state show additional lines in addition to the known diffraction lines of the α-, β- and AlLi phases. These lines were indexed as the reflections belonging to the phase having a composition corresponding to $MgLi_2X$, where X can be Al, Zn, or Cd (the θ-phase). A study of the phase composition at high temperatures made it possible to determine the temperature ranges in which the excess phases exist. It was found that the lines of the θ-phase are retained in the X-ray diffractograms after heating up to 175°C. On further heating, they disappear, which allows us to assume that this phase either decomposes to form an α-phase that is closer to equilibrium or enters the solid solution. The diffraction lines of the AlLi phase are present in the diffraction patterns up to 350°C. The reflections of these phases disappear only at 375°C. It is virtually impossible to observe a homogeneous solid solution of the α- and β-phases. We observed the lines of the θ-phase in all the specimens after quenching from the homogenization temperature.

A study of the phase and the structural constituents of the alloys was also carried out using light microscopy. For this purpose, the usual methods of microanalysis are inapplicable because of the insignificant difference in the etching ability of different phases. Under these conditions, the carrying out of metallographic observations in polarized light was effective since there is a contrast in the colour contours of different phase and structural constituents of the alloy.

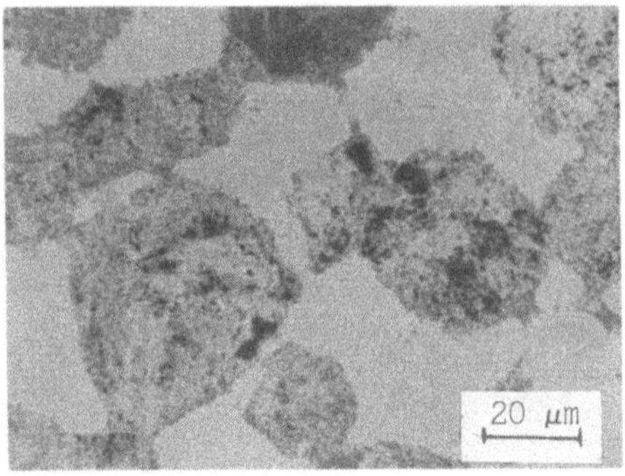

Fig. 4.14. Microstructure of the MA21 alloy in polarized light

The metallographic examinations of the alloy showed that in addition to the α- and β-solid solutions, there are excess phases with different contrast (Fig. 4.14). X-ray microspectral analysis made it possible to qualitatively evaluate the nature of these phases. Thus, it was established that zinc and cadmium dissolve in the AlLi phase.

The inclusion-like dark precipitates form an aluminium-rich manganese-based phase. The aluminium content of these inclusions is as high as 25%.

Within a single grain of β-phase, there are regions with different contrast. This indicates the chemical inhomogeneity of the phase and is probably due to the precipication of the disperse secondary θ-phase particles within the β-grains, the density of inclusions being different in different regions of the β-grains. It was not possible to establish the nature of these inclusions from X-ray microspectral analysis since the size of the particles exceeds the resolving power of this method.

Electron microscopy studies made it possible to confirm the aforementioned assumption and showed that the θ-phase is present in the alloy as finely dispersed precipitates within the grains of β-solid solution (Fig. 4.15). The electron diffraction pattern of the finely dispersed particles consists of rings. Indexing the circular lines showed that they belong to the θ-phase. The point reflections of the electron diffraction pattern that are located between the rings were indexed as the matrix-reflections on the bcc β-phase. However, it was possible to reveal the correspondence between the given reflections and the individual structural elements only by observing the object in dark-field image (Fig. 4.15b). This method was useful for identifying the phases in the alloy.

According to the results of electron diffraction microscopy, the AlLi phase in the MA21 alloy is distributed in the form of isolated and clustered inclusions within the grains of the β-phase at the interphase grain boundaries.

Thus, based on numerous studies of the structure of the MA21 alloy in the original hot pressure-forged state, the presence of the following phases was reliably established: the magnesium-based α-solid solution, the lithium-rich β-solid solution, the AlLi-phase, and the θ-phase, having a formula $MgLi_2X$ where X stands for aluminium, zinc or cadmium.

Let us now examine the phase composition of the MA21 alloy after various types of treatments and attempt to establish the correlation between the variation

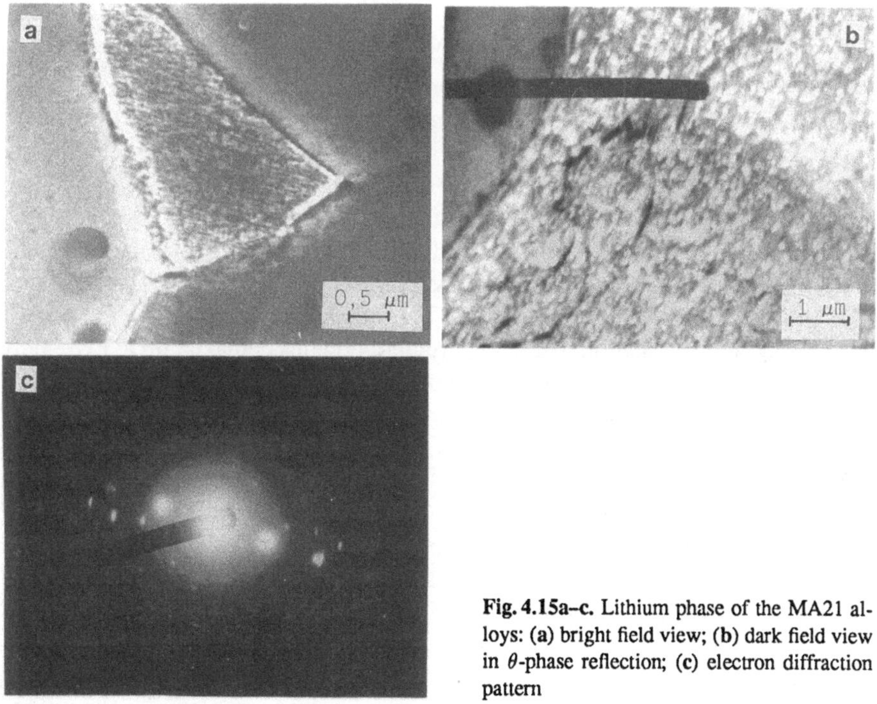

Fig. 4.15a–c. Lithium phase of the MA21 alloys: (a) bright field view; (b) dark field view in θ-phase reflection; (c) electron diffraction pattern

of the properties and the structural changes. No qualitative change in the phase composition is observed as a function of the type of the treatment. However, there is a distinct difference in the quantitative content of the excess phases θ and AlLi, and in their morphology and distribution in the alloy.

A comparative analysis of the integral intensity of the (200) lines of the AlLi phase and the (400) lines of the θ-phase showed that the content of the AlLi phase depends on the quantity of the θ-phase in the MA21 alloy. After heat treatment, superplastic deformation and strengthening, a small amount of the AlLi phase corresponds to a high θ-phase content in the material. A reverse dependence is observed in the original hot-forged bar.

Based on these data, it can be concluded that the θ-phase has a transitional structure and that during the decomposition of the β solid solution, the phase transformations in the alloy occur according to the following scheme: β solid solution $\rightarrow \theta \rightarrow$ AlLi which agrees with the previously proposed scheme [4.3].

A comparative analysis of the variation of the mechanical properties (Table 4.7) and the phase composition of the MA21 alloy (Fig. 4.13) gives a basis for assuming that its thermal strengthening is due to precipitation of the finely dispersed θ-phase from the solid solution during heat treatment; the increased level of the mechanical properties depends on the high content of this phase. In fact, in the original (hot pressure-forged) state, the alloy has a low θ-phase content, and its properties are significantly inferior to those obtained after SP deformation and quenching, which abruptly increases the θ-phase content. The aforementioned assumption is also supported by the fact that during the annealing of the quenched specimens, a distinct correlation between the reduction of strength and reduction in the integral intensity of the diffraction lines of the θ-phase can be observed.

Natural aging for a period of 24–30 h led to a certain increase in the integral intensity of the lines belonging to the θ-phase. In this case, the level of strength increased by 10–20 MPa. At longer durations of aging (300–500 h), the alloy showed overaging followed by a deterioration of its strength. In addition, the microstructural changes occurring in the alloy during SP deformation affect its strength. Let us examine these changes in more detail.

During SP deformation, the microstructure remains equiaxial up to extremely high strains or transforms into an equiaxial structure during the deformation process if the initial microstructure of the material is nonequiaxial (Chap. 2). In the MA21 alloy, grains initially elongated in the direction of pressure-forging become equiaxial during deformation and the nonuniformity of the grain size disappears; furthermore, the mutual displacement of the grains of the phases relative to each other due to intense grain boundary sliding leads to the redistribution of the grains and to the formation of a eutectoid-type structure (Fig. 4.16). The microstructural changes coexist with the equalization of the chemical composition of the phase

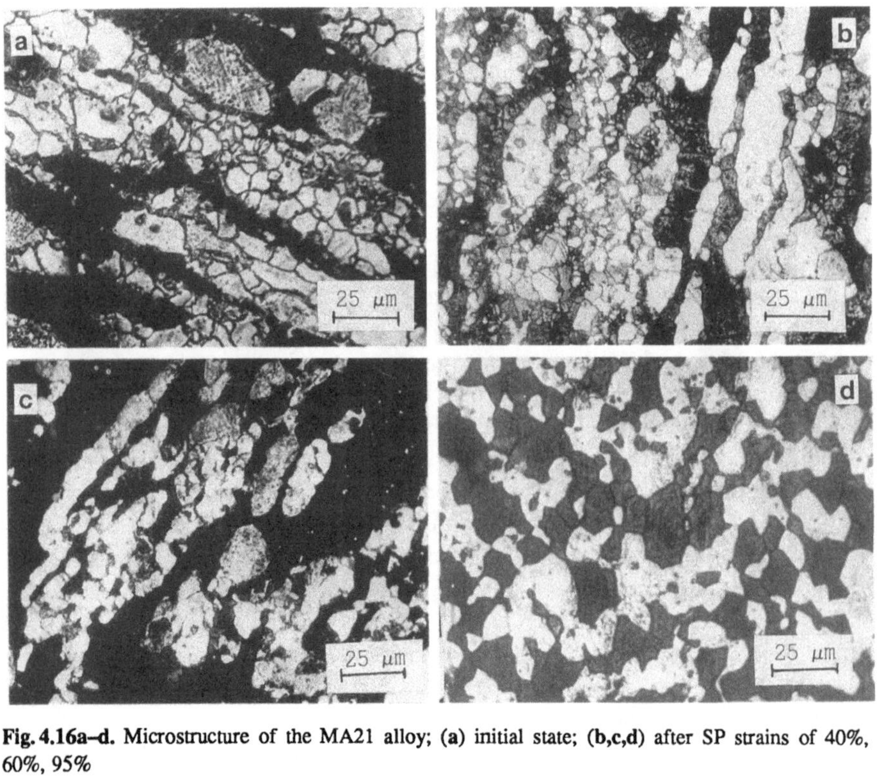

Fig. 4.16a–d. Microstructure of the MA21 alloy; **(a)** initial state; **(b,c,d)** after SP strains of 40%, 60%, 95%

Table 4.8. Mechanical properties of the MA21 alloy as a function of SP strain level

ε_{SPD} [%]	σ_u [MPa]	$\sigma_{0.2}$ [MPa]	δ [%]
40	263	200	17.7
60	272	213	11.5
80	286	220	16.4
95	317	262	13.2

constituents (the α-and β-solid solution) and, at the same time, no substructure is formed during the SP deformation, i.e., the alloy assumes a more equilibrium state. Probably, during the process of SP deformation, such structural changes form one of the main factors leading to an additional increase of the strength characteristics and stabilization of the mechanical properties.

This conclusion is also confirmed by data obtained during a combined study of the variation of the mechanical properties and the microstructure of the alloy that was deformed up to different degrees of strains under superplastic flow. In fact, at $\varepsilon = 40\%$, one observes relatively small microstructural changes as compared to the original material. Thus, it is not surprising that the properties of the alloy virtually do not differ from those of the material subjected to the strengthening heat treatment (Table 4.8). With an increasing degree of SP deformation, the structural nonuniformity gradually disappears and at strains exceeding 80% a uniform microduplex-type microstructure with a grain size of 15 μm develops in the alloy. This is accompanied by a corresponding increase of the strength characteristics. At $\varepsilon = 80\%$, the strength is 15–20 MPa higher than that obtained at $\varepsilon = 40\%$ (Table 4.8). The best combination of strength and ductility is achieved in the alloy after subjecting it to the maximum SP deformation ($\varepsilon = 95\%$). In this case, the contribution of SP deformation to the increase of strength amounts to 45–60 MPa as compared to the properties of the material deformed up to $\varepsilon = 40\%$. In addition, superplastic deformation facilitates a reduction of the anisotropy of the properties of the MA21 alloy. In spite of the fact that the MA21 alloy is distinguished by a high degree of isotropy of the strength characteristics in the hot pressure-forged bars in their initial state ($\sigma_u = 231$ MPa and $\sigma_{0.2} = 167$ MPa in the longitudinal direction and $\sigma_u = 223$ and $\sigma_{0.2} = 161$ MPa in the transverse direction), it exhibits distinct anisotropy of ductility, i.e., in the longitudinal and transverse direction, $d = 18\%$ and 12%, respectively. Deformation under SP conditions retains a high isotropy of strength ($\sigma_u = 248$ MPa and $\sigma_{02.} = 220$ MPa in the longitudinal direction, and $\sigma_u = 280$ MPa and $\sigma_{02.} = 213$ MPa in the transverse direction) and sharply decreases the anisotropy of ductility, i.e., in the longitudinal and transverse direction, the relative elongations are 17% and 18%, respectively. As was indicated earlier, this is apparently due to the increased microstructural nonuniformity of the MA21 alloy and the specific features of the crystallographic texture during SP deformation.

Thus, the experimental results show that the improvement in the multitude of mechanical properties obtained after deforming in the regime of SP flow in addition to the strengthening achieved by quenching is due to the microstructural changes occurring during superplastic deformation.

However, it may be noted that, at the present time, highly systematic studies have not yet been carried out for understanding the physical nature of the instability phenomenon of the properties of the magnesium-lithium alloys.

In order to establish the specific relationship between the structural changes and the stability of the mechanical properties of the alloy, we shall examine specific features of the microstructural changes, phase composition, and fine structure of the MA21 alloy subjected to superplastic deformation and strengthening heat treatment.

Electron microscopy made it possible to establish the specific features of the fine structure of the alloy which are not revealed by the conventional optical microscopic analysis (Fig. 4.17). It was established that, following superplastic deformation and quenching, finely dispersed precipitates of the θ-phase are distributed uniformly throughout the β-grains interior. No coarse particles in the grain and interphase boundaries were observed.

After the strengthening heat treatment, the distribution of the precipitates of the θ-phase within the β solid solution is highly nonuniform. There is a certain amount of coarse precipitates within the grains and in the near-boundary zones.

The obtained data convincingly demonstrate that SP deformation has an effect on the decomposition of the supersaturated β-solid solution and on the morphol-

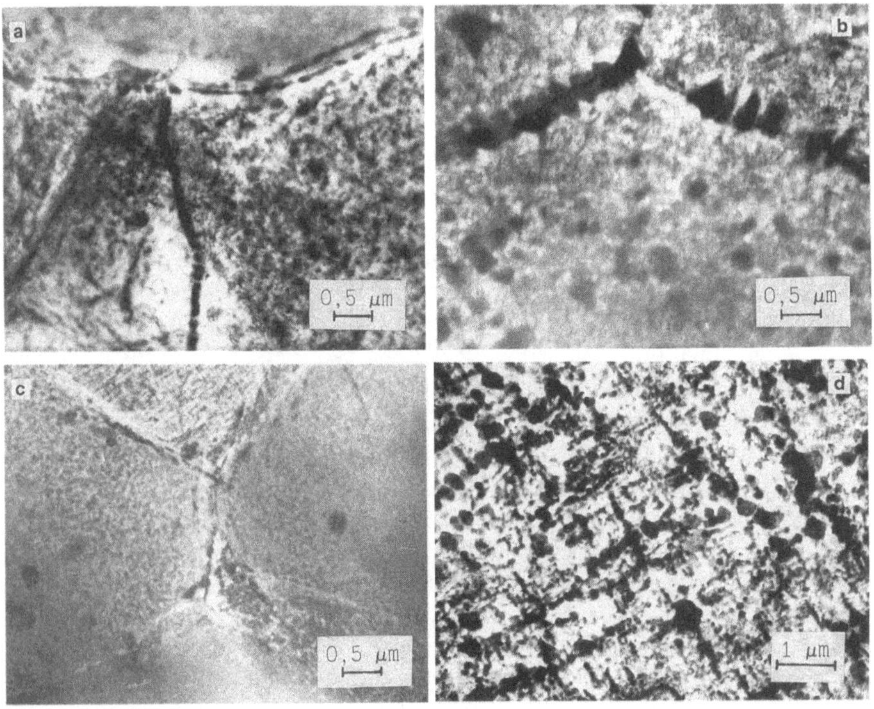

Fig. 4.17a–d. Fine structure of the MA21 alloy after SP deformation **(a,c)** and heat hardening **(b,d)**: **(a,b)** after natural aging for 3 days; **(c,d)** for 3 months

ogy of the precipitated particles. One can expect that the formation of a uniform equilibrium structure is one of the possible reasons for the observed improvement and stablization of the mechanical properties of the magnesium-lithium alloys after superplastic deformation.

The data show that the strengthening of the MA21 alloy after SP deformation and subsequent quenching is different from that after heat treatment. In the former case, the main role was played by the microstructural changes combined with the changes of the structure during quenching, whereas in the latter case, dispersion hardening due to precipitation of the θ-phase from the solid solution was the most important factor.

Below is a more detailed examination of the processes occurring in the alloy during long-term holding or service that finally lead to the alloy's softening after treatment by conventional commercial regimes.

It is known that the properties of multi-phase alloys at room temperatures and elevated temperatures depend on the properties of the phase constituents, i.e., in the case of the MA21 alloy, they are determined by the properties of the α and β solid solutions. It is important to identify the phase in which aging occurs. In order to solve this problem, microhardness measurements were made in the α- and β-phases of the MA21 alloy during the process of holding. The studies were carried out on the alloy deformed in the regime of SP flow and after the strengthening heat treatment. Figure 4.18 shows the effect of the heat treatment and SP deformation on the microhardness of the α- and β-constituents of the alloy as a function of time. These data indicate that during the process of long-term holding (one year), the microhardness of the α-constituent virtually

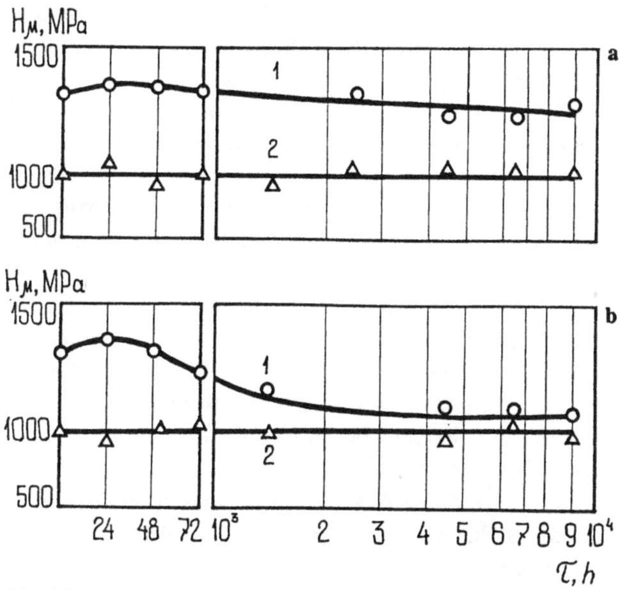

Fig. 4.18a,b. Variation of the microhardness of the MA21 alloy β (Curves *1*) and α-phases (Curves 2): (a) after SP deformation; (b) after heat hardening treatment

does not change, i.e., in the MA21 alloy held at room temperature, aging does not occur in the α-phase.

The microhardness of the β-constituent of the alloy significantly depends on the prior treatment. As the duration of holding increases, the initially similar values of microhardness (H_μ = 1350 MPa), after heat hardening and deformation under SP flow, followed by quenching, deviate considerably. It is interesting to note that in the case of the thermally strengthened alloy, the microhardness of the β solid solution gradually decreases during the first 6 months and, thereafter, it remains virtually unaltered. The microhardness of the β-constituent deformed in the SP regime is a slowly varying function of the duration of holding up to one year at room temperature.

Thus, one can assume that during heat treatment, the strengthening effect observed in the ($\alpha + \beta$) magnesium-lithium alloys is due to strengthening of the β-phase, and the softening of this phase observed during the process of holding the alloy at room temperature leads to instability of its properties. The data of electron microscopic analysis confirmed the aforementioned assumption; a satisfactory correlation was found between the variation of the microhardness and the changes occurring in the structure during the process of prolonged holding. It was established that during the process of long-term holding (three months) following the strengthening heat treatment, coarsening of the particles of the θ-phase occurs in the β-constituent of the alloy, and new precipitates form along the interphase boundaries and the grain boundaries (Fig. 4.17) and give additional reflections in the electron diffraction patterns. Indexing the diffraction patterns showed that these reflections belong to the AlLi phase. A comparative analysis of the changes occurring in the microstructure and the mechanical properties shows that the properties of the alloy deteriorate with time due to significant coarsening of the precipitates of the θ-phase, i.e., during to overaging of the alloy. After processing the alloy in the SP regime, the microstructure is characterized by a high degree of uniformity of the θ-phase precipitates during the initial stages and virtually does not change by an increase in the duration of aging (up to one year) at room temperature.

The difference in the nature of the structural changes occurring during the process of long-term holding after SP deformation and heat treatment is most probably due to chemical inhomogeneity of the β solid solution. In fact, superplastic deformation, caused by an intense activation of the diffusional processes and grain boundary sliding, facilitates a significant homogenization of the chemical composition of the phase constituents of the alloy (Table 4.9), whereas heating prior to quenching (during strengthening heat treatment) does not lead to a significant redistribution of the alloying elements because of the low diffusion rates in the magnesium alloys [4.5]. Thus, observations of the structural state of the alloy during long-term holding showed that softening of the magnesium-lithium alloys after the heat hardening is due to overaging of the the θ-phase and its transformation into the equilibrium AlLi phase.

Thus, the results given in this sectin lead to the following generalizations. Processing of the magnesium alloys under SP conditions is an efficient method

Table 4.9. Content of the alloying elements in the structural constituents of the MA21 alloy

Treatment	Content of the alloying elements [wt.%]											
	α-phase			$\alpha - \beta^{interface}_{region}$			β-phase			$\beta - \beta^{interface}_{region}$		
	Al	Zn	Cd	Al	Zn	Cd	Al	Zn	Cd	Al	Zn	Cd
SP deformation	1.7	1.5	3.2	1.8	1.6	3.2	2.2	2.4	4.2	2.3	2.2	3.8
Strengthen-ing heat treatment	1.7	1.6	3.4	1.8	2.8	4.2	0.94	1.8	4.7	2.8	2.6	4.6

for improving the multitude of mechanical properties. Superplastic deformation treatment (SPDT) is employed with heat magnesium alloys which cannot be hardened. This process makes it possible to obtain isotropic properties in magnesium alloys by producing a textureless state and homogeneous microstructure after SP deformation. Superplastic deformation-cum-heat treatment (SPDHT) is a method of processing applied to heat magnesium alloys which can be hardened. Based on the studies conducted on the MA21 alloy, SP deformation has an effect on the development of transformations in the alloy during subsequent quenching and aging. SPDHT permits one to obtain not only the isotropy of the mechanical properties, but a high level of the room temperature mechanical properties (stable as a function of time) due to the elimination of the chemical and structural nonuniformity in the course of SP deformation.

5. Superplasticity of Commercial Aluminium Alloys

Superplasticity of aluminium alloys was first studied in the Al-33%Cu eutectic alloy [5.1]. After an appropriate treatment, i.e., hot processing, the alloy acquires a microduplex structure and a relative elongation of 2000% at 500°C. Later, superplasticity was observed in other eutectic and near-eutectic aluminium-based alloys, viz., Al-Si, Al-Cu-Mg, etc. [5.1][1].

In view of the poor performance of these alloys, attempts were made at developing new less-alloyed systems with improved mechanical properties and which were superplastically deformable [5.2, 3][2]. Simultaneously, studies of the transformation of well-known commercial alloys into the superplastic state were carried out.

All aluminium alloys in which superplasticity was observed can be divided into two groups: (i) heat hardenable and (ii) heat unhardenable alloys that include both deformable and cast alloys. In recent years, SP deformation of wrought aluminium alloys has become rather important because of possible commercial application of the SP effect. Conventional methods of mechanical working used with this group of alloys might be able to be replaced by processing employing deformation under the conditions of SP flow.

A number of reviews on the SP deformation of aluminium alloys have been published [5.1, 4, 5]. Establishment of the relationship between the chemical composition, phase constitution, and structure of the alloys on the one hand and the characteristics of their superplasticity and performance on the other hand are of considerable theoretical and practical significance. This chapter is a summary of results obtained by the authors and data available in the literature.

---

[1] UK patent No. 1363477 (1974)

[2] UK patent No. 1387586 (1975)

5.1 Effect of Structure, Conditions of Deformation, and Chemical Composition on the Superplasticity of Aluminium Alloys

In order to establish the relationship between the parameters of the SP deformation of aluminium alloys and their chemical composition, phase constitution, and conditions of deformation, a special study was carried out on a number of commercial and model alloys (Table 5.1). The chemical compositions of the tested alloys are typical of the major kinds of aluminium alloys. Among them are the widely used wrought commerical alloys, such as the heat hardenable V96Ts, V93, AK6, AK4-1, and 1420, heat unhardenable AMg6; two-component 1(1.58 Mg), 2(4.1 Cu), and 3(0.5 Zr); three-component 4(1.58 Mg, 0.5 Zr) and 5(4.1 Cu, 0.5 Zr); and commercial-grade (technological purity) aluminium A99. The model alloys contain identical atomic concentrations of magnesium and copper (2 at.%).

Regardless of the differences in the chemical compositions, all the alloys have a common feature, i.e., all of them have a matrix structure which is typical of the wrought aluminium alloys.

Table 5.1. Parameters of superplasticity of tested aluminium alloys. The numerator designates the microcrystalline alloy, the denominator designates the coarse-grained alloy

Alloy (content of the alloying elements, wt.%)	$\bar{d}$ [μm]	t [°C]	$\dot{\varepsilon}_{opt}$ [s^{-1}]	δ [%]	σ_{50} [MPa]	m
AK6 (2.4Cu; 0.68Mg;0.9Si; 0.71Mn)	9–10	515	4×10^{-4}	$\frac{270}{75}$	$\frac{3}{13}$	$\frac{0.55}{0.20}$
AK4-1 (2.3Cu; 1.6Mg; 0.9Fe) 1.1Ni	9–10	530	4×10^{-4}	$\frac{230}{-}$	$\frac{9}{-}$	$\frac{0.45}{-}$
V96Ts (7.56Zn; 2.26Cu; 2.75Mg) 0.15Zr	5–6	470	1×10^{-3}	$\frac{850}{180}$	$\frac{4}{13}$	$\frac{0.60}{0.23}$
AMg6 (6.4Mg; 0.63Mn)	9–10	420	6×10^{-5}	$\frac{410}{150}$	$\frac{4}{12}$	$\frac{0.45}{0.20}$
1420 (5.1Mg; 1.8Li; 0.12Zr)	5–6	450	4×10^{-4}	$\frac{\geq 700}{220}$	$\frac{5}{20}$	$\frac{0.55}{0.22}$
4. (1.58Mg; 0.5Zr)	6–7	500	6×10^{-3}	$\frac{400}{-}$	$\frac{1}{-}$	$\frac{0.33}{-}$
5. (4.1Cu; 0.5Zr	6–7	450	6×10^{-3}	$\frac{500}{-}$	–	$\frac{0.42}{-}$
		500	6×10^{-3}	$\frac{950}{-}$	$\frac{4}{-}$	$\frac{0.53}{-}$

Hot press-forged bars of commercial alloys were used as a starting material. The preforms were subjected to preliminary treatment in order to create a microcrystalline structure with the following average grain sizes: 9–10 μm in AK6 and AK4-1; 5–6 μm in V96Ts and 1420; 4–5 and 9–10 μm in AMg6; and 6 μm in aluminium and model alloys. For comparison, coarse-grained samples of the alloys with a grain size of 35 μm and of aluminium with a grain size of 50 μm were also examined.

Though the aluminium was originally microcrystalline, it displayed no SP flow because the stability of its structure was low, which was manifested in the rapid grain growth during heating and straining [5.6]. In this case, the deformation-induced grain growth exceeds that induced by heating alone at the same temperature.

The low structural stability is also characteristic of the binary model alloys Nos 1 and 2 and of the complex commercial alloy V93 (6.07 Zn, 1.02 Cu, and 1.82 Mg). The rapid grain growth makes superplasticity in these alloys practically impossible. The remaining experimental commercial alloys are of a fairly high structural stability. They exhibit all the characteristic features of superplasticity, viz., a high strain rate sensitivity of the flow stress (m varies from 0.42 to 0.6); no strengthening during deformation; large elongations to fracture (δ = 230–950%) at low flow stresses ($\sigma_{50} = 3 - 9$ MPa) (Table 5.1). In the alloys having a coarse-grained structure, the flow stress was found to be 3–4 times as high, the maximum strains were significantly lower, and the strain rate sensitivity was rather small ($m = 0.2 - 0.25$).

5.1.1 Effect of Structure and Conditions of Deformation. Strain Rate

Figure 5.1 shows the typical strain rate dependence of δ, σ_{50}, and m in the microcrystalline and coarse-grained experimental alloys.

When the structure is microcrystalline (d = 1–10 μm), the strain rate dependence of δ and m of all the alloys (including AMg6 with a grain size of 5 μm) are maximum in the range of $(0.5–2.0) \times 10^{-3}\,s^{-1}$. The range of the optimal strain rates in the alloys studied by other investigators (Table 5.2) is approximately the same ($10^{-3} - 10^{-4}\,s^{-1}$).

The $\ln\sigma - \ln\dot{\varepsilon}$ plots are of a characteristic sigmoidal form. The ternary model alloys also exhibited SP behaviour. In tests where the strain rate was maintained at $10^{-3}\,s^{-1}$ at 500°C, high values of the relative elongation were obtained, in particular, in the Al-Cu-Zr alloy.

The effect of grain size on the parameters of superplasticity in aluminium alloys is similar to that in other materials, i.e., the smaller the grain size, the lower the flow stress, whereas δ and m as well as the optimum strain rate increase (Chap. 1). This proves that the refinement of grains is necessary and practical. A considerable effect has been achieved by creating a submicron crystalline structure ($d < 1$ μm) in these alloys. In this case, the optimum rate of SP deformation shifts towards higher values. For example, the 7475–0.7 Zr alloy prepared by means of powder metallurgy and having a grain size of less than 1 μm exhibits

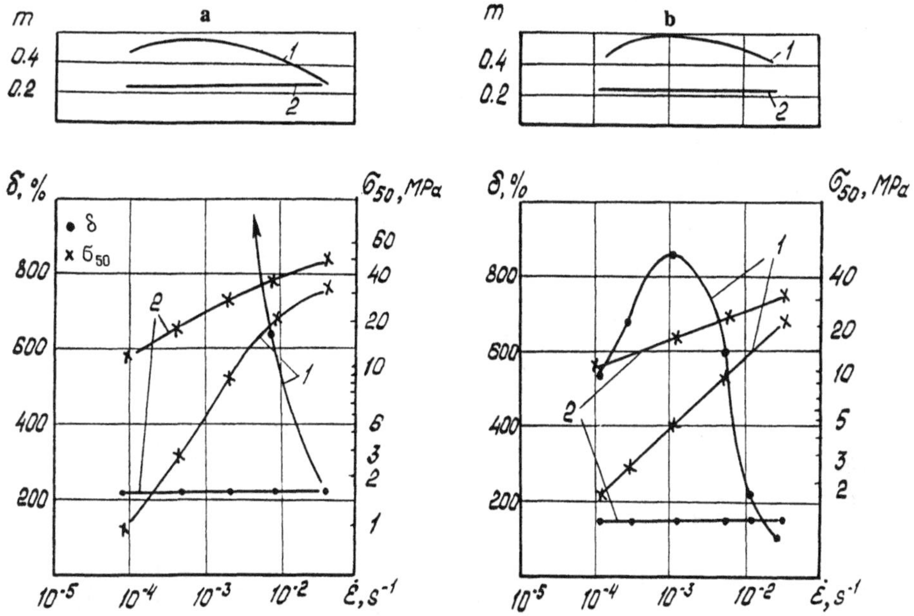

Fig. 5.1a,b. Strain rate dependence of the 1420 alloy at 450°C (a) and the V96Ts alloy at 465°C (b); (*1*) microcrystalline structure; (2) coarse-grained structure. *Points* δ, *crosses* σ_{50}

the highest SP parameters at $\dot{\varepsilon} = 10^{-2} - 10^{-1}\,s^{-1}$ [5.19] and the mechanically alloyed 90211 alloy having a grain size of 0.5 µm displays the highest superplasticity at $\dot{\varepsilon} = 2.5\,s^{-1}$ [5.18] (Table 5.2). Such mechanical behaviour of the alloy is referred to as high-strain-rate superplasticity.

Temperature of Deformation. The effect of the temperature of deformation on superplastic flow in aluminium alloys has been extensively studied [5.1, 4]. It was shown that as the temperature increases, the optimal strain rate shifts towards higher values, the flow stress decreases, and δ and m increase. For example, in a "Supral" type alloy, with rising test temperature, m increases from 0.41 at 450°C up to 0.58 at 540°C and, simultaneously, the flow stress decreases [5.1].

In order to obtain the highest superplasticity in aluminium alloys, they must be strained at the maximum possible temperatures. The increase of the temperature of deformation is limited by a temperature at which rapid grain growth starts and by the temperature of the intergranular oxidation of alloys with stable microstructures.

In alloys with a low or medium content of alloying elements, e.g., AMg6 and AK6, the temperatures of the maximum superplasticity exceed the temperature of the limiting solubility of the main alloying elements, such as zinc, magnesium, silicon, and copper, at which the precipitates containing these elements go into solution. However, at these high temperatures dispersoids such as aluminides of transition metals are not dissolved. This provides for the stabilization of the alloy microstructure.

Table 5.2. Parameters of superplasticity of aluminium alloys

System	Alloy (content of the alloying elements, wt.%)	Grain size [μm]	t_{SPD} [°C]	$\dot{\varepsilon}_{opt}$ [s^{-1}]	δ [%]	σ [MPa]	m	Ref.
Al-Mg	Ingot metallurgy 5456 (4.5Mg; 0.75Zr)	20	275	4×10^{-5}	150	40	0.35	[5.1]
	1570 (5.8Mg; 0.32Sc; 0.30Mn)	–	450	$1 - 5 \times 10^{-3}$	500–700	–	0.4	[5.7]
Al-Mg-Cu	Neopral (5.0Mg; 0.6Cu; 0.7Mn; 0.15Cr)	6	540	6×10^{-4}	—	0.7	0.9	[5.7]
Al-Cu	Supral (6.0Cu; 0.5Zr)	5–6	420–480	$1.5 - 3 \times 10^{-3}$	800–1200	–	0.4–0.5	[5.1]
	1200 (6.4Cu; 0.27Mn; 0.13Ti; 0.05Zr)	–	450	1×10^{-3}	150	20	0.42	[5.8]
	1201 (6.3Cu; 0.27Mn; 0.2Zr; 0.1V)	–	450	3×10^{-3}	–	30	0.45	[5.9]
Al-Cu-Mg	2004 (Supral 100) (6.0Cu; 0.25Mg; 0.4Zr)	–	540	10^{-3}	1800	–	–	[5.10]
	2024 (4.5Cu; 1.5Mg; 0.5Mn)	–	475	–	170	–	0.29	[5.11]
Al-Si-Cu	2091 (2.0Si; 2.2Cu; 1.6Mg; 0.08Zr)	12	500	8×10^{-4}	400	–	–	[5.12]
	1440							

At the temperature of SP deformation, the structure of alloys with a content of alloying elements exceeding the limiting solubility of the main alloying elements also exhibits a small quantity of equilibrium phases coexisting with a highly alloyed solid solution and aluminides of transition metals. These equlibrium phases, S-phase based on Al_2CuMg in the V96Ts alloy; θ-phase based on $CuAl_2$ in Supral S-phase based on Al_2MgLi or γ-phase based on $Mg_{17}Al_{12}$ in the 1420 alloy [5.1], result from a eutectic or peritectic reaction.

A decrease in the temperature of deformation leads to the reduction of the superplastic parameters. However, the extent of this reduction is different in different alloys. For instance, when the temperature of deformation of the AK6

Table 5.2. cont.

Al-Si-Cu	(2.4Si; 1.5Cu; 0.9Mg; 0.15Zr; 0.1Be)	2–3	520	3.6×10^{-3}	400	6	0.5	[5.13]
	8090 (2.5Si; 1.2Cu; 0.7Mg; 0.12Zr)	–	500	10^{-3}	600	–	–	[5.12]
Al-Zn-Ca	8050 (5Zn; 5Ca)	1–2	550	2×10^{-2}	640	0.2	0.5	[5.7]
Al-Zn-Mg-Cu	7075 (5.6Zn; 2.5Mg; 1.5Cu; 0.2Cr; 0.17Fe; 0.08Si)	10–15	510	10^{-4}	600	–	–	[5.14]
	1950 (6.0Zn; 2.3Mg; 0.4Mn; 0.15Cr; 0.5Fe; 0.5Si)	9–10	470	10^{-3}	500	6	0.4	[5.15]
	7475 (5.7Zn; 2.1Mg; 1.5Cu; 0.2Cr; 0.08Fe; 0.04Si)	14	516	2×10^{-4}	1200	0.8	0.75	[5.16]
	1973 (5.9Zn; 2.4Mg; 1.9Cu; 0.2Zr; 0.14Fe; 0.10Si)	11	500	2.4×10^{-4}	360	–	–	[5.17]
Al-Cu-Mg	Power metallurgy 90211 (4.4Cu; 2.0Mg; 1.1C)	0.5	475	2.5	520	30	0.26	[5.18]
Al-Zn-Mg-Cu	7075 + 3.8Mn	–	475	2×10^{-1}	200	–	0.3	[5.1]
	7475 + 0.7Zr	1	520	10^{-1}	600	5.8	0.55	[5.19]

alloy decreases from 515 to 465°C, the maximum elongation decreases from 270 to 130% at $\dot{\varepsilon} = 4 \times 10^{-4}\,s^{-1}$, which is accompanied by a significant increase in the flow stress. Similarly, the maximum ductility of the AK4-1 alloy rapidly decreases from 230 to 130% with the temperature of superplastic deformation decreasing from 525 to 450°C. On the other hand, in the AMg6 alloy, large elongations are retained even when the temperature decreases from 520 to 400°C [5.1].

The difference in temperature conditions of deformation of the considered alloys is significant. The homologous temperatures of AK6 and AK4-1 alloys are higher than those of the AMg6 alloy, although it is in the latter that the parameters of superplasticity decrease to a lesser extent. Such a difference in the variation of the parameters of superplastic flow is, apparently, related to the fact that in the AK6 and AK4-1 alloys the temperature decrease is accompanied by a transition towards a region below the temperature of solubility of the excess phases, whereas in AMg6, both temperatures exceed the temperature of the dissolution of the Al_3Mg_2-based excess β-phase. In the Al-Zn-Mg alloy containing zirconium, the ductility rapidly increases with heating to a temperature above those of maximum solubility of the zirconium-free secondary phases [5.1].

Based on the temperature dependence of the superplastic parameters, temperatures of quenching can be recommended providing for the SP deformation of the heat hardenable aluminium alloys. They ensure maximum heating employed for processing semifinished products made of the wrought aluminium alloys. When large strain levels are required during SP deformation, as in the case of pneumatic forming, it may be advisable to employ lower temperatures, because during prolonged deformation, even thermally stable alloys can exhibit a significant deformation induced grain coarsening. For instance, in a Supral type alloy, after 400% elongation at an initial optimum strain rate of $3 \times 10^{-3}\,s^{-1}$, the average grain size was 5.8, 7.0, and 9.2 µm at 450, 480, and 580°C, respectively, and the maximum strains at these temperatures were 960, 740, and 420%, respectively.

For the Supral alloy, with its varying ductility, strain rate sensitivity and flow stress, a temperature of 480°C was chosen as the optimum temperature of SP deformation [5.1], which is slightly lower than the temperature of the alloy quenching (500°C).

Selection of the maximum possible temperatures of the SP deformation of aluminium alloys has a number of practical advantages, for example, the possibility of combining heating required for the deformation of a preform with heating to the temperature of the alloy quenching and the possibility of straining at high rates since the range of the optimum rates of SP deformation shifts to higher values.

However, the decrease in temperature is also of practical importance because it makes it possible to utilize less expensive materials for making die-tools and prolongs service life. The submicron crystalline structure provides for a considerable decrease in the temperature of SP deformation (Chap. 1). Submicron-grained materials have the same SP parameters as materials with a grain size of 8–11 µm, but at a temperature which is 200–250°C lower. This phenomenon is referred to as "low-temperature superplasticity" [5.20].

5.1.2 Effect of Chemical Composition

The features of the SP flow of aluminium alloys with a matrix structure discussed above make it possible to establish a specific relationship between their chemical and phase composition and the possibility of their developing SP flow.

According to data presented in Table 5.1, SP parameters are different in the standard and model alloys. In alloys containing zirconium additives (e.g., V96Ts, 1420) SP parameters are the highest as indicated by the combined analysis of all the parameters taking their maximum and minimum values into account. Alloying with manganese (AK6 and AMg6) is less efficient. The AK4-1 alloy containing iron and nickel has the lowest thermal stability and, because of the accelerated grain growth during SP deformation, displays the lowest ductility and coefficient m and the highest flow stress.

Among the model aluminium alloys, the highest parameters of superplasticity are obtained in the alloy containing copper along with zirconium. Addition of magnesium to the Al-0.5%Zr alloy leads to significantly lower values of δ and m (Table 5.1). (In both alloys, alloying was accomplished using identical atomic contents of the copper and magnesium additions).

Effect of Transition Metals. A comparative study of the chemical compositions of the commercial and model alloys shows that the presence of a certain amount of transition metals (which makes recrystallization extremely difficult) is an essential condition for obtaining and stabilizing a microcrystalline structure. In the absence of these elements, not only the binary Al-Cu and Al-Mg model alloys, but also the highly alloyed V93 lack thermal stability (just as pure aluminium) and are not capable of deforming superplasticity. On the other hand, all alloys containing transition metals exhibit all symptoms of superplasticity in spite of their large differences with respect to other alloying elements (Table 5.1). To a large extent, this is due to the possibility of developing a fairly stable microcrystalline structure by means of an appropriate treatment.

Transition metals have a multiple effect on the structure evolution. The effect is seen during the solidification of the ingots, their annealing and during the final operations of the thermomechanical treatment, (during which the final microcrystalline structure is developed).

According to [5.1, 13], among the transition metals, zicronium and scandium have the most significant effect on the formation of microcrystalline structure, and a combined alloying with zirconium and scandium is the most favourable [5.7].

The data obtained on the Al-1.58%Mg alloy with 0.3%Zr and without zirconium illustrate the specific features of structure evolved during solidification and during subsequent processing.

An addition of 0.3%Zr to the aluminium-magnesium alloy leads to a significant grain refinement of the microstructure during slow cooling of the ingot in a hot mold (Fig. 5.2). This structural change is due to the formation of Al_3Zr intermetallic compounds during the initial stages of solidification which have a modifying effect during the subsequent crystallization of the alloy. Primary aluminides of other transition metals also have a similar effect on the structure of the aluminium alloys. A characteristic feature of this alloy is that in addition to the binary aluminide of zirconium, a ternary T-phase $Al_xMg_yZr_z$ is also formed [5.1].

Accelerated crystallization (during solidification of the ternary alloy in a cold mold) leads to an abrupt coarsening of the grains instead of grain refinement from the increased degree of supercooling and the corresponding increase in

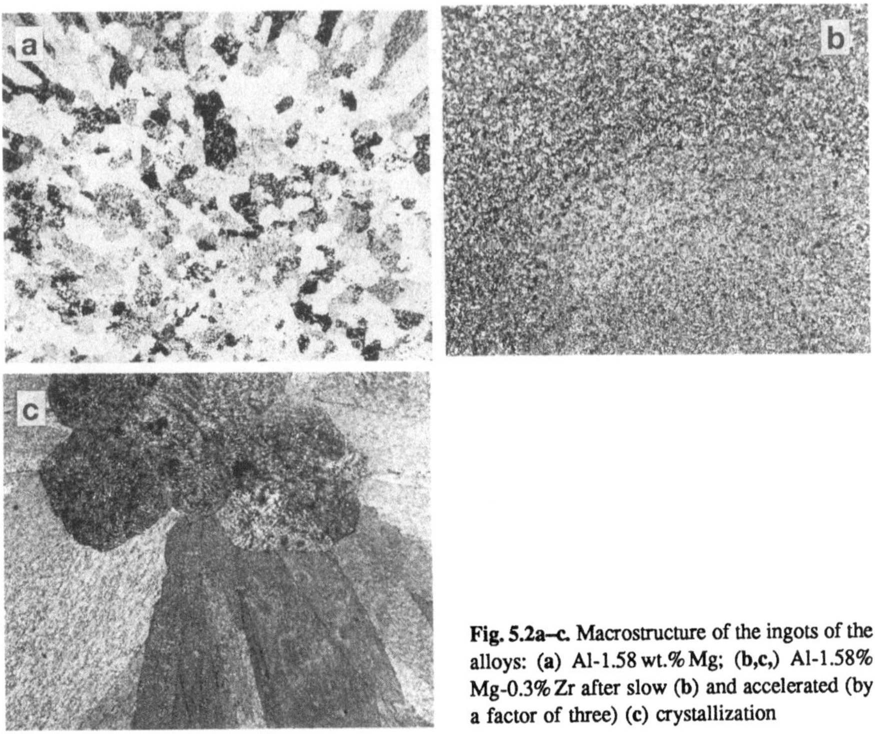

Fig. 5.2a–c. Macrostructure of the ingots of the alloys: (**a**) Al-1.58 wt.% Mg; (**b**,**c**,) Al-1.58% Mg-0.3% Zr after slow (**b**) and accelerated (by a factor of three) (**c**) crystallization

the number of crystallization nuclei (Fig. 5.2). In fact, grain coarsening occurs because fast cooling suppresses the formation of primary Al_3Zr crystals losing their modifying effect. In this case, zirconium remains in the solution which increases the degree of its supersaturation.

A similar relationship was found between the solidification rate, the macrograin size, and the degree of supersaturation of the solid solution in a Supral type alloy [5.1]. It was shown that a uniform microcrystalline structure and high indices of superplasticity can be obtained in this alloy only in the presence of a significantly high supersaturation of zirconium in the solid solution. Therefore, a criterion was suggested for evaluating the quality of the alloy, viz., the size of the micrograins must exceed 100 μm. This fact implies that during the solidification of the ingot, the rate of crystallization must be sufficiently high in order to ensure a suppression of the formation of primary intermetallic compounds and their modifying effect, and to simultaneously achieve the required supersaturation of the solid solution with zirconium.

The addition of zirconium affects not only the macrostructure, but also the microstructure of the alloy. At high rates of crystallization, the alloy shows a distinct dendritic structure. In view of such microstructural features of the alloys, a criterion was suggested[3] for evaluating the required degree of supersaturation of

[3] UK Patent No. 1387586 (1975); US Patent No. 3876474 (1975)

the solid solution with zirconium in order to attain high parameters of superplasticity, viz., during solidification of the Supral ingot, the dendrite cell size must not exceed 30 μm. Refinement of the dendritic structure is due to the increased rate of crystallization which simultaneously facilitates the required supersaturation of the solid solution.

The ingots of aluminium alloys are usually subjected to annealing in order to increase the engineering ductility and to improve the quality of the hot-deformed semifinished products. Annealing induces the decomposition of the solid solution (supersaturated with zirconium) to form highly dispersed precipitates of a metastable modification of Al_3Zr, i.e., in this case, annealing stimulates heterogenization in addition to facilitating homogenization.

The above structural changes which occur during annealing of the ingot of the Al-Mg-Zr alloy are typical. In alloys containing transition metals, nonequilibrium primary intermetallic compounds (compounds based on the main alloying elements including copper, magnesium, and zinc) and other types of compounds such as $CuAl_2$, Al_2CuMg, Mg_2Si, Al_3Mg_2 and $MgZn_2$ go into solution. Simultaneously, secondary phase (metastable modifications of aluminides of the transition metals, e.g., Al_3Zr, Al_6Mn, Al_7Cr, Al_3Fe, Al_3Sc) are precipitated from the solution supersaturated with respect to the transition metals.

The final evolution of the microcrystalline structure occurs during the heat-strain treatment of the ingots or hot-deformed semifinished products and is induced by recrystallization processes. The results of the studies conducted on the Al-1.58%Mg and Al-1.58%Mg-0.3%Zr make it possible to discuss the specific features of the effect of the transition elements on the structure evolution processes.

The ingots of the alloys annealed at 450°C for 24 h were subsequently subjected to cold extrusion up to a strain of 60%. The specimens cut out from the strips were then subjected to recrystallization annealing. After heating at 550°C for 1 h, the aluminium-magnesium alloy assumed a polyhedral structure with an average grain size of 1000 μm, whereas the alloy containing zirconium had an average grain size of 30 μm. In both cases, structure evolution occurred as a result of the primary recrystallization process and collective recrystallization following it. Zirconium raises the temperature at which primary recrystallization starts (Table 5.3). The effect of zirconium on the collective recrystallization was evaluated by comparing the average growth rates of the recrystallized grains. It was established that at an identical initial grain size of 7 μm and a dwell of 5–15 s at 600°C, the average rate of grain growth in the zirconium-free alloy is 10–15 times higher than that of the zirconium-bearing alloy. These data show that the additional zirconium makes the occurrence of primary recrystallization extremely difficult, facilitates grain refinement, suppresses collective recrystallization, and, thereby, stabilizes the fine-grained structure.

Additions of other transition metals have an analogous effect, but their effectiveness is different. This effect is related mainly to the presence of dispersed secondary precipitates of the aluminides of the transition metals which remain undissolved in the alloy at high temperatures.

Table 5.3. Effect of zirconium on the recrystallization of alluminium alloys

Alloy	Temperature at start of recrystallization [°C]	Duration of primary recrystallization at 500°C,[c] [s]
Al-1.58%Mg	260	< 5
Al-1.58%Mg-0.3%Zr[a]	375	~300
Al-0.3%Zr	500	incomplete
Al-0.5%Zr	560	incomplete [5.3]

[a]During slow crystallization of the ingot (in a hot mold)
[b]Duration of annealing is 2 hours
[c]Determined by the time of the disappearance of the original fibre structure; heating of thin (1 mm) samples in a saltpeter bath at 500°C

The advantage of alloying aluminium materials with zirconium (as compared to additions of other transition metals) stems from the fact that during the decomposition of the supersaturated solution, this element forms the most finely dispersed secondary precipitates of the excess Al_3Zr phase, which do not coarsen up to 500°C [5.1].

It is known that the maximum microstructural stability is achieved in the presence of the optimum degree of dispersion of the products of decomposition. For instance, in Supral and in the Al-Mg-Zr and Al-Mg-Zn-Zr alloys, the microstructure having an average grain size of 5 µm is stabilized by the Al_3Zr particles measuring approximately 0.1 µm in diameter [5.1]. Both higher and smaller degrees of dispersion are less effective.

The volume concentration of the dispersed products of decomposition and the uniformity of their distribution across the body of the grains depends on the extent of alloying. For example, when alloying with zirconium, about 0.5%Zr is required for obtaining high indices of superplasticity [5.1]. Usually aluminium alloys are alloyed with Zr up to 0.15 wt.%. But even at 0.2%Zr, a deterioration of the parameters of superplasticity may occur because of the difficulty in obtaining a uniform fine-grained structure.

A characteristic feature is that the presence of other transition elements in the alloy does not compensate for the deficiency of the zirconium content. Thus, the AA2219 alloy (6%Cu, 0.3%Fe, 0.3%Mn, 0.22%Zr, and 0.2%Sn) which is close to Supral with respect to the mechanical properties and contains the same amount of copper and a high total content of the transition metals (0.8%) (but having a lower zirconium content) does not exhibit symptoms of superplasticity [5.2].

Effect of Alloying Elements Other than Transition Metals. Alloying of aluminium with only transition metals is insufficient. Without such elements as copper, magnesium, and zinc, aluminium alloys containing only transition metals cannot acquire microcrystalline structure. For example, in a binary aluminium-zirconium (0.3%) alloy, it is practically impossible to obtain such a structure, and, hence, to obtain superplasticity. In spite of considerable cold pre-straining by hydraulic extrusion (85%), the alloy acquires a polygonized structure during subsequent heating. At 500°C only partial recrystallization take place. In the Al-0.5%Zr alloy recrystallization starts at 560°C due to the very finely dispersed particles of Al_3Zr. The elongation of this alloy is rather small. The addition of magnesium (1.58 wt.%) or copper (4.1%) into the Al-0.3%Zr alloy makes it possible to obtain a structure with an average grain sizes of 7 μm by means of recrystallization annealing and to convert the alloys into an SP state (Table 5.1).

According to [5.3], the magnitude of the relative elongation increases with increasing zinc content in the Al-0.5%Zr alloy. At 850°C, in an alloy containing 3, 5, and 10%Zn, it amounts to 360, 580, and 620%, respectively. Such an increase of the parameter of superplasticity with increasing zinc content is probably associated with some increase in the homologous temperature of deformation due to a decrease in the solidus temperature. It was established that in the absence of zinc, the Al-0.5%Zr alloy recrystallizes only at 560°C and a small relative elongation is observed.

It should be noted that the presence of zinc, copper, and magnesium in the solid solution affects superplasticity. It is also significant that their presence facilitates the development of microcrystalline structure due to primary recrystallization. In the absence of copper, magnesium and zinc, in an alloy containing only zirconium, recrystallization is suppressed and polygonization occurs predominantly. The investigation of model alloys (Table 5.1) indicates that in the presence of copper, aluminium alloys exhibit higher indices of superplasticity than in the presence of magnesium. The nature of the effect of copper and magnesium of the solution on superplasticity can be related to the operation of two factors. With identical atomic contents, copper decreases the temperature of incipient melting to a greater extent than magnesium and thus, at the same test temperature (500°C), the homologous temperature of the Al-Cu-Zr alloy is significantly higher than that of the Al-Mg-Zr alloy. For the binary alloys of aluminium containing 2 at.%Cu and 2 at.%Mg, a temperature of 500°C corresponds to 0.92 and 0.85 T_m, respectively. Moreover, the specific features of the copper and magnesium atoms probably also have an effect on the structure of the grain boundaries (Chap. 2). This is indicated by the fact that at approximately equal homologous temperatures (0.85 T_m), the Al-4.1%Cu-0.5%Zr alloy shows a higher rate of dissocation of the lattice dislocations that are trapped by the grain boundaries and correspondingly higher indices of SP flow than the Al-Mg-Zr alloy (Table 5.1).

When analyzing the effect of copper on the parameters of superplasticity, it should be taken into account that a part of copper exceeds the solubility limits (even when the copper content of the alloy is relatively low) and forms a second (excess) phase $CuAl_2$. According to *Grimes* et al. [5.3], the particles of this phase in Supral play a significant role in suppressing recovery, and lead to fine recrystallized grains at the zirconium-depleted near-boundary zones. Because of this, the structural uniformity of the alloy and, consequently, the uniformity of its deformation increase.

According to electron microscopic investigations [5.21], the $CuAl_2$ particles play an important role in hindering the grain growth from the moment of attaining a grain size approximately equal to the interparticle spacing. Such a grain size is obtained as a result of coarsening during superplastic deformation up to large strains (200–300%).

Based on the established relationships between the chemical and the phase compositions of the aluminium alloys and their superplasticity, it is possible to predict their superplastic behaviour and, to a first approximation, the composition of new superplastic aluminium alloys. With regard to the main alloying elements, most of the industrial aluminium alloys meet the requirements for transforming them into the SP state. However, not all alloys contain transition metals that ensure effective suppression of recrystallization. Thus, only the alloys containing a sufficient quantity of one or more transition metals (zirconium, manganese, chromium) in addition to the main alloying elements are suitable for practical processing under the conditions of superplastic deformation.

It is advisable to correct the composition of the commercial alloys intended for superplastic forming by introducing Zr instead of other transition metals and employing high solidification rates when casting the ingots. The combined alloying with zirconium and scandium is even more advantageous since it can significantly decrease the cooling rate during the solidification of ingots [5.13]. Such a correction of the alloy compositions substantially simplifies subsequent processing aimed at obtaining a microcrystalline structure and stipulates the increase in the maximum ductility and decrease in the flow stress during SP deformation.

5.2 Processing Methods for Obtaining Microcrystalline Structure in Commercial Aluminium Alloys

In order to obtain a microcrystalline structure in aluminium alloys, special processing is required. In practice, only two methods are applicable to the hot deformed semi-finished products of the industrial alloys: (i) recrystallization annealing after cold or warm deformation or (ii) hot deformation accompanied by dynamic recrystallization. In both cases, a microcrystalline structure is developed due to the formation of a large number of recrystallization centers in the deformed matrix of the alloy.

It is rather difficult to create a microcrystalline structure in aluminium alloys using static or dynamic recrystallization because of the high stacking fault energy which causes intensification of the recovery processes and, as a consequence, the rate of nucleation of the recrystallization centers decreases during recrystallization annealing and hot deformation [5.22]. Therefore, when using a simple processing scheme involving deformation and recrystallization annealing, it is possible to obtain a microcrystalline structure only in a limited number of cases by subjecting the alloy to large deformations prior to recrystallization annealing. Thus, in order to obtain a structure having an average grain size of 9 μm in the hot press-forged AMg6 alloy, it was necessary to give a few passes of cold rolling with a strain of 70% or more before conducting recrystallization annealing [5.1]. In a case of a hot press-forged semifinished product of the V96Ts alloys, a microcrystalline structure was obtained by employing multipass rolling with a total strain of 90% [5.1]. In order to obtain a microcrystalline structure in the alloys of the Al-Zn-Mg system, one must use deformation by cold rolling at 80% strain [5.4].

In order to achieve an average grain size of 10 μm in the hot press-forged preforms of the AK6 alloy, cold press-forging with 70% deformation was insufficient (with subsequent recrystallization annealing carried out in an electric furnace at 520°C). The required grain refinement was achieved using special treatments.

Working with such high strains at ambient temperature or at warm temperatures (up to 200°C) is technologically unsuitable since it requires high energy and labour consumption and leads to rapid wear of the tools and their destruction. Thus, it is applicable for obtaining a microcrystalline structure only in small preforms, but even in this case, it may not be effective (as was found in the case of AK6). Consequently, most of the aluminium alloys require complex treatments.

The special treatments employed for the structure refinement in the industrial wrought aluminium alloys include: prior regulated heterogenization, rapid heating when carrying out recrystallization annealing treatment, powder and granular technologies, etc.

Prior Regulated Heterogenization. This is a method of controlling the structure of aluminium alloys underlying the intermediate thermo-mechanical treatment (ITMT) offered for processing heathardenable aluminium alloys[4]. The common feature for all kinds of ITMT is the acquisition of a certain quantity of particles of excess equilibrium phases of special dispersion in the alloy matrix. In aluminium alloys having a matrix-type structure, this is achieved by the precipitation of the secondary phases during the decomposition process of the supersaturated solid solution.

The required degree of refinement and volume fraction of the products of decomposition are obtained by means of heterogenization treatment in experimentally established modes. The only undisputable fact is that in the case of aluminium alloys, the temperatures at which final evolution of the most favourable heterogeneous structure must occur lie below the solvus, but above the temperature at which the excess phase is precipitated in a highly dispersed state (the precipitation hardening temperature). The heterogenization temperature must be approximately in the 250–400°C range. In order to control heterogenization, one can employ not only a thermal effect (heat treatment), but also a deformation effect that effectively influences the morphology and the kinetics of decomposition of the supersaturated solid solutions [5.23].

---

[4] US Patent Nr. 4092181 (1978)

Methods of attaining regulated heterogenization required for producing a microcrystalline structure in aluminium alloys are examined in [5.2, 24, 25]. According to the patents[5], dispersion-strengthened aluminium alloys are first brought into a homogeneous condition by solution treatment, and thereafter they are overaged at a temperature exceeding the precipitation hardening temperature. They are then plastically deformed at the overaging temperature or below it and are further subjected to recrystallization annealing. The secondary excess phase precipitating during overaging facilitates an increase in the number of primary recrystallization centers and restricts grain growth.

Controlling the recrystallized structure by carrying out regulated heterogenization of Supral forms the basis for working out the composition and the method of processing the alloy. Thus, after rapid cooling, the alloy is annealed at 370°C. During annealing, the particles of zirconium aluminide, having a particle size of 10–30 μm, precipitate from the solid solution.

Large cold or warm deformation (70–80%) is required for obtaining a microcrystalline structure after heterogenization. To avoid this limitation, a method called two-stage deformation-heat treatment (TSDHT) was suggested[6]. This is based on the formation of a microcrystalline structure via the combination of static and dynamic recrystallizations. To induce static recrystallization, at the initial stage of treatment the strain level is relatively small (50% or less) and the temperature is 20–300°C. The subsequent recrystallization annealing is combined with heating to prepare the second stage of deformation. At this stage, the treatment is carried out under conditions providing for hot deformation during which formation of a microcrystalline structure is completely due to dynamic recrystallization. This is accomplished under the temperature-strain rate regimes of deformation corresponding to superplasticity. Consequently, the second stage of treatment can be combined with the formation of products using superplastic deformation. In this case, hot deformation and subsequent formation of microcrystalline structure take place during the initial stage of straining before SP deformation starts.

TSDHT can be used with or without heterogenization annealing. At present, the former method is virtually the only efficient method of obtaining microcrystalline structures in high aluminium alloys, e.g., in V96Ts where the development of recrystallization is impeded [5.25].

However, ITMT with prior regulated heterogenization cannot be considered as an absolute and universal method for obtaining microcrystalline structure in the heat hardenable aluminium alloys. This treatment is ineffective for the alloys with low thermal stability exhibiting rapid coarsening of grains due to normal grain growth (AK4-1 and V93 are examples of such alloys). The stability of structure is ensured by highly dispersed thermally stable aluminides of transition metals inhibiting grain boundary migration in the alloy.

---

[5] US Patent No. 4092181 (1978); French Patent No. 2388893 (1978)

[6] USSR Patent No. 1126625 (1984)

Thus, in general, the formation of a microcrystalline structure in the aluminium alloy requires a "bimodal" distribution of secondary excess phases during processing [5.25]. The first mode (the size of the particles of aliminides of transitions metals is 10–30 μm) results from the decomposition of the solid solution oversaturated with transition metals. The second mode [the size of the particles of secondary excess phases, i.e., intermetallic compounds of the basic alloying elements, must be equal to or exceed a critical size d_ε (Chap. 3)] is yielded by the special heterogenization treatment mentioned above.

In the case of insufficient alloying of the system, the required volume fraction of the excess phase cannot be achieved. Apparently, this is the reason why it is not possible to obtain the required degree of grain refinement in the AK6 alloy due only to different heterogenizing treatments.

Heterogenization is ineffective in the alloys in which secondary excess phases are not sufficiently stable. Because of the coarsening of the particles of the excess phase during the process of heating up to the temperature of the recrystallization annealing operation, it is not possible to obtain the desired grain refinement.

Employing Fast Heating. It was previously shown (Chap. 3) that an increase in the heating rate during recrystallization annealing is accompanied by grain refinement. This is due to the increased nucleation rate of the recrystallization centers. A study was conducted on the AK6 alloy for establishing the effectiveness of employing high heating rates for obtaining microcrystalline structure in aluminium alloys. Preforms measuring 38 mm in diameter and 44 mm in length were subjected to induction and furnace heating. In the 300–450 °C range at which recovery processes actively occur in the AK6 alloy (but recrystallization of the alloy does not start), the average heating rates were 7 K/min and 25 K/s during furnace and induction heating operations, respectively. Increasing the heating rate led to a decrease of the average grain size from 14 to 9.5 μm.

Employing fast heating is of great advantage after low and medium strains, since, in this case, deformation is nonuniform and, consequently, the incubation periods necessary for nucleating the recrystallization centers are different in different areas of the sample. Fast heating leads to elimination of the difference in the incubation periods. This makes it possible to use lower deformation before recrystallization annealing and, thereby, offers a significant technological advantage when developing a microcrystalline structure in the aluminium alloys.

Employment of Powder and Granular Technologies. Microcrystalline structures can be created in semifinished products using powder and granular technologies [5.18, 19, 26]. The latter includes the following processes: (i) preparation of powders and granules; (ii) their cold or hot compacting; (iii) degassing; (iv) hot pressing; and (v) forming (e.g., forging, rolling, or pressing) of hot-pressed semifinished products.

This method of manufacturing semifinished products offers vast possibilities for controlling their structure by means of substantial suppression of intracrystalline segregation and eliminating coarse inclusions of primary phases. This is achieved by high and ultra-high rates of solidification during the quenching of small volumes of metal from the liquid state. Powders with sizes of 10–500 μm are prepared by spraying the melt with a jet and cooling it in a stream of inert gas. Granules (particle size 1–3 mm) are crystallized during immediate contact with the cooling medium, and the spraying may be accomplished in a variety of ways [5.26].

Superplasticity of powdered and granulated aluminium alloys is possible in a number of systems, viz., Al-Si of eutectic and proeutectic compositions [5.27], Al-Zn-Mg-Cu (alloys of the 7XXX series) [5.28], etc.

The possibility of obtaining a microcrystalline structure using the above described technologies stems from a number of factors. Severe grain refinement in microingots is inherited in semifinished products, which increases the number of nucleation centers during subsequent hot treatment, specifically ITMT.

However, this method of preparing microcrystalline aluminium alloys, liable for superplastic deformation, is not universal. One of the reasons for this is that powder and granular alloys exhibit a tendency toward premature fracture due to the presence of oxides and a large content of oxygen and hydrogen which make the alloys brittle and induce a rapid formation of voids during SP deformation [5.29]. For example, in the granulated V96Ts alloy voids appear during tensile SP deformation at 50% strain. Semifinished products made of this alloy were manufactured from granules (0.5–2.0 mm in diameter) by means of hot pressing of brickets at 430°C with the strain degree of 94%. High strain rate sensitivity of the flow stress, and low resistance to deformation are observed in the alloy. However, its maximum ductility never exceeds 160% because of the voids.

The possibility for overcoming the above described limitations is different in diverse alloys. The solution of this problem is impeded by the lack of systematic data on the effect of microingot structure on the structure of the final product, and by the imperfection of the technology used.

Other methods of producing microcrystalline structures in the semifinished products of aluminium alloys, including fractional deformation [5.4] and others [5.1] are also often employed.

5.3 Effect of Superplastic Deformation on the Structure and Properties of Aluminium Alloys

The properties of the aluminium alloys depend on the initial microstructure, the structural changes occurring during heating and holding at the temperature of SP deformation, during deformation, and during subsequent heat treatment. In order to identify the 'true' effect of SP deformation alone, it is necessary to evaluate experimentally the effect of each of these factors on the properties of the aluminium alloys. With this aim, we processed part of the specimens according to the following scheme: heating to the temperature of SP deformation, holding at this temperature for a duration equal to that required for accomplishing SP deformation, and subsequent standard heat treatment. Such a treatment is termed as "deformationless" treatment (DLT).

There are two types of treatments designed for aluminium alloys on the basis of SP deformation. They can be referred to as superplastic deformation-cum-heat treatment (SPDHT), and superplastic deformation treatment (SPDT). SPDHT is applicable only for heat hardenable superplastic alloys that contain transition metals stabilizing the microstructure and, hence, thermal stable at the quenching temperature. Figure 5.3 explains the purpose and the sequence of operations of deformation and thermal effects on the alloys during this treatment. SPDT is intended for heat unhardenable alloys, and its scheme differs from SPDHT by the absence of quenching and aging operations (5–7 in Fig. 5.3).

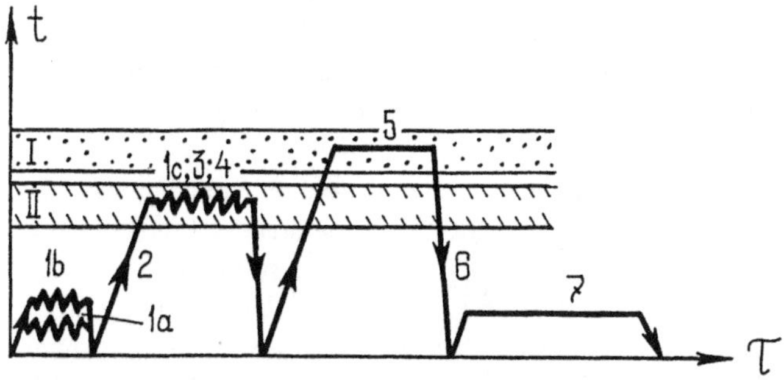

Fig. 5.3a–c. Scheme of superplastic deformation heat treatment (SPDHT) of aluminium alloys that can be hardened by heat treatment: (I) temperature range of dissolution of the excess phases not containing transition metals; (II) temperature range of the maximum engineering ductility; (*1*) prior deformation (**a** cold deformation; **b** warm deformation; and **c** hot deformation) to obtain a microcrystalline structure; (2) heating; (*3*) recrystallization annealing; (*4*) final shaping under the conditions of superplasticity; (*5*) heating prior to quenching; (*6*) rapid cooling (quenching); (*7*) aging

The efficiency of treatments incorporating SP deformation can be determined by comparing the structure and the properties of the alloys after these treatments with those of the alloys subjected to conventional treatment.

Such a study was carried out both on the heat hardenable alloys V96Ts, AK6, 1420, and AK4-1, and on the heat unhardenable alloy AMg6[7]. In order to deform under the SP conditions, a microcrystalline structure having an average grain size of 8–10 μm (5 μm in the case of V96Ts) was obtained in the alloys. For the conventional treatment, we used alloys in the form of hot press-forged rods with an initial coarse-grained structure. In all the cases, preforms were deformed by upsetting with identical strains (reductions) of 50%. During conventional treatment, the alloys were deformed using hydraulic or mechanical presses at a strain rate commonly used for mechanical working of metals: $10^{-2} - 10^{-1}$ and $10^{-1} - 10^{0}\ s^{-1}$, respectively. SP deformation was accomplished using a specially

[7] The study was carried out in collaboration with V.G. Trifonov on the alloys V96Ts, AK6, and 1420 and in collaboration with Yu.B. Timoshenko on the AMg6 and AK4-1 alloys

modernized 100-ton hydraulic press under isothermal conditions maintaining a strain rate equal to $\dot{\varepsilon}_{opt}$ and an optimum temperature for each alloy. An approximately constant rate was maintained in the automatic regime of operation of the press, and the temperature during deformation was held to within 5°C. The forgings were water quenched (20°C) or air cooled (the 1420 alloy). In the case of SPDHT, this operation was carried out immediately after completing the deformation. The loss of time during withdrawl of the forged preforms from the tool amounted to 10–15 s. After quenching (cooling), all specimens were subjected to aging according to the typical regimes for each of the alloys.

Structural Changes. Transmission electron microscopic and metallographic studies established that the fine structure obtained after deformationless treatment and the fine structure of the superplastically deformed alloys are qualitatively similar. However, the microstructure changes significantly during the treatments; namely, additional grain growth occurs in the transverse and longitudinal directions at different rates in different alloys. Figure 5.4 shows the curves of grain size (grain area) distribution for three states of the alloy AK6: 1 – for the original preform having a microcrystalline structure, 2 – after heating without deformation, 3 – after SPDHT. The shape and the location of the distribution curves are typical of the microstructural changes occurring during collective recrystallization [5.22]. A comparison of curves 1 and 2 shows that heating without deformation leads to a change in grain size due to usual (static) collective recrystallization. A comparison of curves 2 and 3 makes it possible to evaluate the effect of SP deformation proper on the structural changes. In this case, the grains grow larger and faster than in the case where heating alone is employed (Chap. 2).

In most of the studied alloys, anisotropy of growth was observed; the grains grow to a greater extent in the direction of the maximum tensile deformation. This leads to the development of metallographic texture and the shape factor attains a value of about 1.1–1.4. Such a feature of SP deformation-induced structural

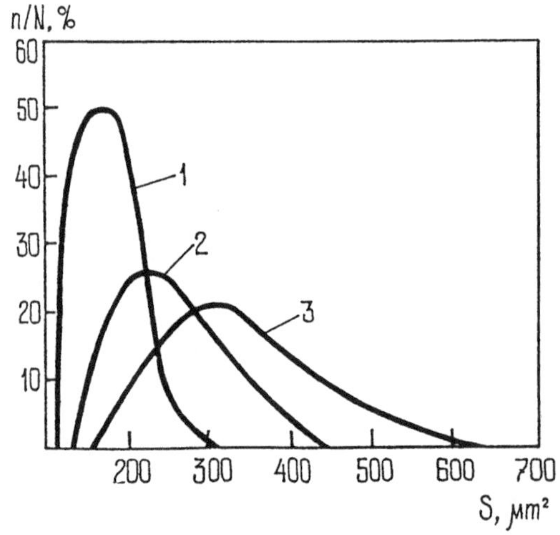

Fig. 5.4. Grain size distribution curves of the AK6 alloy (n/N is the ratio of the number of grains of a given area S and the total number of grains): (*1*) original preform having a microcrystalline structure; (2) after deformationless treatment (DLT) at 515°C; (*3*) after superplastic deformation heat treatment (SPDHT) at $\dot{\varepsilon}_{opt}$, $\varepsilon = 50\%$ and 500°C

change on the mode of deformation is greater in the tensile mode than under compression. The anisotropy of growth depends on the composition of the alloy; at identical strains, the shape factor of different alloys is not the same.

The specific features of the effect of SP deformation on the structure of alloys include the observed increase in the uniformity in those cases where the initial fine-grained structure contains individual coarse grains or their clusters. Deformationless treatment does not eliminate this nonuniformity. During SP deformation of the V96Ts and 1420 alloys, a deformation-induced growth of fine grains constituting the main part of the structure occurs. In the coarse grains, fragmentation occurs and new fine grains form at their boundaries. As a result, the uniformity of the micro- and macro-structures increases. The weak dependence of the alloy structure on the strain level during SP deformation also contributes to its uniformity. As a result, a uniform fine-grained structure is retained throughout the preform regardless of the local strain.

Thus, SP deformation helps to solve one of the most complex problems of physical metallurgy with respect to aluminium alloys, namely, the elimination of the nonuniformity of grain size distribution in the finished product. It is difficult to solve this problem under the conditions of conventional treatment, because it is impossible to eliminate the nonuniformity of strains within the entire volume of the semifinished product.

Among the peculiarities of the effect of SP deformation on the structure are the significant spreading of crystallographic texture and the formation of voids during tensile deformation.

As in grain growth and its anisotropy, the intensity of void formation depends on the composition of the alloy, which was also observed in the alloys based on other elements [5.1].

When tensile straining is substituted with upsetting, neither electron microscopy nor metallographic analysis reveal any voids. This indicates that the void formation in the aluminium alloys subjected to SP deformation under nonuniform compression is suppressed, the latter being characteristic of most methods of materials forming. Hence, porosity has no effect on the mechanical properties of the alloys.

However, intensive void formation occurs at large strains during the superplastic forming of products out of sheets and thin-walled preforms, the two-axial tensile state of stress being characteristic of this process. In this case, the void formation can essentially deteriorate the material strength and ductility.

The essential changes of structure take place in aluminium alloys during SP deformation, which is not the case during conventional deformation.

Mechanical Properties. When the changes in the structural states of the alloys are well-pronounced, the efect of SP deformation on the mechanical properties (viz., σ_u, $\sigma_{0.2}$, and δ) is negligible. Few variations of mechanical properties under significant changes in the structure are characteristic of aluminium alloys. For instance, a low sensitivity of the mechanical properties to the structural changes can be noticed by comparing the structure and the properties of a number of alloys after high-temperature thermomechanical treatment and conventional treatment [5.23] and in some cases by comparing the strength of the alloys with the recrystallized and nonrecrystallized structures [5.30]. The relationship between the strength and ductility (Table 5.4) depends on the strain rate during conventional treatment, i.e., the higher the strain rate, the faster the softening during subsequent heating prior to quenching under the conditions of conventional treatment. When deforming at a low rate ($10^{-1} - 10^{-2}\,s^{-1}$), heating of the alloys prior to quenching is not accompanied by recrystallization. However, when the strain rate increases ($10^{-1} - 10^{0}\,s^{-1}$) during subsequent heating at the quenching temperature, the alloys were completely recrystallized.

The results of mechanical testing agree with the effect of strain rate on the state of the alloys. When deforming at a high strain rate, the strength of the alloys after conventional treatment was significantly inferior to those obtained after the treatments involving SP deformation (Table 5.4). When deforming at a low strain rate, the alloys AK6, V96Ts, and 1420 assume higher strength after the conventional treatment. After the treatments based on superplastic deformation, the ductility of the alloys is more independent of the strain rate at which deformation during conventional treatment was carried out. In the medium-strength alloys having relatively high ductility (AMg6 and AK6), the increase of δ due to SPDHT is insignificant. On the other hand, in the alloys having a low ductility (medium-strength 1420 and high-strength V96Ts), the increase of ductility is significant after SPDHT (δ increases by 1.5–2 times).

The treatment involving SP deformation also causes a decrease in the anisotropy of mechanical properties. For example, in the case of AK6, the ratio of the yield stress in the transverse and the longitudinal directions was equal to 1 and 0.96 after SPDHT and conventional treatment, respectively.

Table 5.4. Effect of the mode of treatment on mechanical properties at room temperature

Alloy	Treatment	σ_u [MPa]	$\sigma_{0.2}$ [MPa]	δ [%]
AK6	DLT[a]	445	352	15
	SPDHT	432	338	15
	CT[b]	411	319	–
	CT[c]	483	363	14
AMg6	DLT	396	196	26.5
	SPDT	399	204	28
	CT[b]	366	175	26
V96Ts	SPDHT	626	585	5.5
	CT[c]	663	623	–
1420	DLT	457	296	–
	SPDHT	462	301	14
	CT[c]	471	317	8

[a]Deformationless treatment
[b]Conventional treatment ($\dot{\varepsilon} = 10^{-1} - 10^{0}\,s^{-1}$)
[c]Conventional treatment ($\dot{\varepsilon} = 10^{-2} - 10^{-1}\,s^{-1}$)

Table 5.5. Effect of treatment mode on the intercrystalline corrosion resistance of the V96Ts alloy

Treatment	Intensity of corrosion damage[a]	Depth of corrosion [mm]		Amount of flowed out hydrogen
	[%]	Maximal	Average	[cm^3/mm]
CT	62	1.16	0.67	0.160 ± 0.013
SPDHT	55	1.16	0.56	0.062 ± 0.003
DLT	40	0.66	0.53	–

[a]Intensity of corrosion is the ratio of the sum length of the damaged regions to the total length of the specimen edge

Corrosion Properties. The effect of the treatment method on the resistance to corrosion depends on the nature of the alloy. For instance, in the AMg6 alloy that is resistant to intergranular corrosion (IGC), no difference in the magnitude of corrosion is observed after SPDT and conventional treatment. In the V96Ts alloy that exhibits satisfactory resistance to IGC, this property significantly increases after SPDHT.

The tests on IGC showed that a decrease in the intensity and depth of corrosion as well as a decrease (more than twice) in the hydrogen outflow are observed after processing under an SP condition (Table 5.5). According to a 10-grade scale, the tendency for the stratifying corrosion was 5 and 8 after SP deformation and conventional treatment, respectively. Similar data on the increasing resistance to stratifying corrosion after SPDHT were obtained for the D16 alloy [5.31]. Corrosion cracking tests performed under a fixed tensile load and periodically affected by a solution of NaCl in distilled water indicated that at $\sigma = 245$ MPa the time before fracture after SPDHT is twice as long as after conventional treatment.

The difference in the effect of treatment on the V96Ts alloy resistance to corrosion is due to the peculiarities of its microstructure [5.32]. The structure refinement increases the resistance of the alloy to intergranular corrosion; hence, the resistance to corrosion after SPDHT is stronger than that after the conventional treatment, but weaker than that after the deformationless treatment. Strain induced grain growth during SP deformation deteriorates corrosion-resistance.

Aging Kinetics. The influence of SP deformation on the structural state of the alloys is also revealed by the fact that the kinetics of alloy aging after deformation differs from that after treatment without SP deformation. For example, tests conducted on the as-quenched AK6 alloy, after 10 s from the moment of quenching in water at 20°C, indicated that after heating without deformation the alloy hardness was higher than after SP deformation, while the electrical conductivity was lower. However, after annealing, in particular, during the first 5 hours, the rate of natural aging after SP deformation was found to be significantly higher, which was proved by a greater decrease in the electrical conductivity and a more rapid increase in the alloy hardness. A difference in the kinetics of the decomposition of the solid solution was observed even during the artificial aging of the AK6 and 1420 alloys.

The Nature of the Effect of Treatment on the Properties of the Alloys. Based on the specific features of the structural state of the alloys subjected to the treatments incorporating SP deformation, one can explain the differences in the properties as compared to those after conventional treatment. Since SP deformation is not accompanied by work hardening, the higher level of the strength obtained after SPDHT (as compared to the level of properties obtained after conventional treatment which imparts a recrystallized structure to the alloys) can be attributed to the Hall-Petch strengthening due to the difference in the grain sizes after the treatments. That is why the microcrystalline structure obtained after SPDHT ensures a greater resistance of the alloys to plastic deformation than the structure obtained after conventional treatment (the alloys AK6 and AMg6 at $\dot{\varepsilon} = 10^0 - 10^{-1}\,s^{-1}$; Table 5.4). In the case of conventional treatment, with a low strain rate, 'structural strengthening'[8] can occur. Its magnitude exceeds the strengthening effect due to the Hall-Petch hardening. Thus, the strength of the AK6, V96Ts, and 1420 alloys observed after conventional treatment using a strain rate of $10^{-2} - 10^{-1}\,s^{-1}$ are superior to those obtained after SPDHT (Table 5.4).

The higher ductility of the alloys subjected to SP deformation can be attributed to the uniform and fine-grained structure obtained in the alloys deformed under superplastic conditions. In the V96Ts and 1420 alloys that exhibit a low ductility after conventional treatment, the elimination of structural strengthening during SPDHT makes an additional contribution to the alloy ductility.

The above discussed peculiarities of the effects of treatments based on superplasticity and conventional processing on the mechanical properties are valid in the case the forming procedure under the bulk-stressed condition. As mentioned above, cavitation causes a significant decrease in ductility and strength during superplastic forming of intricately shaped products, requiring large two-axial tensile strains. Forming with counter-pressure is the most efficient for eliminating this limitation of SP deformation. In aluminium alloys this is accomplished using a hydraulic-static pressure of about 2–4 MPa to suppress cavitation.

The decrease of anisotropy of the mechanical properties during the treatment incorporating SP deformation is caused by the elimination of metallographic texture (grain elongation in the direction of working) and spreading of the crystallographic texture (grain orientation) which lead to anisotropy in the aluminium alloys [5.33].

The analysis of the effect of SP deformation on the alloy structure and physical and mechanical properties indicates that treatments involving superplastic

---

[8] 'Structural strengthening' is one of the unique features of aluminium alloys; it refers to the strengthening of the heat treated, deformed, semifinished products or components due to the retention of an unrecrystallized (polygonized) structure after quenching. This type of strengthening is possible if the quenching temperature is below recrystallization temperature

deformation can be considered as new methods of thermo-mechanical treatment of aluminium alloys. These differ from other known thermo-mechanical treatments not only in their technological schemes, but also in the specific structural features and properties they induce in the alloys.

5.4 Effect of Grain Size on the Structural Strength of Aluminium Alloys

Ensuring a high enough structural strength of alloys in final products after SPDT and SPDHT is one of the least studied aspects of the problem of the commercial application of SP deformation. Structural strength of alloys is essential for the safe performance of products under certain service conditions. The reliability of products is more often determined by the 'safe damage' of material, i.e., by the safe operation of components with cracks. This is assessed by the resistance to crack growth (RCG). In cases when the crack appearance in components is inhibited or detection is difficult, the reliability is determined by the resistance to the crack formation (RCF), i.e., by the 'safe resource'.

The effect of SPDT and SPDHT on the properties of aluminium alloys manifests itself mainly in the effect of microcrystalline structure. However, this refers only to the processes of bulk forming, but not to pneumatic sheet-metal forming. The problem of studying the effect of SP treatment on constructive strength is basically reduced to the comparison of the crack resistance of alloys with coarse-grained and microcrystalline structures.

In order to minimize the effect of other structural factors (except the grain size), the AMg6 alloy having a matrix-type structure with relatively small quantities of primary and secondary excess phases was examined. The RCG and RCF parameters were evaluated at ambient temperature under static, impact, and cyclic loading conditions. The alloy was tested in three completely recrystallzed states with the banded distribution of excess phases particles. These states differ only in their average grain sizes: 35 μm in a coarse-grained state, and 11 and 8 μm in two microcrystalline states produced by cold deformation, namely, by flattening hot-pressed rods followed with recrystallization annealing[9].

Table 5.6 shows the parameters of the mechanical behaviour of the alloy. Imparting a microcrystalline structure to the AMg6 alloy leads to the Hall-Petch strengthening and increase of ductility under quasi-static tensile deformation. As a result of transition from coarse-grained structure to microcrystalline (8 μm), the fracture energy increases by almost 10%.

In order to establish a relationship between the increase of the fracture energy and one of the fracture stages (nucleation or growth of a crack), tests were conducted according to Kahn's method (tensile straining of flat samples with side notches). It was established that the growth of the fracture energy is basically

[9] Experiments performed by M.V. Markushev

Table 5.6. Mechanical properties of the AMg6 alloy (samples with longitudinal or LT orientation)

$\bar{d}$ [μm]	σ_{02} [MPa]	σ_u [MPa]	δ [%]	a_f [a] [kJ/m^2]	a_g [kJ/m^2]	KCT [kJ/m^2]	KCU [kJ/m^2]	$\sigma_u^{0.3}$ [b] [MPa]	$\sigma_u^{0.05}$ [MPa]	σ_{-1} [MPa]
35	157	352	23.2	143	165	280	460	448	423	150
11	162	354	24.6	155	170	250	430	450	415	165
8	165	362	25.1	173	180	210	420	–	–	165

[a] a_f and a_g are specific energies required for crack formation and growth evaluated according to Kahn's method.
[b] σ_u^r is an ultimate tensile stress of cylindrical samples with notches (r in mm is the radius of the notch)

due to the increasing of resistance to crack formation (Table 5.6), whereas the energy of crack growth increases insignificantly.

Crack resistance under impact loading was estimated via bending the samples with introduced fatigue cracks (KCT) and with notches of a radius of 1 mm (KCU). The variation of KCT (Table 5.6) indicates that the resistance to crack growth essentially decreases with the decreasing grain size. This is also the cause of the decrease of KCU, characterizing the sum energy spent on the crack nucleation and growth. The crack formation energy is equal to the difference between KCU and KCT, and it increases with the structure refinement, as in the case of static tests.

A similar effect of grain size on the impact strength is observed in other aluminium alloys, viz., V95 [5.34], 7050 [5.35], etc.

It should be noted that the microcrystalline AMg6 alloy is more sensitive to the notch. This follows from the estimation of the effect of grain size on the ultimate tensile strength (σ_u^r) of cylindrical samples with stress concentrations having the tip radia of 0.30 and 0.05 mm. Table 5.6 shows that σ_u^r decreases with the increasing notch sharpness, and this decrease is more acute in the microcrystalline alloy. This fact apparently, confirms the earlier conclusion that, in a microcrystalline structure, resistance to crack growth decreases.

A greater sensitivity to the sharpness of the stress concentrator of the microcrystalline AMg6 alloy is also observed during the impact loading, which is confirmed by the impact toughness data (Table 5.6).

The results of low-cycle fatigue tests ($R = 0.1$ and $v = 0.5$ Hz) of the longitudinal samples show that the endurance of the alloy decreases with the decreasing grain size (Fig. 5.5). This is due to the acceleration of the fatigue crack growth rate in the microcrystalline alloy (Table 5.7), since the resistance to the crack formation increased with the grain refinement. Figure 5.6 indicates that the number of cycles (N_1) required for the formation of a crack 1 mm long is considerably larger in microcrystalline than in a coarse-grained alloy. Since the rate of the crack growth is higher in the presence of a microcrystalline structure, the increase of N_1 can be explained only by a higher resistance to crack nucleation.

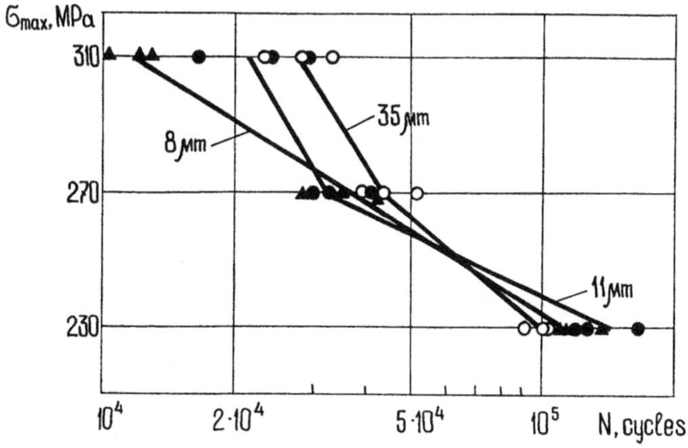

Fig. 5.5. Fatigue curves of the AMg6 alloy with various grain size; (○ – 35 μm, ● – 11 μm, ▲ – 8 μm)

Table 5.7. Effect of grain size of the AMg6 alloy on the fatigue crack growth rate (LT samples, $R = 0.1$, $\nu = 11$ Hz)

ΔK [MPa · $m^{1/2}$]	Crack growth rate [10^{-7} m/cycle]		
	$d = 35\,\mu m$	$d = 11\,\mu m$	$d = 8\,\mu m$
10.1	1.44	1.47	1.53
11.5	2.61	2.90	3.17
13.4	4.22	5.05	5.35
15.2	5.24	6.80	7.06

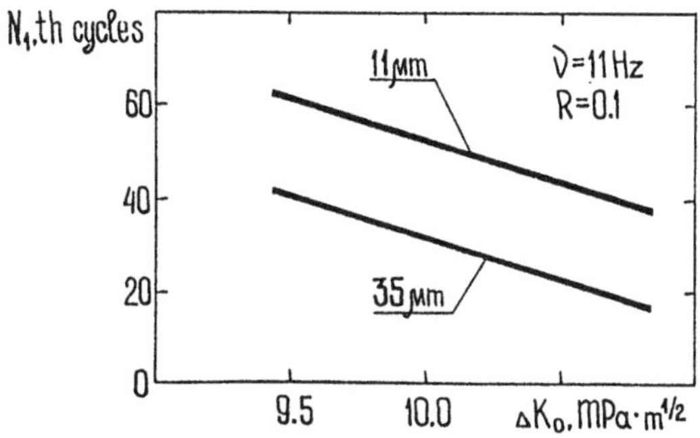

Fig. 5.6. Effect of grain size on the number of cycles required for the formation of a fatigue crack 1 mm long

In the region of high-cycle fatigue, the microcrystalline structure provides a longer endurance of the alloy and, consequently, increases its fatigue limit (Table 5.6). A similar change of the fatigue characteristics with the grain refinement in aluminium alloys was reported in [5.36, 37].

Thus, it was clearly established that the effect of microcrystalline structure on the crack resistance is related to the growing resistance to the crack formation and to the decreasing resistance to the crack growth. Therefore, it is possible to explain the change of the structural strength parameters. For example, the low resistance to low-cycle fatigue, durability, K_{1c} and KCT, and high sensitivity to sharp stress concentrators stem from a lower resistance to crack growth, whereas the increasing fatigue limit and endurance during high-cycle fatigue is due to the enhanced resistance to crack formation time of crack nucleation constituting about 70–90% of the fatigue durability [5.36]).

The relationship between the resistance to crack nucleation and crack growth depends on the nature of the alloy. When the resistance to crack nucleation is strong, the fatigue strength can also be strong in the region of low-cycle fatigue, as was established in the AK4-1 alloy by the authors.

Conclusion. The use of microcrystalline aluminium alloys for the production of principle parts and the superplastic forming of the latter provide for the homogeneity of their structure and mechanical properties throughout their volume, reduce the anisotropy of properties, and improve their durability due to increasing resistance to various kinds of corrosive fracturing and high-cycle fatigue.

The microcrystalline aluminium alloys are useful for ensuring the reliability of important components calculated by the safe resource which limits the service life to a time required for the nucleation of cracks. To ensure the safe damage, the application of these materials must be preceded with the assessment of their reliability, taking into account the low resistance to crack growth in microcrystalline aluminium alloys.

6. Superplasticity of Titanium Alloys

Among the hcp metals, polycrystalline titanium has the highest ductility. This is because slip along the prismatic and the pyramidal planes is possible in α-titanium, whereas in other hcp metals it occurs essentially along the basal planes. In addition, twinning plays an important role during the plastic deformation of α-titanium [6.1].

The alloys based on bcc β-titanium also possess a significant ductility. Plastic deformation of the β-phase occurs by slip as well as twinning. Its ductility depends to a large extent on the content of the alloying elements.

Most of the titanium alloys exhibit a significantly lower ductility than pure titanium. Due to their high flow stress and severe work hardening, it is extremely difficult to subject them to cold deformation. When the temperature is raised, the deformation resistance of titanium alloys decreases significantly and ductility increases [6.2]. However, when conventional hot working operations are employed, the nonuniformity and inhomogeneity of deformation lead to the formation of the so-called "zones of intense flow" in the alloys which, in turn, lead to a severe macro- and microstructural heterogeneity. The formation of these zones is also facilitated by the low thermal conductivity of the alloys. To avoid nonuniformity of deformation and structure on the one hand and the inherently low ductility of the alloys on the other requires multistage processing involving a number of intermediate heating operations. Heat treatment subsequent to hot deformation does not always make it possible to achieve complete elimination of the microstructural nonuniformity in the semifinished products and to obtain the required combination of the mechanical properties. However, it is extremely important that structural materials possess the proper combination of properties such as strength and ductility, fatigue resistance, high-temperature strength, fracture toughness, etc., which depend on the structural state. Thus, let us examine the specific features of structure evolution in titanium alloys.

The phase constitution and the microstructure of the titanium alloys vary as a function of the alloying elements. The microstructures of titanium alloys are based on solid solutions of α- and β-titanium. The quantitative ratio of these phases in the annealed condition forms the basis of classification of titanium alloys into α and β alloys, pseudo-α and pseudo-β alloys, and the two-phase $\alpha + \beta$ alloys. The variation of the quantitative ratio of the α- and β-phases is significantly affected by the alloying elements (those classified as α-stabilizers such as aluminium and β-stabilizers such as molybdenum, vanadium, chromium, iron, etc.) and also by the heat treatment parameters. When cooling from certain initial temperatures, it becomes possible to obtain the metastable phases β, α' (α'') at room temperature. One of the characteristic features of the α and $(\alpha+\beta)$-alloys is that they exhibit an abrupt coarsening of the microstructure during their transition into the β-region. This process is not as severe as in the high alloy β-alloys.

Significant microstructural changes occur even during cooling of the alloys. When cooling from the β-region, a lamellar microstructure is developed with the β grains in the α and $(\alpha+\beta)$-alloys as a result of polymorphic transformations. The degree of dispersion (fineness) of the colonies of plates significantly depends on the alloying elements. On annealing in the β-region, one observes considerably finer particles of α-phase in the highly-alloyed pseudo-β alloys than in the case of α- and $(\alpha+\beta)$-alloys [6.2]. The lamellar microstructure exhibits extraordinary stability and is extremely difficult to chagne by heat treatment.

There is a wide variety of microstructural states in alloys subjected to treatments in different phase regions. However, one can distinguish two basic morphological types of precipitates, viz., globular and lamellar.

For the sake of convenient presentation of the titanium structures and for retaining uniqueness of terminology, we shall use the following terms in this chapter: the microcrystalline structure (globular, equiaxial) and the coarse-grained (lamellar) structure.

Depending on the type of the microstructure, titanium alloys possess different combinations of mechanical properties. The alloys having a microcrystalline structure exhibit better strength, impact strength, and fatigue limit than the alloys having coarse-grained lamellar microstructure [6.2]. At the same time, the two-phase alloys with a lamellar structure show superior characteristics of high-temperature strength (heat resistance) and fracture toughness.

Treatment of titanium alloys under superplastic conditions is of significant interest for technology. Such a treatment makes it possible to diminish the resistance to deformation as well as to increase the ductility of titanium alloys. At the same time, it is impotant to examine the effect of SP deformation on the microstructure and different mechanical properties of titanium alloys. The chemical composition of the alloys under study as well as the parameters of SP deformation are given in Table 6.1.

6.1 Plasticity and Superplasticity of the Titanium Alloys in the Single-Phase Region

Retention of the microcrystalline structure of the titanium alloys is difficult in the α- as well as in the β-regions because of the rapid grain growth occurring during heating. It would appear that attainment of high ductility of the alloys is impossible as in the case of commercial-grade aluminium and magnesium (Chaps. 4, 5). However, results indicate high ductility of alloys in the presence of coarse grains in the initial state [6.3–6]. In the α-alloy VT1-00 (commercial-grade

Table 6.1. Characteristics of superplastic deformation of titanium alloys

Alloy Type	Chemical composition [mass %] Mo	V	Sn	Zr	Al	Fe	C	N	O	Si	H	Ti	Alloy grade	Initial grain size [μm]	Temperature of the trans-formation $\alpha+\beta \rightarrow \beta$ [°C]	Conditions of SP deformation t [°C]	$\dot{\varepsilon}$ [s^{-1}]	σ [MPa]	δ [%]	m
α	–	–	–	–	0.31	0.25	0.05	0.035	0.10	0.07	0.08	balance	VT1-00	10[a]	890	600	1.3×10^{-3}	80	100	0.26
β	11.4	–	5.5	4.2	–	0.08	0.015	0.012	0.10	–	0.01	balance	VT30	200	745	900	3.3×10^{-3}	55	200	0.3
$\alpha+\beta$	–	3.75	–	2.0	6.0	0.30	0.02	0.05	0.20	0.15	0.015	balance	VT6	1.0	990	900	3.1×10^{-3}	16	1000	0.54
$\alpha+\beta$	3.5	–	–	1.7	6.6	0.11	0.02	0.01	0.087	0.27	0.002	balance	VT9	2	1000	950	2.7×10^{-3}	29	1800	0.6
$\alpha+\beta$	3.2	–	–	–	6.0	0.4	0.1	0.05	0.15	0.3	0.015	balance	VT8	0.1	1000	600	7×10^{-4}	230	480	0.33

titanium), superplasticity was observed at a grain size of 20 µm [6.6] and 40 µm [6.5]. It was also established that in commercial grade titanium RC-70 (grain size of 18 µm) and in the α-alloys Ti-5%Al-2.5%Sn (18 µm) superplasticity takes place in the temperature range of polymorphic transformation $\alpha \rightleftarrows \beta$, but not in the α-region [6.3]. Most unusual is the fact that in the β-alloys, viz., Ti-8%Mn, Ti-15%Mo, and Ti-3%Al-11%V-13%Cr, a superplastic condition was noted in the β-region with d = 200–500 µm [6.4]. There is evidence that superplasticity is observed in another β-alloy, namely, β-III at a grain size of 30 µm [6.7].

Let us examine the effect of the initial microstructure and the temperature and the strain rate of deformation on the ductility of a single-phase alloy and the microstructural changes under the conditions of high ductility. Commercial-grade titanium VT1-00 (Table 6.1) was used [6.8] and was studied in the cold-rolled (E = 50%) and annealed (850°C, 2 h) conditions. After annealing, the grain size was found to be approximately 150 µm.

Figure 6.1 shows the effect of the deformation temperature on the mechanical properties of the VT1-00 alloy in both structural states. The temperature dependence of the relative elongation has a nonmonotonic nature: initially δ increases, thereafter it decreases abruptly, and again increases. With increasing test temperature, the ductility of the cold worked alloy increases more rapidly. At a temperature exceeding 850°C, the relative elongation is virtually independent of the initial structural state of the alloy.

The variation of the flow stress with the temperature of deformation also shows a nonmonotonic, but less distinct, trend. The flow stress of the alloy de-

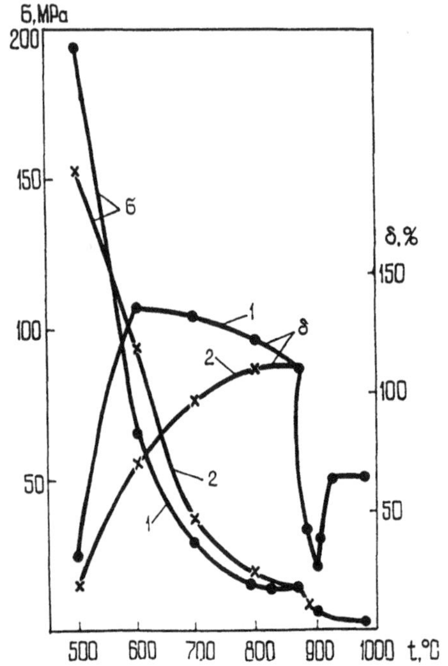

Fig. 6.1. The dependence of flow stress σ_{50} and relative elongation δ on the deformation temperature of the cold-rolled (*1*) and annealed (2) alloy VT1–00 at $\dot{\varepsilon} = 1.3 \times 10^{-3}\,s^{-1}$

creases rapidly with increasing temperature. At a temperature exceeding 850°C, the flow stress is independent of the initial structure of the alloy, and, at a temperature exceeding 875°C, the variation of the flow stress as well as ductility is abnormal. The flow stress and the relative elongation rapidly decrease in the same temperature range. A subsequent increase of temperature does not have a considerable effect on the flow stress.

The complex nature of the temperature dependence of δ and σ of the alloy is due to structural changes occurring during the process of heating. Recrystallization occurs upon heating the cold rolled alloy to 600°C and the microstructure of the alloy shows a grain size of 10 µm. A subsequent increase of temperature continuously increases the grain size, and at 850°C the grain size does not virtually differ from that of the annealed alloy (d = 150 µm). The $\alpha \rightarrow \beta$ transformation of the alloy takes place in the temperature range between 885–890°C [6.2]. At these temperatures, the microstructure of the alloy consists of two phases ($\alpha + \beta$). The grain size of the α-phase is approximately 150 µm and the β-phase is distributed along the boundaries of the α-phase. Upon heating the alloy in the β-region, coarsening of the microstructure takes place. At 925 and 1000°C, the grain size d = 300 and 800 µm, respectively. The formation of the fine-grained microstructure after heating to 600°C sharply improves the ductility and reduces the flow stress. However, a subsequent increase of temperature does not significantly affect the elongation.

The following observations were made during a study of the strain rate dependence of the VT1–00 alloy at different temperatures. At 600°C, the relative elongation significantly depends on the strain rate. At a strain rate of $6.6 \times 10^{-2}\,s^{-1}$, the elongation is equal to 50%, whereas at $\dot{\varepsilon} = 3.3 \times 10^{-3}\,s^{-1}$, it is 140%. A subsequent decrease in the strain rate leads to some decrease of the elongation. The magnitude of the coefficient m also depends on the strain rate. On decreasing the strain rate to $3.3 \times 10^{-3}\,s^{-1}$, m increases from 0.1 to 0.26; however, a subsequent decrease in the strain rate does not lead to a significant change in m. The tests conducted at other temperatures in the α-region did not reveal a substantial effect of the strain rate on the ductility,and m was found to be approximately 0.2.

Thus, the ductility of the VT1–00 alloy can be increased in the 600–700°C range due to prior grain refinement. However, in this case, the indications of superplasticity are weak. This is probably due to the absence of microstructural stability in the alloy.

It is interesting that at high temperatures in the α-region the ductility of coarse-grained titanium is fairly high. What are the reasons for the enhanced ductility of the coarse-grained titanium at higher temperatures in the α-region? A detailed study of the mechanical behaviour and structural variations of annealed titanium during compression in the temperature range of 550–850°C was carried out.

The increase of σ with the increasing strain was observed (Fig. 6.2). The lower the temperature and the higher the strain rate, the more severe is the hardening during deformation. These results were obtained by *Galeev* and *Valiakhmetov*.

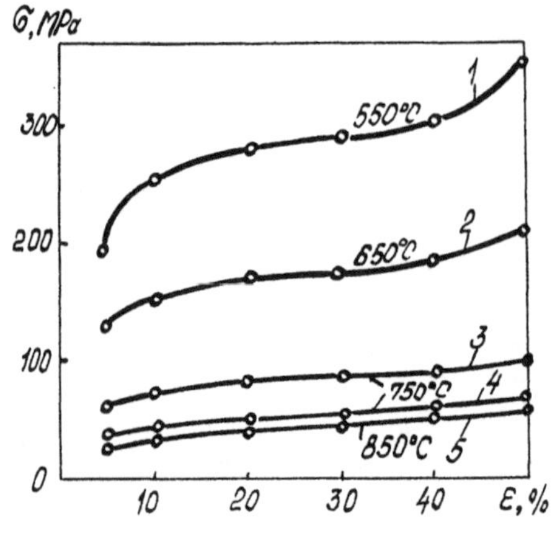

Fig. 6.2. Flow stress σ and deformation of VT1-00 titanium at different temperature and test rates: Curves (*1,2,3,5*) $\dot{\varepsilon} = 2.2 \times 10^{-3}\,s^{-1}$; curve (*4*) $\dot{\varepsilon} = 2.2 \times 10^{-4}\,s^{-1}$

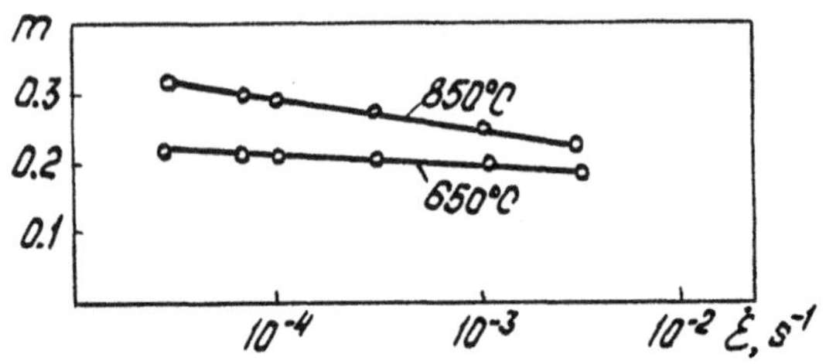

Fig. 6.3. Dependence of the strain-rate sensitivity coefficient m for VT1-00 titanium at different temperatures

The coefficient m measured by abruptly changing the strain rate by 50% while straining with an initial rate of $2.2 \times 10^{-3}\,s^{-1}$, proved to be dependent of the temperature-rate conditions of deformation (Fig. 6.3). At 650°C m hardly changes with the strain rate, and its value never exceeds 0.2. At 850°C the strain rate dependence of m is observed and it attains the value of 0.31 at $\dot{\varepsilon} = 5 \times 10^{-5}\,s^{-1}$.

The study of the microstructure of the deformed samples of the alloy showed that the common feature of hot deformation is the development of dynamic recrystallization and formation of new grains; both the number and size of these grains increase with the increasing temperature and decreasing strain rate (Table 6.2).

The analysis of microstructural evolution at $t = 750°C$ and $\dot{\varepsilon} = 2.2 \times 10^{-3}\,s^{-1}$ showed that during an early stage of plastic flow some portions of grain bound-

Table 6.2. Average recrystallized grain size d_r and recrystallized volume V_r in α-titanium after deformation up to $\varepsilon = 80\%$ at different temperatures

$\dot{\varepsilon}$ [s^{-1}]	2.2×10^{-3}				2.2×10^{-4}
t [°C]	550	650	750	850	750
d_r [μm]	0.8	2.9	4.8	8.8	8.5
V_r [%]	39	58	65	76	77

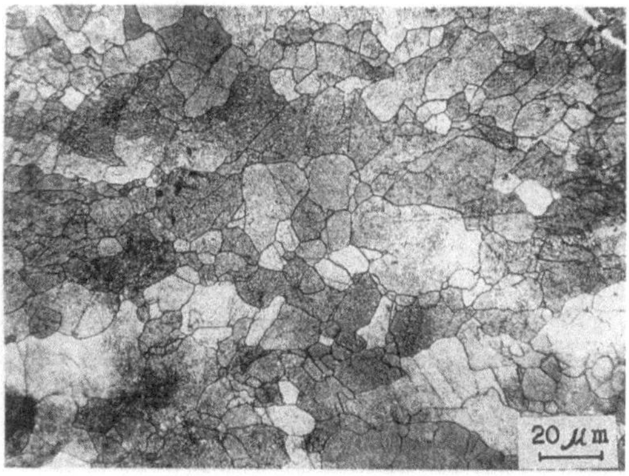

Fig. 6.4. Microstructure of titanium VT1–00 after deformation at $t = 750°C$ and $\dot{\varepsilon} = 2.2 \times 10^{-3}\,s^{-1}$ up to $\varepsilon = 80\%$

aries are meandering, and a substructure is observed in the grain interior. At $\varepsilon = 15\%$ the appearance of first recrystallized grains, sized 2–10 μm, is observed in the vicinity of the initial boundaries. When the strain level increases to 50%, the initial grains become elongated in the direction of deformation, and the quantity of recrystallized grains increases. At $\varepsilon = 80\%$ the recrystallized grains, sized from 2 to 15 μm, constitute about 65% of the total volume (Fig. 6.4).

Thus, dynamic recrystallization develops during the hot deformation of commercial-grade titanium, which ensures favourable conditions for superplastic flow. Due to the realization of the SP flow at temperatures of the top parts of the α-region, titanium has improved parameters of ductility.

No symptoms of superplasticity are observed in commercial-grade titanium in the phase transformation interval which contradicts the data of [6.3, 5]. For example, the elongation of the samples decreases instead of increasing. It should be noted that in spite of the low values of δ of the VT1–00 titanium, the cross section area reduction ψ is as high as 100%. The low value of the sample relative elongation in combination with the high value of ψ indicate that there is a loss in the stability of plastic flow during the initial stage of deformation. Structural studies showed that in the $(\alpha + \beta)$-region of the VT1–00 alloy, a coarse-grained two-phase microstructure is created hindering the development of superplastic flow; consequently, the deformation of the alloy develops according to the usual mechanisms.

In the β-region of the VT1–00 alloy, the indications of superplastic flow are not observed (although some increase in the relative elongation is observed at $t > 900°C$). It is difficult to determine the reasons for the low elongation levels of the alloy in the β-region, because the microstructure is subjected to significant changes during the process of cooling as a result of phase transformation. In order to study the specific features of hot deformation in the β-region, it is necessary to have an alloy in which the high-temperature state is retained after quenching. Thus, the VT30 alloy (Table 6.1) was studied in the hot rolled and annealed states [6.9]. After annealing at 900°C for a period of 6 h, the grain size of the β-phase was 200 μm.

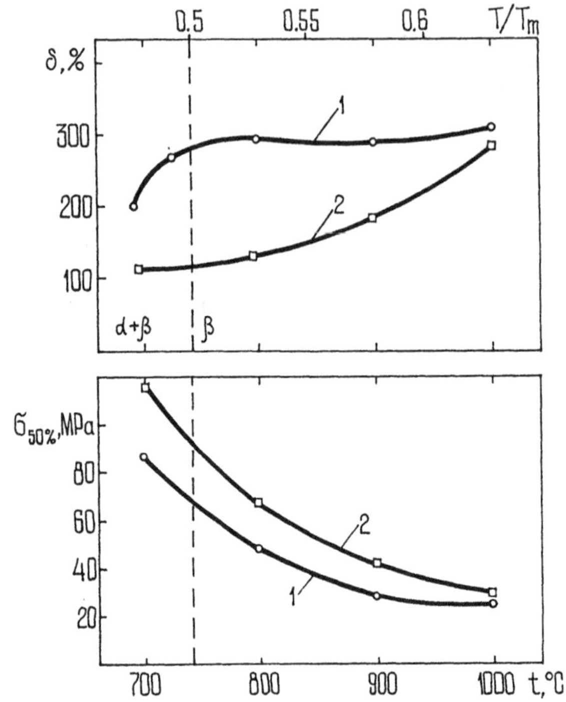

Fig. 6.5. Dependence of flow stress σ and relative elongation δ on the deformation temperature of the hot-rolled (*1*) and annealed (2) VT30 alloy at $\dot{\varepsilon} = 1.3 \times 10^{-3}\,s^{-1}$

The dependence of the flow stress of the alloy on the deformation temperature is monotonic (Fig. 6.5): the higher the temperature, the lower the flow stress. The flow stress of the hot rolled and the annealed alloys differs throughout the experimental temperature range up to 1000°C, at which stage the ductility of the alloy does not depend on its initial state. In the 800–1000°C range, the ductility of the annealed alloy increases to a greater extent (as compared to the hot rolled alloy): from $\delta = 140\%$ at 800°C up to $\delta = 340\%$ at 1000°C.

It should be noted that the obtained plots do not look abnormal when passing through the point of phase transformations (as was observed in the case of α-titanium).

Let us analyze the nature of the stress-strain curves at different temperatures and strain rates. They are characterized by a common regularity which is clearly manifested by curves obtained at 900°C (Fig. 6.6). At the initial stage of deformation the flow stress increases rapidly and after a deformation of 5–10%, it decreases up to a certain level; thereafter, it varies slowly as a function of the strain. The nature of the $\sigma - \varepsilon$ curves is retained at all test temperatures; however, the lower the strain rate, the lower is the peak value of the flow stress. The stress vs strain rate plots based on these data (the stress corresponding to 50% strain was taken) were used to evaluate the coefficient m. At 800°C, the maximum value of m in the hot-rolled state was equal to 0.3. In the annealed alloy, the maximum of m is displaced towards low strain rates, and its magnitude is virtually the same. Evaluation of the dependence of m on the strain rate at

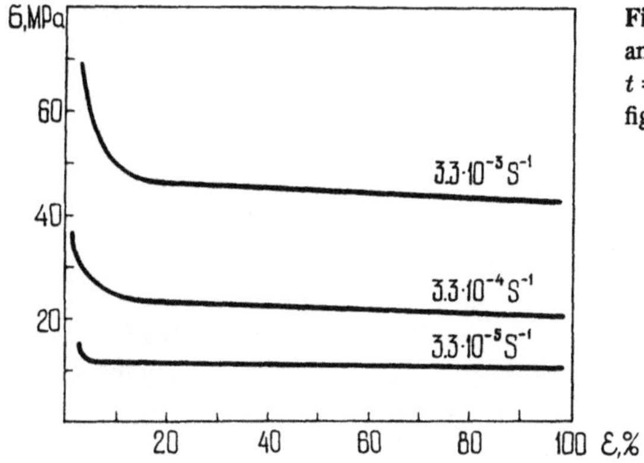

Fig. 6.6. True curves of flow stress σ and deformation ε of the VT30 alloy at $t = 950°C$ and test rates indicated on the figure

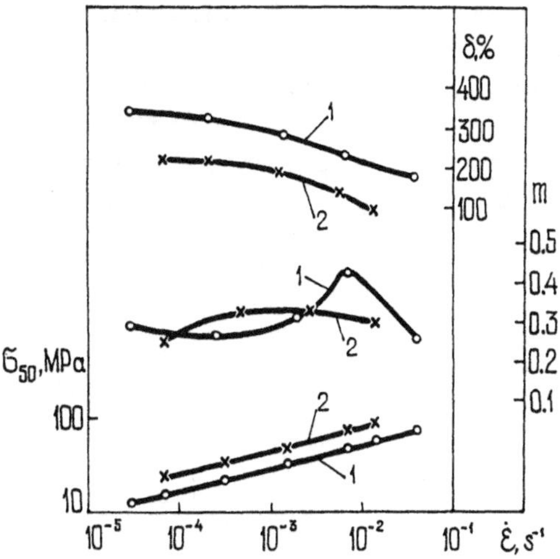

Fig. 6.7. Dependence of flow stress σ (*lower curves*), relative elongation δ (*upper curves*) and the strain rate sensitivity coefficient m (*middle curves*) for the hot-rolled (*1*) and annealed (2) VT30 alloy at $t = 900°C$

900°C gave essentially similar results, but the strain rate dependence is more clearly seen in (Fig. 6.7). At a strain rate of $3.3 \times 10^{-3}\,s^{-1}$, $m = 0.4$ for the initially hot-rolled condition. In the annealed condition, m varies slowly with strain rate and is equal to 0.3. There is virtually no correlation between δ and m as a function of strain rate.

It should be noted that the relative elongation of the alloy in either state (hot rolled or annealed) depends on the strain rate, i.e., the smaller the strain rate, the greater the ductility. A difference in the ductility and the flow stress levels of the alloys in different conditions is retained in the entire range of strain rates. The ductility of the hot-rolled alloy is greater than that of the annealed alloy, and the flow stress follows a reverse trend. It is important to note that in the region of low strain rates and maximum values of m, deformation of the specimens occurs uniformly.

Thus, the data confirm the fact that the VT30 alloy shows the following symptoms of SP flow in the β-region: the strain rate dependence of the coefficient

m which attains values exceeding 0.3; a high relative elongation with uniform deformation of the specimen; and the strain independence of the flow stress at the optimum strain rates. In general, the results agree with previously reported data [6.4]. However, there is a difference in the value of m; when it is in the annealed alloy, in no case does the coefficient m exceed 0.3. However, in the coarse-grained condition, coefficient m was close to unity. Probably this difference is due to the differences in the procedures used for determining m.

The metallographic study of samples deformed at 800–1000°C indicated that complex structural changes occur during the deformation of the VT30 alloy in both states. The typical feature observed is the development of dynamic recrystallization, i.e., formation of new grains and subgrains and the appearance of meandering grain boundaries (Fig. 6.8). It can be seen that the formation of new grains is caused by the migration of local portions of initial grain boundaries. The size of grains and subgrains (revealed metallographically) increases with rising temperature and decreasing strain rate. Symptoms of dynamic recrystallization were observed during the deformation of the alloy in the β-region as well [6.10].

During the deformation of the hot-rolled alloy at 900°C and $\dot{\varepsilon} < 10^{-3}\,s^{-1}$ the initial grains remain equiaxial and their growth is observed. Moreover, the following fact is a contradiction to the usual superplastic behaviour: the grain

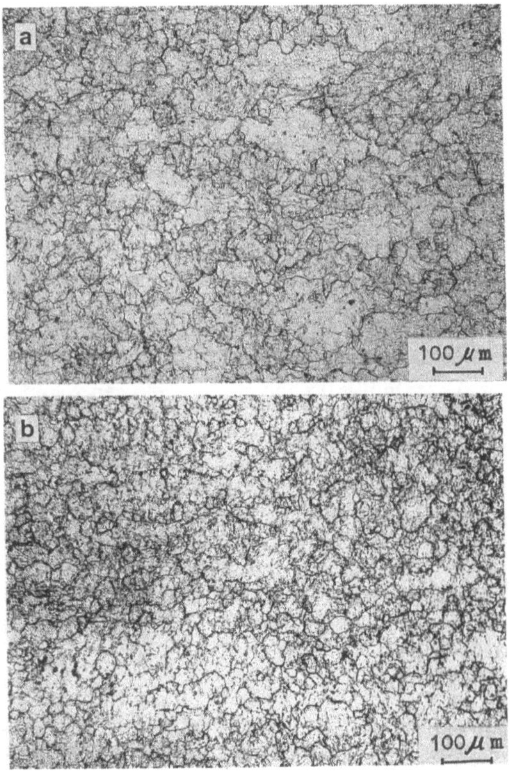

Fig. 6.8. Microstructure of the hot-rolled VT30 alloy after deformation at $t = 900°C$ and $\dot{\varepsilon} = 3.3 \times 10^{-3}\,s^{-1}$ up to (**a**) $\varepsilon = 100\%$ and (**b**) 180%

Table 6.3. Strain rate dependence of the microstructural characteristics of the VT30 alloy and its exposing time at 900°C

State	$\dot{\varepsilon}$ [s^{-1}]	Grain size after deformation [mm]	Grain size after holding for a time equal to the deformation time [mm]	Subgrain size[a] [mm]
Hot-rolled state	3.3×10^{-5}	210	300	15–25
	3.3×10^{-4}	100	150	5–10
	3.3×10^{-3}	–[b]	100	2– 4
Annealed state	6.6×10^{-5}	200	235	10–20
	2.7×10^{-4}	–[b]	210	6–10
	3.3×10^{-3}	–[b]	200	3–10

[a]Subgrain size was measured by means of metallographic studies
[b]Grain size was not measured because of the loss of the contours of the original grains

growth is less rapid during deformation than while holding for a time equal to the time of deformation (Table 6.3). At 100% strain many grains lose their contours as a result of their fragmentation and recrystallization. However, the latter process is apparently suppressed with decreasing strain rate, since at $\dot{\varepsilon}$ $10^{-4}\,s^{-1}$ no recrystallized grains are found. At strains exceeding 100%, microstructure is retained. The evaluation of an average subgrain size at 200 and 300% strain shows that the subgrain size virtually does not change when compared to the size at 100% strain.

At a higher strain rate ($\dot{\varepsilon} = 3.3 \times 10^{-3}\,s^{-1}$) in the annealed alloy, the grains are elongated in the direction of tensile loading, and they become more elongated with the increasing strain level and strain rate. During the hot deformation of the annealed alloy, a non-uniform microstructure is formed, subgrains first appearing near the grain boundaries. In the hot-rolled alloy (under the same conditions), deformation is more uniform and subgrains are formed throughout the volume of the original grains. When the strain increases, the microstructure becomes more heterogeneous. Recrystallized grains, sized 10–50 µm, are formed in the vicinity of grain boundaries (Fig. 6.8a), and at 100% strain, initial grain boundaries lose their contours (Fig. 6.8b). It should be noted that, under intensive etching fine, subgrains with a size corresponding to data presented in Table 6.3 are observed in these grains.

With a further increase of strain rate ($\dot{\varepsilon} \sim 10^{-1}\,s^{-1}$), the grains are elongated in the direction of tensile loading, subgrains and grains being formed predominantly near the grain boundaries. Deformation of the alloy in the β-region at other test temperatures leads to qualitatively similar changes in the microstructure.

The study of the dislocation structure of the alloy produced during deformation at 900°C under conditions corresponding to the maximum values of m showed that during the initial stage of deformation (5–10%) elongated subgrains are formed, and during the steady-state flow ($\varepsilon \sim 20\%$) equiaxial subgrains

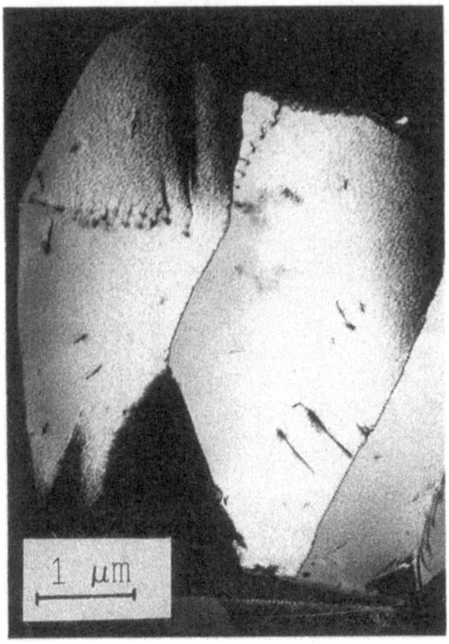

Fig. 6.9. Dislocation structure of the hot-rolled VT30 alloy after deformation up to 100% at $\dot{\varepsilon} = 1.3 \times 10^{-3}\,s^{-1}$ $t = 900°C$

appear. The misorientation and structure of subgrain boundaries depend on the strain level. During the initial stages of deformation, dislocation walls and subgrain boundaries are formed. When the strain level increases, the subgrain boundaries become thinner (Fig. 6.9). In determining the azimuthal misorientation at subgrain boundaries, it was established that the misorientation of adjacent subgrains continuously grows with deformation from a few tenths of a degree up to 4–5°C. It was established that this process was accompanied by a growth of subgrains. Dislocation slip develops in coarse subgrains, and fine subgrains with a misorientation of a few tenths of a degree appear again. According to electron microscopie data, the size of coarse subgrains varies from 8 to 30 μm and are of middle-angle misorientations. In some volumes, disintegrated subgrain boundaries and enhanced density of dislocations can be observed. Similar microstructural changes occur at other test temperatures as well.

Based on the study of the deformation relief, it was established that the symptoms of microstructural changes coexist with the displacement of grains during deformation. It is important to note that the displacement of marker scratches is observed at the boundaries of both the original and recrystallized grains (Fig. 6.10). The magnitude of the grain displacement is nonuniform; in addition, grain boundary sliding is accomplished only along a few boundaries. The displacements were measured along the original grain boundaries and the boundaries of grains formed during deformation. The evaluation of the contribution of grain boundary sliding to the total strain of the hot-rolled alloy at 900°C and $\dot{\varepsilon} = 6.2 \times 10^{-3}\,s^{-1}$ (the conditions corresponding to the maximum value of m) yielded a value of $35 \pm 7\%$. Decreasing the strain rate to $1.4 \times 10^{-4}\,s^{-1}$ decreases the contribution of grain boundary sliding to $18 \pm 4\%$.

The results of the investigation show that high ductility is observed in the VT30 alloy with a coarse-grained microstructure. However, the phenomenology

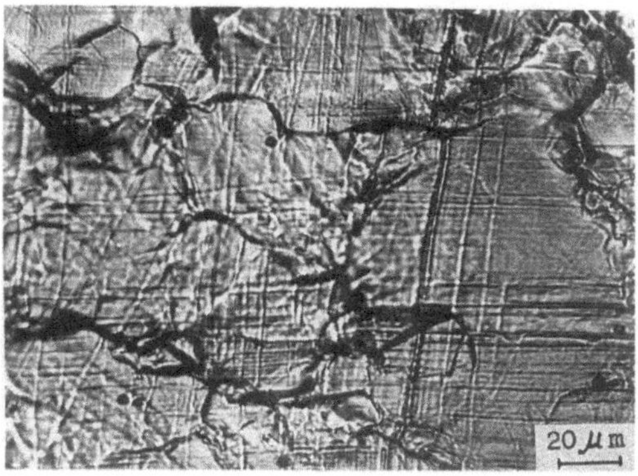

Fig. 6.10. Relief of the surface of the hot-rolled VT30 alloy after deformation at 900°C and $\dot{\varepsilon} = 6 \times 10^{-3}\,s^{-1}$ to $\varepsilon = 130\%$, cooling to room temperature, polishing and subsequent deformation under the same conditions up to $\varepsilon = 20\%$

and nature of this ductility are different from those of conventional structural superplasticity. The data obtained prove that the deformation starts with the multiple slip which results in the formation of a network of subgrain boundaries, primarily in the vicinity of grain boundaries. The prevailing mechanism of their formation is the local migration of the boundaries of the original grains. The formation of high-angle boundaries is probably inhibited by the development of dynamic recrystallization. During deformation, subgrains grow and their misorientations increase.

Grain boundary sliding develops along the boundaries of new and old grains which determines high values of the strain rate sensitivity coefficient ($m = 0.4$) at this stage of deformation. However, the instability of the alloy microstructure during heating in the β-region induces the grain growth which hinders grain boundary sliding. The grown grains start to accumulate dislocations; consequently, a substructure is formed and dynamic recrystallization repeats. The continuous regeneration of subgrain and grain structures and the observed heterogeneity of microstructure manifested in the coexistence of both coarse and fine grains and subgrains is preserved throughout the process of deformation. Though subgrains and grains of the alloy grow during deformation, their average size under steady state flow changes insignificantly, which is possible only if new subgrains and grains are continuously formed. The importance of the subgrain and grain structures formed can also be understood from experiments on a temporary interruption of deformation. The instantaneous interruption of deformation at the steady flow stage has practically no effect on the dynamically established relationship between the size of the new subgrains and grains and, hence, causes no change in the tensile diagram. After a prolonged interruption, the alloy microstructure undergoes a considerable change due to static recovery and recrystallization. This is why further deformation again leads to the appearance of the flow stress peak.

The structural changes observed during the hot plastic deformation of the alloy are due to intensive dynamic recovery and recrystallization. They are accompanied by the development of SP flow. In general, the aforementioned effects are unusual for hot deformation and are probably induced not only by the refining of original microstructure, but primarily by similar rates of dynamic recrystallization and SP flow in this alloy.

Thus, in one-phase α- and β-areas of titanium alloys, features of superplastic flow are observed. In the α-area superplasticity is displayed only when there is a microcrystalline structure. In the β-area superplasticity was displayed in a coarse grained high alloy. In this case, SP flow is characterized by a specific combination of dynamic recrystallization and SP flow. The results obtained in the present research and in [6.7, 11, 12] do not deny a possibility of conventional structural SP which is displayed in the β-area. In [6.7] Sp features were detected in a fine-grained β-III alloy, the β-area temperatures being close to the temperature of polymorphous transformation. However, the microstructure of such an alloy was unstable, the grains grew during deformation and the high values of relative elongation to rupture were obtained probably due to microstructural changes similar to those in the VT30 alloy.

6.2 Superplasticity of Titanium Alloys in the Two-Phase Region

It has been shown that the existence of a microcrystalline structure in the original state of a two-phase titanium alloy Ti-6%Al-4%V has a decisive influence on the alloy superplasticity [6.3]. However, titanium alloys are remarkable because the maximum values of the strain-rate sensitivity coefficient and elongation were observed in the temperature range of phase transformations. Indeed, the development of the $\alpha \rightleftarrows \beta$ polymorphous transformation can substantially alter the alloy microstructure parameters (grain size, phase and chemical composition). During hot deformation, the formation of a microstructure can also be influenced by static and dynamic recrystallization. To distinguish between these effects, two alloys, VT6 and VT9, were systematically studied [6.13]. The choice of these alloys is not accidental; they contain different β-stabilizers, viz., vanadium and molybdenum, but the content of the α-stabilizer (aluminium) is almost identical (Table 6.1). Thus, it is possible to establish the effect of chemical composition of the alloys on superplasticity. The temperature of complete polymorphic transformation (the β-transus) did not differ significantly. The initial state of the alloys was as bars, hot-rolled in the $(\alpha + \beta)$-region. The dependence of the flow stress and the relative elongation on the temperature and the strain rate of these alloys (Figs. 6.11–14) is similar. The results obtained on the VT9 alloy will now be examined in detail.

The temperature dependence of the elongation is characterized by a distinct maximum in the 900–950°C range (Fig. 6.13). In this case, the relative elongation exceeds 1800% (the ultimate values could not be recorded because of the limited traverse of the cross head of the testing machine). Either a decrease or an increase of temperature leads to a sharp decrease of the elongation of the alloy. At a

Fig. 6.11

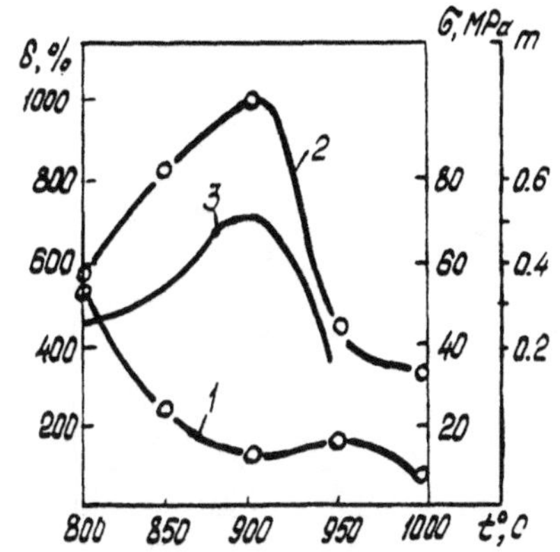

Fig. 6.12

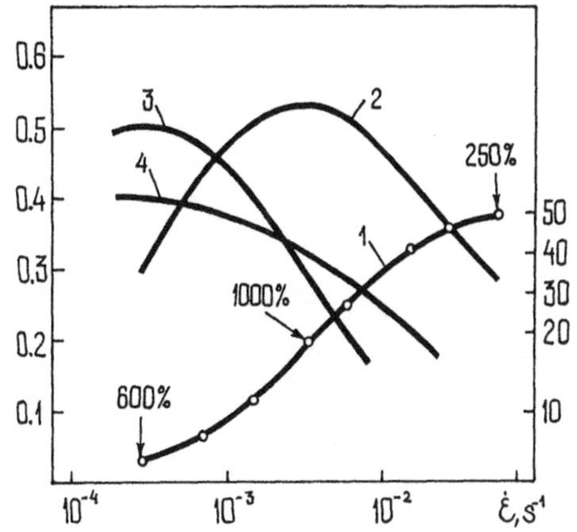

Fig. 6.11. Dependence of flow stress σ (*1*), relative elongation δ (*2*) and strain-rate sensitivity coefficient m (*3*) on the deformation temperature of the VT6 alloy at $\dot{\varepsilon} = 3.1 \times 10^{-3}\,s^{-1}$

Fig. 6.12. Dependence of the strain-rate sensitivity coefficient m (*2,3,4*) and flow stress σ (*1*) of the VT6 alloy on the deformation temperature: Curves (*1,2*) 900°C; (*3*) 850°C; (*4*) 950°C

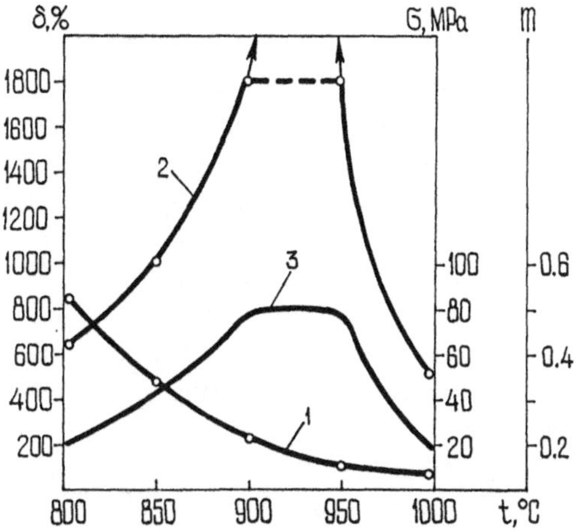

Fig. 6.13. Dependence of flow stress σ, (*1*) relative elongator δ (*2*) and strain-rate sensitivity coefficient m (*3*) on the deformation temperature of the VT9 alloy at $\dot{\varepsilon} = 3.1 \times 10^{-3}\,s^{-1}$

temperature close to the temperature of the $(\alpha+\beta) \rightarrow \beta$ transformation, ductility remains rather high ($\delta = 500\%$).

The relative elongation and coefficient m of the alloy shows a significant dependence on the strain rate (Fig. 6.14a,b). There is a correlation between the maximum values of the relative elongation and the coefficient m. At 900°C, the maximum of m (0.58) is observed at $\dot{\varepsilon} = 1.2 \times 10^{-3}\,s^{-1}$, and at $t = 950$°C, m is 0.54 which corresponds to $\dot{\varepsilon} = 3.1 \times 10^{-3}\,s^{-1}$.

The nature of the true stress-strain curves of the VT9 alloy is typical of superplastic flow. The dependence of the mechanical properties of the VT6 alloy

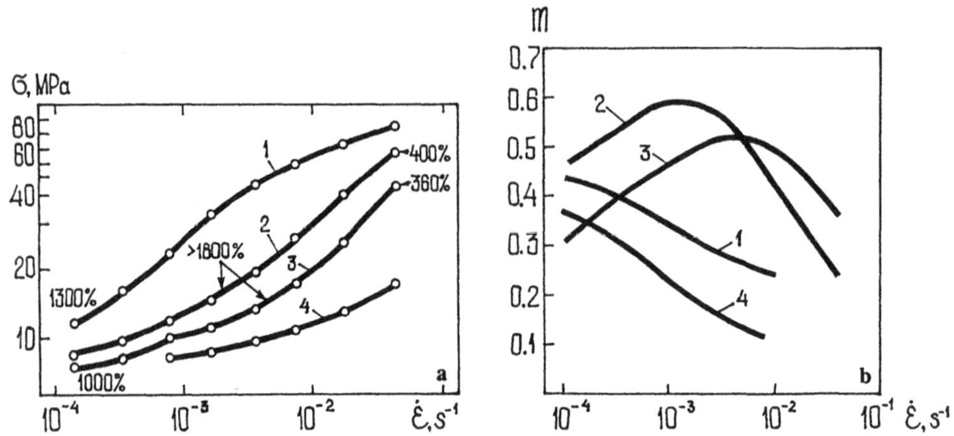

Fig. 6.14. Dependence of flow stress σ (a) and strain-rate sensitivity coefficient m (b) of the VT9 alloy at different temperatures: Curve (*1*) 850°C; (2) 900°C; (*3*) 950°C; (*4*) – 1000°C

on the temperature of deformation and the strain rate resembles that of the VT9 alloy. In the case of VT6, at the optimum temperature of SP and $\dot{\varepsilon} = 3.1 \times 10^{-3}\,\mathrm{s}^{-1}$, the relative elongation was 1000% and $m = 0.54$. Thus, the elongation maxima of both alloys are observed at temperatures corresponding to the $(\alpha+\beta)$-region and at temperatures that are 50–100°C below the temperature of the $(\alpha + \beta) \rightarrow \beta$ transformation. Similar dependence of the elongation and flow stress on the deformation temperature and strain rate has been observed in other two-phase titanium alloys [6.6, 14, 15].

However, there is a distinct difference between the ductility levels of the VT6 and VT9 alloys: the elongation of the VT6 alloy is less and its temperature region corresponding to SP deformation is narrower. In order to explain the obtained results, we shall examine the data of microstructural studies. In this case, it should be noted that the microstructure of the titanium alloys not only depends on the prior treatment of the material, but also varies during the process of heating up to the deformation temperature.

The original structures of the VT6 and VT9 alloys are microcrystalline and consist of fine α- and β-grains that are elongated in the direction of rolling. The VT9 alloy shows a more distinct metallographic texture than the VT6 alloy. We shall examine the changes occurring in the microstructure and the phase composition of the VT9 alloy during the processes of heating and holding.

In the original state, it has a nonrecrystallization microstructure. The first indications of recrystallization are observed after heating to a temperature exceeding 800°C; after holding at 850°C for 60 min, the initially elongated grains of the α- and β-phases show considerable coarsening and rounding, and after heating up to 900°C and holding for a short holding time, 10 min, the alloy is completely recrystallized.

Figure 6.15 presents the data concerning the changes occurring in the grain size of the α- and β-phases at 900–1000°C. It can be seen that a significant

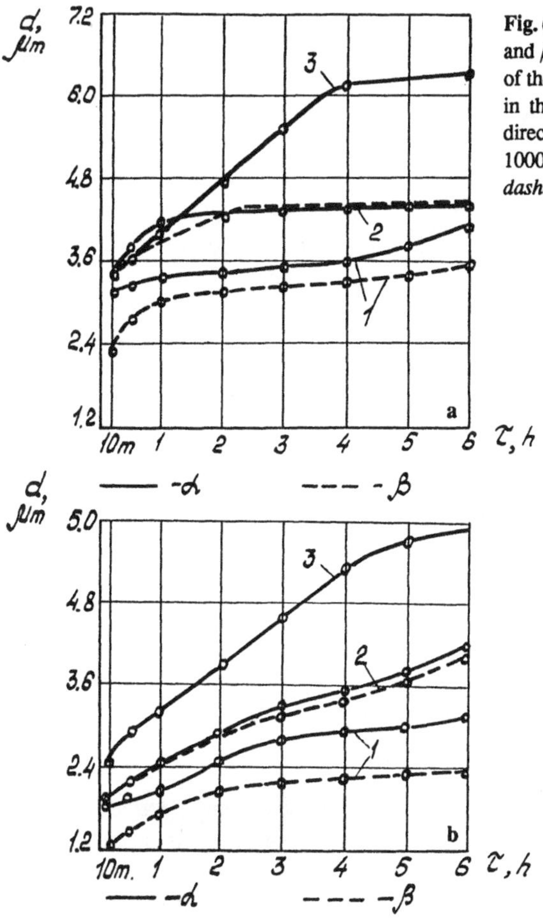

Fig. 6.15. Variation of the grain size of α- and β-phases of the VT9 alloy as a function of the temperature and the time of exposure in the longitudinal (**a**) and transverse (**b**) directions: Curve (*1*) 900°C; (*2*) 950°C; (*3*) 1000°C. *Solid curves* are for the α-phase, *dashed curves* for the β-phase

coarsening of both phases occurs with increasing temperature as well as with increasing duration of heating. The most rapid grain growth is observed during the initial stages of heating (30–60 min) and at a temperature exceeding 950°C. Upon heating the alloy to 1000°C, the grains of the β-phase exhibit abrupt coarsening. The original extent of elongation of ten grains of the phases decreases with increasing temperature and with increasing duration. It should be noted that at 900 and 950°C, a significant reduction in the extent of elongation of the grains of the phases occurs only after a period exceeding 2–4 h.

The recrystallization processes causing microstructural changes in the titanium alloys during the heating operation are accompanied by phase transformations.

The results of phase analysis of the VT9 alloy subjected to heating and exposing are given in Table 6.4. These data indicate that there is a significant change in the phase composition of the alloy at relatively small changes in temperature. The most striking change in the phase composition is observed during the

Table 6.4. Dependence of the β-phase content in the VT9 alloy on the heating temperature and exposing time. (Quantitative analysis of phases was carried out metallographically to an accuracy of 1.5%)

Temperature [°C]	Content of the β-phase [Vol. %]		
	Exposing time [min]		
	10	30	120
900	23	30	35
950	41	49	51
1000	63	75	75

Table 6.5. Size of grains of α- and β-phases and phase composition of VT6 alloy during exposure to heat. Numerator is for grains size of α-phase, denominator is for grains size of β-phase

Tempera-ture [°C]	Size α- and β-grains [µm]			Volume of β-phase [%]		
	Time of treatment [min]			Time of treatment [min]		
	10	30	120	10	30	120
950	1,1/0,7	2,4/1,1	3,6/1,7	21	37	38
950	3,5	6	9	42	50	57

initial stages of exposing. Apparently, the occurrence of phase transformations during the process of holding is due to the extremely slow redistribution of the alloying elements present in titanium during the heating process, their time being controlled by a relatively low value of the diffusion coefficient of molibdenum [6.1].

Quantitatively similar changes in the microstructure and the phase composition of the VT6 alloy occur during the process of heating (Table 6.5). These data show that when heated in the upper portion of the $(\alpha+\beta)$-region, the microstructure and the phase composition of the two-phase titanium alloys are subjected to considerable changes due to the development of recrystallization and phase transformations. Thus, the mechanical properties of the hot worked two-phase titanium alloys are determined by the microstructure formed during the process of heating prior to deformation. Evolution of a two-phase microcrystalline structure during the process of heating facilitates the attainment of the maximum ductility. At a test temperature exceeding the optimum temperature, the ductility decreases because of the substantial coarsening of the grains of the phases (Fig. 6.15, Table 6.4, 5). At temperatures below the optimum, the alloy plasticity decreases, because the diffusion processes are insufficiently active.

Let us analyze the difference in superplastic behaviour of the alloys. First of all, we shall discuss the reasons underlying the difference in the temperature dependence of the flow stress. In contrast to VT9, the flow stress of the VT6 alloy does not decrease after exhibiting an abrupt reduction on attaining 900°C, and it increases upon increasing the temperature up to 950°C.

The differences in temperature dependence of the properties must be associated with the specific features of the microstructural changes occurring during the process of heating. This assumption is confirmed by the data presented in Table 6.5 and Fig. 6.15, and the aforementioned facts: when heated at a temperature exceeding 900°C, the grains of the α- and β-phases of the VT6 alloy coarsen to a greater extent than those of the VT9 alloy. For comparison it should be indicated that, while in the VT9 alloy at 900°C at the beginning of deformation the grain size of the α-phase is 2.2 μm, the grain size of the VT6 alloy α-phase is 1.1 μm at the same temperature. However, at 950°C, the grain size of the α-phase of VT9 and VT6 is 2.5 and 3.5 μm, respectively. In the 900–950°C range, the volumetric ratio of the phases existing in the alloys just before the deformation begins did not differ significantly. Apparently, the lower stability of the microstructure of the VT6 alloy results in lower values of the SP flow parameters m and δ as compared to those of VT9, and superplasticity is observed in a narrower temperature range.

Since the grain sizes in both alloys are similar, the differences in the properties can be attributed to the effect of chemical composition on the stability of the microstructure and the structural changes occurring during superplastic flow.

Just as in the superplastic deformation of other alloys, in the optimum temperature range of SP, the $\sigma - \dot{\varepsilon}$ curves of the two-phase titanium alloys have three distinct regions of SP flow (Chap. 1). Table 6.6 gives results dealing with the structural changes occurring in the alloys within each of these regions of the SP curve.

Grain growth is the most characteristic structural change occurring in the VT6 and VT9 alloys during SP deformation. In this case, a decrease in the strain rate and an increase in the strain level lead to an increase in the grain size. During deformation, grain growth is more active than when held at a test temperature

Table 6.6. Grain sizes of the α- and β-phases and phase composition of the VT6 and VT9 alloys after deformation. The numerator designates grain size in the tensile direction, and the denominator indicates the grain size in the perpendicular direction

	VT6				VT9				
t [°C]	$\dot{\varepsilon}$ [s^{-1}]	d [mm] (ε = 200%)		β-phase [vol.%] (ε = 200%)	$\dot{\varepsilon}$ [s^{-1}]	d [mm] (ε = 200%)		β-phase [vol.%]	
		α-phase	β-phase			α	β	ε = 50%	ε = 200%
900	3.1×10^{-4}	6.8/6.0	5.0/4.8	43	1.2×10^{-4}	5.8/4.3	4.2/3.1	37	40
	3.1×10^{-3}	4.5/4.0	3.5/3.1	42	1.2×10^{-3}	4.0/3.6	2.8/2.5	33	35
	3.1×10^{-2}	2.5/2.3	1.3/1.2	35	1.2×10^{-2}	3.0/2.0	1.8/1.2	27	31
950					3.1×10^{-4}	6.4/4.2	6.5/4.2	50	53
					3.1×10^{-3}	4.5/4.2	4.6/4.2	48	50
					3.1×10^{-2}	4.0/2.3	3.2/2.1	41	44

for a time equal to the time of deformation. At high strain rates corresponding to the III region of SP flow, grain growth is virtually absent. Similar microstructural changes have been repeatedly observed during superplastic deformation of titanium alloys [6.14, 15]. However, the change in the grain size during SP deformation is not identical for different alloys. In particular, in the VT6 alloy, it is more rapid than in the VT6 alloy. After a deformation of 200%, the average grain size of the α-phase of the VT6 alloy tested at $\dot{\varepsilon} = 3.1 \times 10^{-4}\,\mathrm{s}^{-1}$ was 6.4 μm, whereas it was equal to 5 μm in VT9 tested at the same temperature and at the same strain level, but maintaining a lower strain rate ($1.2 \times 10^{-4}\,\mathrm{s}^{-1}$).

Thus, a greater microstructural stability of the VT9 alloy as compared to the VT6 alloy is apparently the reason for a wide range of the SP temperature and strain rates and for the higher ductility of the VT9 alloy.

The dependence of the shape of the grains of the phases existing in these alloys on the strain rate is rather complex (Table 6.6). At all investigated strain rates, no change in the form of the grains of the α- and β-phases was observed. At 900–950°C VT9 alloy shows the elongation of the grains of both phases in the direction of tensile loading at low and high strain rates; at the optimum strain rates, there is a significant reduction in the extent of elongation. The reasons underlying the change occurring in the grain shape during SP deformation were discussed in Chap. 2.

An interesting observation was made with regard to the change in the phase constitution of the alloys when studying the microstructural changes occurring during SP deformation. Table 6.6 shwos that the amount of the β-phase increases during the process of SP deformation. The constitutional changes depend on the strain rate. The most significant change in the phase composition was observed during the alloy deformation in SP regions I and II. In SP region III, the change in the contents of the α- and β-phases with an increasing degree of deformation was less significant. Furthermore, the data presented in Table 6.6 show that phase transformations occur more rapidly during the initial stages of deformation.

The change in the contents of the α- and β-phases must be related to the fact that the diffusional processes are intensified during superplastic flow (Chap. 2). Raising the temperature at which deformation is carried out does not seem to lead to the attainment of phase equilibrium (because of the relatively low diffusional mobility of the β-stabilizing elements). For example, for obtaining 49% β-phase at 950°C, the VT9 alloy requires a holding time of 30 min; on the other hand, virtually the same quantity of the β-phase is obtained after subjecting it to 50% deformation within 2.5 min at a strain rate of $3.1 \times 10^{-3}\,\mathrm{s}^{-1}$.

The presented data confirm the previously made suggestions [6.3, 16] regarding the possible occurrence of phase transformations during SP deformation of titanium alloys. However, the changes in the phase composition mainly occur during the initial stage of deformation.

A study of the deformation mechanism during SP flow of titanium alloys shows that its features are similar to those of other alloys (Chap. 1) in the state of structural superplasticity. This is indirectly confirmed by an experiment conducted on the VT9 alloy, i.e., after subjecting to a strain of 25% at 950°C

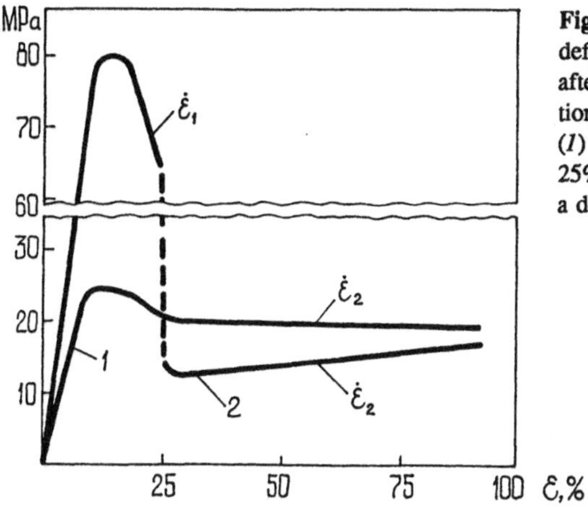

Fig. 6.16. True curves of flow stress σ – deformation ε of the VT9 alloy, obtained after tension at 950°C and at a deformation rate of $\dot{\varepsilon}_2 = 3.1 \times 10^{-3}\,s^{-1}$: Curve (*1*) initial state; (2) after pre-straining to 25% at $\dot{\varepsilon}_1 = 3.1 \times 10^{-1}\,s^{-1}$ followed by a decrease to $\dot{\varepsilon}_2$

and at a strain rate (exceeding the optimum strain rate) equal to 500 mm/min ($\dot{\varepsilon} = 3.1 \times 10^{-3}\,s^{-1}$) that corresponds to the optimum value for SP.

Figure 6.16 shows the $\sigma - \varepsilon$ plots characterizing the flow stress of the alloy without prior deformation and after a prior strain of 25%. It can be seen that there are significant differences in the tensile stress-strain curves. After being subjected to prior deformation at a high strain rate, the flow stress levels decrease abruptly. However, during a subsequent increase of the degree of deformation, the curves tend to come closer. A detailed explanation for the variation of flow stresses as a function of the changes in the strain rates was given in Chap. 2. In spite of the specific features of the changes occurring in the phase composition of the titanium alloys, this experiment indicates the existence of a close relationship between the two basic mechanisms of superplastic flow, viz., grain boundary sliding (GBS) and intragranular dislocation slip (IDS). The effect of diffusion creep is discussed in [6.17] reporting on the enrichment of the α-phase longitudinal boundaries with molibdenum in the Ti-4%Al-4%Mo-2%Zr-0.5%Si.

Thus far we have dealt with the phenomenon of superplasticity in the two-phase titanium alloys with an ultrafine-grained microstructure. Let us compare these data with the mechanical properties of the titanium alloys having a lamellar microstructure[1].

Figure 6.17 shows the variation of σ, δ, and m on the strain rate at 950°C for the VT9 alloy with a lamellar microstructure that was obtained by quenching the alloy from the β-region (from 1020°C). It can be seen that within a certain range of strain rates, the alloy with a platelike microstructure shows an increased strain rate sensitivity of the flow stress. The maximum value of m (0.36) is observed at $\dot{\varepsilon} = 3.1 \times 10^{-4}\,s^{-1}$. A characteristic feature is that the parameter m determined at a strain of 10% is independent of the strain rate. Over the entire range of strain rates, m = 0.2–0.25. The relative elongation increases with decreasing strain rate

[1] The experimental data were obtained in collaboration with R.Y. Lutfullin.

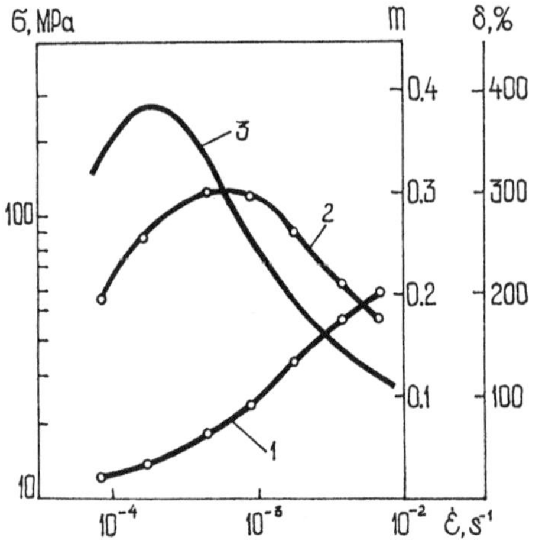

Fig. 6.17. Dependence of flow stress σ (*1*), relative elongation δ (2) and strain-rate sensitivity coefficient m (*3*) of the VT9 alloy with lamellar structure at 950°C (σ and m are calculated for $\varepsilon = 40\%$)

and attains its maximum value (290%) in the range of the maximum values of m. The tensile deformation of the specimens occurs with the formation of the neck, and this deformation induces a high degree of elongation. For the purpose of comparison, it can be noted that under the same conditions of temperature and strain rate, it is possible to obtain an elongation exceeding 1800% in a fine-grained alloy. The optimum strain rate of an alloy with a platelike structure is an order of magnitude less than that of an alloy with an equiaxial microstructure. In addition, in the presence of the platelike microstructure, strain rate dependence of the flow stress can be observed only after attaining a strain not less than 40%.

The nature of the stress-strain curves of the specimens of the VT9 alloy originally having lamellar and equiaxial microstructures is also significantly different. The flow stress levels of the alloys with a platelike structure abruptly decrease after attaining a strain of 5–10%; and at strains exceeding 40%, they vary slowly. Such a dependence is a characteristic for the entire experimental range of strain rates. In this case, during the initial stage of deformation, the flow stress of an alloy with a platelike microstructure is 2.5–3 times higher than that of the alloy with an equiaxial microstructure. However, at $\varepsilon > 40\%$, the difference in the magnitude of flow stress decreases significantly. At $\dot{\varepsilon} = 3.1 \times 10^{-4}\,\mathrm{s}^{-1}$ and $\varepsilon = 40\%$, the flow stresses of the alloy with a microcrystalline structure is equal to 10 MPa, whereas under the same conditions, the alloy with a platelike structure has a flow stress of 14 MPa.

Thus, the titanium alloys initially having a lamellar microstructure exhibit an unusual variation of mechanical properties, i.e., the symptoms of SP flow are observed after a certain stage of deformation, viz., strain rate dependence of the relative elongation and the coefficient m and stabilized deformation in the neck. The superplastic behaviour is due to the transformation of the lamellar microstructure into equiaxial, discussed in detail in Chap. 3.

Thus, the two-phase titanium alloys exhibit the symptoms of superplastic flow in the presence of either a fine-grained or a lamellar microstructure. When heating these alloys to a temperature near that of the $(\alpha+\beta) \rightarrow \beta$ transition, phase transformations and recrystallization processes occur in them and, consequently, the conditions required for obtaining a microcrystalline structure are created either during the process of heating or during the process of isothermal deformation; in the latter case, the lammalar microstructure is transformed into an equiaxial microstructure.

The specific features of the two-phase titanium alloys include accelerated phase transformations during the process of superplastic deformation. This process does not play a decisive role in the development of SP deformation since it is observed during the initial stage of deformation, but it does facilitate evolution of an microcrystalline structure which is essential for SP flow. In the two-phase titanium alloys, the optimum temperature conditions of SP deformation depend on the temperature of the $(\alpha+\beta) \rightarrow \beta$ transitions. Upon passing into the one-phase region, an abrupt grain growth and a decrease in the ductility of the alloys are observed. Consequently, the temperature of the $(\alpha+\beta) \rightarrow \beta$ must be treated as the upper limit for realizing SP deformation under these conditions. The lower limit of the temperature range of SP is determined by the temperature corresponding to the beginning of recrystallization. For instance [6.18], in the Ti-6%Al-4%V alloy, subjected to 80% cold straining, recrystallization occurs at 560–670°C and symptoms of SP flow [6.19] are observed beginning at 685–725°C (at 725°C: $\dot{\varepsilon} = 10^{-5} - 4.7 \times 10^{-4}\,\mathrm{s}^{-1}$, $\delta > 250\%$, and $m = 0.3$–0.5) which is approximately 300°C less than the temperature of the $(\alpha+\beta) \rightarrow \beta$ transition.

The initial microstructure has a significant effect on the characteristics of the SP flow of the two-phase titanium alloys. The refinement of the globular microstructure of titanium alloys increasing the ductility and reducing the flow stress considerably widens the ranges of the temperature and the strain rate corresponding to SP deformation. Thus, in the VT14 alloy (Ti-5%Al-3%Mo-1%V), for example, the decrease of grain size from 3.5 to 1 μm shifts the optimum of superplasticity at 875°C to higher strain rates of 1.5 order of magnitude [6.20]. As a result, at $\dot{\varepsilon} = 1.2 \times 10^{-1}\,\mathrm{s}^{-1}$, the relative elongation to rupture increases from 100 to 850% and the flow stress decreases from 140 to 50 MPa. When a submicrocrystalline structure is created in the alloys, the lower limit of the SP temperature abruptly drops [6.21]. Table 6.7 shows how the grain sizes of the phases affect the superplasticity of the VT8 titanium alloy. An alloy with a submicrocrystalline structure displays superplastic features at 600°C, while in an alloy with a grain size of 5 μm this happens at 700°C. The alloy relative elongation at 600°C attains 480%. The maximum value of the coefficient m is

Table 6.7. Mechanical properties of the VT8 alloy with submicrocrystalline structure

d [μm]	t [°C]	$\dot{\varepsilon}$ [s^{-1}]	σ [MPa]	δ [%]	m
5	800	5×10^{-4}	80	650	0.5
5	700	5×10^{-4}	300	185	0.28
0.1	600	7×10^{-4}	230	480	0.33

0.33. Hence, obtaining the submicrocrystalline state in the titanium alloys allows a lowering of the SP temperature by 400°C or more below the temperature of the $(\alpha + \beta) \rightarrow \beta$ transition.

In addition to the grain sizes of the phases, the SP parameters are substantially influenced by the structure homogeneity. In [6.22] six casts of the Ti-6%Al-4%V alloy were examined. They were different in the concentrations of the alloying elements, the size, shape, and distribution of grains, and in their crystallographic texture. The values of the flow stresses of samples of different heats, e.g., in region I of superplasticity, differed by an order of magnitude. The maximum values of the strain rate sensitivity coefficient varied from 0.6 to 0.9, and the relative elongation to rupture varied from 230 to 675%. The authors concluded that the realization of superplasticity is affected by the microstructure heterogeneity: the α-phase grain pile-up, the nonuniformity of grain sizes, and the metallographic structure. Since the main mechanism of SP deformation is grain boundary sliding, the existence of coarse grains in the phases or their pile-ups slows down its development and, thus, diminishes ductility in heats with heterogeneous microstructures. This is also confirmed in [6.23]. The best superplastic characteristics were manifested by the Ti-6%Al-4%V alloy samples which had a comparatively large average grain size, but a smaller difference in the sizes of coarse and fine grains. This indicates that to attain superplasticity the alloy grains must not only be equiaxial, but similar in size.

It should be noted that analyzing the probable role of crystallographic texture in the difference of the mechanical properties of the casts under consideration, the authors concluded that this role is negligible [6.22]. However, data (Chap. 1) indicate that in titanium alloys crystallographic texture exerts a considerable influence over the SP parameters.

There is one more factor affecting the SP deformation parameters, i.e., the quantitative ratio of phases. In titanium alloys it depends both on the temperature of heating and on the alloying level. The structure is most stable when the volume ratio of phases is 1:1.

However, the optimal parameters of superplasticity do not always meet these conditions. In alloys VT6 (Table 6.1), Ti-6%Al-4%V, and Ti-6%Al-2%Sn-4%Zr-2%Mo [6.24] at an optimal temperature, the volume ratio of phases was $\alpha : \beta = 3 : 2$. Thus, during the alloy heating when the volume ratio of phases were 1:1, the grain size substantially coarsened as compared to lower temperatures and the ductility remarkably decreased.

Medium-alloyed titanium alloys, containing about 10–20% of the β-phase in the equilibrium state at ambient temperature, exhibit the highest parameters of superplasticity. This is apparently due to a favourable combination of certain conditions important for superplasticity, namely, the stability of microstructure during deformation which provides for the rise of the homologous temperature of deformation.

The parameters of superplasticity essentially depend on the chemical composition of the alloys. This is confirmed by the comparison of data obtained on the VT6 and VT9 alloys. The difference in their mechanical properties is determined by their content of different β-stabilizing elements, viz., vanadium in the VT6 alloy and molybdenum in the VT9 alloy. The diffusion coefficient of vanadium in titanium is larger than that of molybdenum [6.1], which probably makes the VT6 alloy more inclined to grain growth during SP deformation, thus decreasing its ductility. This is indirectly confirmed by the fact that the substitution of molybdenum by vanadium in titanium alloys [6.14] deteriorates the parameters of superplasticity. At the same time the alloying of the Ti-6%Al-4%V alloy with yttrium inhibits grain growth during SP deformation because of the formation of dispersed particles and thus considerably improves the ductility of the alloy [6.25].

It should be noted that when titanium alloys contain β-stabilizing elements with a diffusion coefficient exceeding those of the elements of the original compositions, the superplastic properties abruptly increase at lower temperatures. This problem was solved in [6.26] by adding iron, nickel, and cobalt to the Ti-6%Al-4%V alloy, the diffusion mobility of these elements being higher than that of vanadium. At the same time the temperature of superplasticity of the alloy was 150°C lower.

The hydrogenization of titanium alloys also significantly decreases the temperature of superplasticity [6.27]. In processing titanium alloys, it is important to ensure their high performance. This must be considered when alloying is performed with the objective of improving their superplastic parameters. To fulfil this task, experiments were first mathematically modeled [6.28]. A new titanium alloy GT-9 was developed, which exhibits a relative elongation of 770%, and a flow stress of 73 MPa, and satisfactory performance at t = 750°C and $\dot{\varepsilon} = 6.7 \times 10^{-4}\,\mathrm{s}^{-1}$.

6.3 Methods of Obtaining Microcrystalline Structure in Titanium Alloys

As was shown earlier, obtaining a superplastic state is considerably easier in titanium alloys than in other materials. Superplasticity of titanium alloys can be realized even without special prepration of the microstructure, for example, during hot deformation of a coarse-grained β-alloy and the two-phase alloys with a lamellar microstructure. This is due to the specific features of the microstructure of the titanium alloys during the process of heating and hot deformation which transforms the lamellar and coarse-grained microstructure into a microcrystalline structure (Chap. 3). The ductility characteristics obtained in the alloys having such a microstructure are fairly high (δ = 200–300%), and a difference in the flow stresses of the coarse-grained and the microcrystalline materials is observed essentially during the initial stages of deformation and the difference becomes insignificant during the final stages. This makes it possible to process most of the titanium alloys in the SP condition without prior preparation of the microcrystalline structure.

Refinement of the microstructure of the preforms subjected to heat treatment or deformation in the β-region requires consideration of certain features of the phase transformations occurring in the titanium alloys. During their cooling it is necessary to prevent the formation of the α-phase layer among the boundaries of the β-grains (such a layer causes the uniformity of deformation), and also to try to obtain thin α-phase plates which exhibit the most rapid spheroidization. In order to eliminate the α-phase layer, the preforms are deformed in the $(\alpha + \beta)$-region while being cooled to required temperatures or immediately after supercooling. In both cases, it is possible to increase the uniformity of the microstructure formed during hot working. In this case, the start of deformation before the precipitation of the α-phase is rather efficient since an increase in the dislocation density facilitates the formation of shorter and thinner platelets during cooling. Quenching from the β-region is efficient, too [6.20], since the fine-plated martensite formed while cooling dissociates during the subsequent heating in the $(\alpha + \beta)$-region and yields a fine dispersed mixture of phases and provides for a microstructure with a grain size of less than 1 μm. According to [6.20], the indices of superplasticity of the alloys processed according to the suggested method are significantly superior to those observed after conventional rolling in the $(\alpha+\beta)$-region. For example, in the VT14 alloy tested at 875°C and $\dot{\varepsilon} = 1.3 \times 10^{-1}\,s^{-1}$, the flow stress decreases from 145 to 55 MPa and ductility (elongation) increases from 100 to 900%.

Adherence to the conditions described above makes it possible to obtain a uniform microcrystalline structure in semifinished and final products during their subsequent treatment in the $(\alpha+\beta)$-region. This can be either deformation in the temperature-rate regimes of superplasticity (Chap. 3), or conventional forging or gatorizing [6.29]. The latter is performed in titanium alloys at a temperature which is from 10 to 100°C below the temperature for at least 50% strain. Subsequent hot working will facilitate recrystallization and provide for a microcrystalline structure.

Cold working followed by recrystallization annealing is also useful for creating fine-grained structure in titanium alloys. Such a treatment substantially increases the ductility of titanium VT1–00 (Sect. 6.1). Quenching followed by cold working can accelerate the dissociation of the martensite structure or of the metastable β-solid-solution. The formation of differently oriented precipitates of the α- and β-phases during the aging operation makes it possible to obtain a microcrystalline structure during subsequent heating up to the recrystallization temperatures. For example, *Froes* et al. [6.7] could obtain a significant improvement in the characteristics of superplasticity of the β-III alloy (an analog of the VT30 alloy) by prior quenching from the β-region (solution treatment) followed by cold working and aging. After such a treatment, the grain size is 0.5–1 μm. An alloy with a coarse-grained microstructure is capable of exhibiting the superplastic effect in the β-region only, whereas after such a treatment, the alloy shows high ductility and low flow stress in the $(\alpha + \beta)$-region. At 700–730°C (the temperature of complete polymorphic transformation is 745°C) and in the strain range of $10^{-4} - 6 \times 10^{-5}\,s^{-1}$, the flow stress amounts to 26–67 MPa and the relative elongation amounts to 200–500%.

The prior treatment involving cold rolling up to 59% and annealing makes it possible to decrease the temperature of SP deformation of the Ti-6%Al-4%V alloy due to the refinement of the microstructure [6.7]. The alloy which is usually worked in the $(\alpha + \beta)$-region at a temperature of 925°C shows a low flow stress (30–60 MPa) at 815°C in the strain rate range of $10^{-5} - 10^{-3}\,s^{-1}$.

When employing any method for obtaining a microcrystalline structure, one should consider that not only grain size refinement, but also a sharp crystallographic texture can have a significant effect on the superplastic flow of alloys. For instance, in the VT6 alloy, a grain size of 1 μm was obtained by deforming at 850–900°C with a strain per pass not less than 50%. After this, rolling was carried out at 750°C maintaining reductions of 10–20% with intermediate annealing operations in order to obtain a sharp crystallographic texture. Figure 6.18 shows the dependence of σ, δ, and m of the alloy on the strain rate at 700°C. It can be seen that at such a low temperature of working (with reference to the titanium alloys), VT6 exhibits all the symptoms of SP deformation. In the $10^{-4} - 10^{-3}\,s^{-1}$ range of strain rates, the relative elongation was 600–400% at σ = 60–150 MPa and m = 0.2–0.5. Similar results were obtained in [6.19].

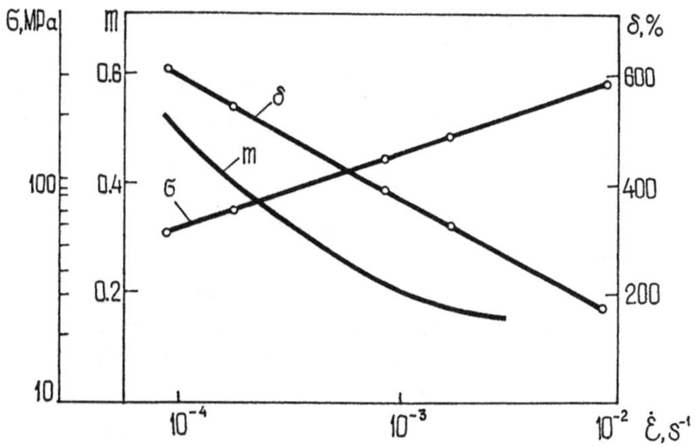

Fig. 6.18. Dependence of flow stress σ, relative elongation δ and the strain – rate sensitivity coefficient m on the deformation rate of the VT9 alloy at 700°C after pretreatment

6.4 The Effect of Superplastic Deformation on the Mechanical Properties of the Titanium Alloys[2]

The studies on magnesium and aluminium alloys (Chaps. 4, 5) showed that working them in the superplastic condition leads to specific structural changes. Thus SP deformation can be used as a method of strengthening and obtaining the desired combination of the mechanical properties of these alloys. In particular, the effect of superplastic deformation on the properties of the titanium alloys is interesting. In contrast to the magnesium and aluminium alloys, the development of the mechanical properties of the titanium alloys is significantly affected by the polymorphic transformations occurring during the process of heating and cooling.

We studied the mechanical properties of the VT9 alloy after superplastic deformation and after heating and exposure at the deformation temperature, but without (actually) deforming, i.e., deformationless treatment (DLT) [6.30]. After processing according to these two schemes, the preforms of the alloys were air cooled. Under such cooling conditions, the microstructure of the alloys was subjected to extreme changes (with reference to the high-temperature structural state), because not only dissociation of the metastable phase, but also a change in the quantity (content) and the size of the primary α-phase occurred [6.2]. The preforms were subsequently subjected to aging according to the standard regime. Thereafter, some of the preforms of the alloy were exposed for a long time at the aging temperature (the thermal stability test). Such a treatment is not equivalent to overaging because during long-term exposing, the alloys are not softened, but strengthened by the dissociation of the metastable phases. It is important to note that thermal stability (heat resistance) is sensitive to the initial structural state of the alloy [6.2]. These facts lead to a number of questions with regard to the effect of superplastic deformation on the mechanical properties of the titanium alloys. First, whether SP deformation accompanied with phase recrystallization affect the alloy properties, or if the effect is identical to that produced by exposing at the temperature of heating. Second, (in case SP deformation affects the properties of titanium alloys) it is important to show the relative efficiency of the strengthening treatment involving SP deformation together with other known modes of processing of these alloys, in particular, as compared to the high-temperature thermomechanical treatment (HTMT) involving quenching from the temperature of the final stage of hot working. Finally, it would be interesting to compare the properties of the alloys subjected to conventional treatment and to SP deformation.

Unfortunately, systematic studies concerning the effect of SP deformation on the properties of the titanium alloys are virtually nonexistent in the published literature. There have been certain studies [6.16, 31, 32] which showed that isothermal deformation conditions facilitate the evolution of a uniform microstructure

[2] The experimental studies were carried out in collaboration with R.Ya. Lutfullin.

in titanium alloys and, thereby, improve the desired set of mechanical properties. However, the investigators did not analyze the strain rate regimes of processing and the data concerning the initial microstructure of the alloys. Therefore, it is not possible to correlate these results with the parameters of SP deformation and to reveal its effect on the properties.

Since the superplastic behaviour of the VT9 alloy has been studied in detail (Sect. 6.3), this alloy can be used for investigating the aforementioned aspects. A positive feature of this alloy is that the temperature corresponding to the optimum superplasticity coincides with the temperature at which heat treatment, conventional treatment, and high-temperature thermomechanical treatment are conducted. This simplifies a comparison of the results of various treatments. Finally, it is easy to obtain a microcrystalline structure or a lamellar microstructure in the alloy.

The VT9 alloy was studied in two conditions, viz., with an initially microcrystalline (globular) structure and with a dispersed lamellar microstructure. The temperature of the complete polymorphic transformation $(\alpha + \beta) \rightarrow \beta$ of the microcrystalline alloy was 1000°C and that of the alloy with the plate-like structure was found to be 990°C. Preforms were deformed by compression under isothermal conditions at strain rates ranging from 10^{-4} up to $10^{-1}\,s^{-1}$ at the following temperatures: 970°C in the case of a fine-grained alloy and 950°C for one with a dispersed lamellar microstructure. The degree of deformation of the preforms amounted to 75% and the duration of holding after heating up to the forging temperature was 10 min. Aging was carried out at 530°C for 6 h. Deformationless treatment was carried out at the deformation temperatures of the alloy at a duration of 1 h, following which the specimens were cooled and were subjected to aging under the same regimes. The cooling rates of all the preform processes according to different regimes did not differ significantly. This was ensured by selecting the appropriate cooling conditions. Heat treatment for stabilizing the alloy was carried out at 500°C for a period of 200 h.

Regardless of the distinct difference in the microstructures of the alloy in the selected conditions, under these thermal regimes, they have fairly close values of the optimum strain rates (region II of SP deformation); i.e., $3.1 \times 10^{-3}\,s^{-1}$ for the microcrystalline alloy and $1.2 \times 10^{-3}\,s^{-1}$ for the alloy with a dispersed lamellar structure. Based on these data, we selected the range of the strain rates and carried out working of the alloy in this range.

Table 6.8 shows the results of the mechanical testing of the alloy. The data show that, as compared to deformationless treatment, working in the superplastic state leads to a superior combination of the mechanical properties of the alloy and increases the ultimate tensile strength by 80–100 MPa with simultaneous retention of a high level of ductility. In the case of the initial lamellar structure, the effect is intensified and the flow stress increases by 100 MPa.

The data also show that the mechanical properties of the alloy significantly depend on the strain rate. With increasing strain rate, the ultimate tensile strength and the yield stress increase, but the thermal stability decreases. This tendency is particularly observed in the alloy with a lamellar microstructure. When deforming

Table 6.8. Mechanical properties of the VT9 alloy with different initial microstructures after SP deformation and conventional treatment. The numerator gives values obtained after deformation and heat treatment with aging at 530°C for 6h, and the denominator indicates values obtained after additional holding of the alloy at 500°C for 200h

Initial structure of the alloy	$\dot{\varepsilon}$ [s^{-1}]	Tensile strength σ_u [MPa]	$\sigma_{0.2}$ [MPa]	δ [%]	ψ [%]
Micro-	1.2×10^{-1}	1230/1240	1040/1120	16.0/14.0	45/30
crystalline	6×10^{-2}	1220/1240	1000/1100	17.0/14.0	49/35
(globular)	2.5×10^{-2}	1210/1230	1030/1100	15.0/16.0	46/34
structure	4×10^{-3}	1200/1210	1030/1100	16.0/15.0	46/38
	CT	1130/1160	1025/1060	16.0/14.0	46/40
	2.5×10^{-4}	1190/1200	1010/1050	18.0/17.0	46/42
Dispersed	6×10^{-2}	1260/1290	1180/1210	8.0/6.0	39/18
lamellar	2.6×10^{-3}	1230/1260	1160/1170	13.0/9.0	44/36
structure	3.6×10^{-4}	1200/1240	1080/1120	15.0/9.0	40/32
	CT	1120/1150	1060/1080	14.0/9.0	40/32

the preforms at a strain rate of region III of SP deformation ($6 \times 10^{-2}\,s^{-1}$), the relative reduction in area decreases by more than 2 times. Working the alloy at a strain rate below the optimum strain rate leads to some decrease in the strength properties. Thus, the obtained data indicate that processing the alloy in region II ensures the best combination of the strength and ductility characteristics in the presence of an initial microcrystalline as well as a dispersed lamellar phase.

In order to deliniate the reasons for the superior strength characteristics of the alloy after SP deformation (as compared to those obtained after DLT), let us examine the structural changes occurring in the alloy after different types of treatments. The alloy with a microcrystalline structure will be examined in greater detail.

After deformationless treatment (exposing at 970°C for 1 h), the structure of the alloy is characterized by a nonuniform distribution of the equiaxial grains of the α- and β-phases. The average grain size of the α-phase is 2.5 µm. The nature of the structural change occurring during the process of upsetting as a function of the strain rate is similar to that observed during tensile deformation (Table 6.9). The grains of the α- and β-phases coarsen substantially at a strain rate of region I of SP deformation. At the strain rates of region I and II, the microstructure of the alloy is more uniform that in the case of deformationless treatment. In addition, during hot working of the alloy the quantity of primary α-phase changes depending on the strain rate. The structure of the alloy deformed at a strain rate of $1.2 \times 10^{-1}\,s^{-1}$ shows 59% primary α-phase, whereas after deforming at the optimum strain rate, the primary α-phase is 50% (time of deformation is 3 min). Similarly, after deformationless treatment, the alloy contains 50% α-phase, i.e., decreasing the strain rate during hot working facilitates attainment of the equilibrium state to a greater degree. As noted in Sect. 6.1, the changes

Table 6.9. Parameters of the structure of the initially microcrystalline VT9 alloy after upsetting at 970°C

Strain rate [s^{-1}]	Grain size of the primary α-phase [μm]	Content of the primary α-phase [vol.%]	Lattice parameter of the β-phase [nm]
2.5×10^{-4}	3.4	51	0.3263
4×10^{-3}	2.3	50	0.3259
2.5×10^{-2}	2.1	50	0.3248
6×10^{-2}	1.9	57	0.3246
1.2×10^{-1}	1.8	59	—

occurring in the phase composition are apparently related to the redistribution of the alloying elements which takes place slowly during the process of heating and exposing, but becomes significantly accelerated (as a result of activation of the diffusion process) when the strain rate is decreased during SP deformation. This assumption is also supported by the variation of the lattice parameter of the β-phase (Table 6.9). The smaller the strain rate of the alloy, the larger is the lattice parameter and, consequently, the greater the degree of depletion of the alloying elements from this phase. It is necessary to note that the lattice parameter of the β-phase changes most significantly after deforming in regions I and II (Table 6.9); in this case, as the strain rate decreases, its value approaches the lattice parameter of the β-phase of a heat-treated specimen (0.3269 nm). It should be noted that straining in region II eliminates the initial texture [6.13].

Similar dependence of the structural changes on the strain rate of SP deformation is observed in the VT9 alloy with an initially dispersed lamellar structure [6.13]. However, the initial nonuniform structure is not eliminated by working the alloy in region III. After deforming in region II, the microstructure of such an alloy consists essentially of a homogeneous mixture of the equiaxial grains of the α- and the β-phases. A study of the microstructure of the alloy using diffraction microscopy confirmed the data obtained in the metallographic studies. When deforming the alloy in regions II and I, the lamellar microstructure transforms into an equiaxial microstructure, and the structure of the alloy shows only a few (individual) dislocations. When deforming the alloy in region III, the alloy retains its structural inhomogeneity because of the incomplete transformation of the lamellar structure into an equiaxial structure; furthermore, a substructure (particularly distinct in the α-phase) forms.

Thus, data of structural analysis confirms the existing difference in the states of the alloy after SP deformation and DLT that was revealed during the study of the mechanical properties. By comparing the microstructure of the alloy and its properties after different treatments, one can conclude that the structural heterogeneity existing in the original condition of the alloy or additionally developed during the course of deformation at high strain rates in region III of SP deformation exerts a significant influence on its mechanical properties. The obtained improvement in the strength characteristics after SP deformation (as compared to

DLT) is due to the elimination of structural nonuniformity and the transformation of the lamellar microstructure into an equiaxial microstructure. In this case, due to the accelerated phase transformations and recrystallization, SP deformation facilitates a more equilibrium state of the alloy.

Apparently, the increased structural and chemical homogeneity of the alloy leads to an increased uniformity of dissociation of the metastable phases during the cooling process as well as during aging. However, it is difficult to confirm this experimentally in the titanium alloys because of the high degree of dispersion (fineness) of the structure. The validity of this assumption was established during a study of the effect of superplastic deformation on the properties of the magnesium alloys (Chap. 4).

The data obtained on the VT9 alloy show that the specific effect of SP deformation on the structure and the properties of the alloy is not only retained, but intensified during the process of subsequent phase recrystallization. Such a conclusion is based on the fact that after the SP deformation, as well as after DLT of the aluminium and magnesium alloys, there is no significant difference in the properties as in the case of the VT9 alloy. Finally, the extensive investigations conducted on the structural states of the alloy make it possible to identify the reasons underlying thermal instability of the alloy, which is observed when deforming the alloy with a plate-like microstructure in region III (Table 6.8). After such a treatment, the microstructure of the alloy is characterized by the maximum structural and chemical inhomogeneity because of the incomplete transformation of the lamellar microstructure into an equiaxial microstructure and the incomplete redistribution of the alloying elements during deformation (Table 6.9). Apparently, these are the factors that cause a change in the nature of aging of the alloy during the process of holding at 500°C for a period of 200 h which, in turn, leads to a loss of thermal stability of the alloy.

It must be taken into account that grain growth during SP deformation can be rather rapid; therefore, the mechanical properties of an alloy (in particular, the strength) are inferior to those of the initial state of the alloy and of that produced by deformation at high strain rates. However, in this case, an increase of other parameters (e.g., creep limit) is observed [6.33].

That the aforementioned features of the effect of SP deformation on the structure and the mechanical properties of the VT9 alloy are of universal nature was confirmed by the studies conducted on the VT3-1 (Ti-6%Al-2.5%Mo-2%Cr), VT8, and VT23 (Ti-5.5%Al-2%Mo-4.5%V-1%Cr) alloys.

Let us examine the relative effects of processing under superplastic conditions and other strengthening methods of the titanium alloys, viz., high-temperature thermomechanical treatment (HTMT) and conventional treatment (CT) [6.13].

It is known that high levels of various mechanical properties are obtained during HTMT of titanium alloys. This is due to the high defect density developed during the high-temperature deformation process and inherited during the phase transformation occurring when quenching the alloy. The conventional treatment is usually performed under conditions differing from those maintained for conducting HTMT. CT is carried out in several stages with intermediate reheating operations until the required degree of deformation is attained, and the preforms are air cooled and subjected to subsequent heat treatment.

The studies were conducted on the VT9 alloy with a dispersed lamellar microstructure. HTMT and conventional treatment were performed on a press. The regimes (parameters) of HTMT were heating the preforms up to a temperature of 950°C, $\dot{\varepsilon} = 5\,s^{-1}$, $\varepsilon = 60\%$, and the time loss for discharging the reforms from the furnace before water cooling was 10 s. The regimes of conventional treatment were heating the preforms up to a temperature of 950°C for 1 h, air cooling, and aging at 530°C for 6 h. The total degree of deformation was 75%.

The regimes of SP deformation and the equipment were described earlier. It is necessary to note that some of the preforms were water quenched after deformation (the time lag before quenching was 25 s). After HTMT and SP deformation with quenching, tempering was carried out at 570°C for 2 h. In order to compare the mechanical properties, DLT was conducted along with HTMT and SP deformation.

Table 6.10 shows the mechanical properties of the VT9 alloy after HTMT, SPD, and DLT. It can be seen that the ultimate tensile strength and the impact strength of the superplastically deformed alloy are the same as those obtained in the alloy subjected to HTMT. There is no significant difference in the anisotropy of mechanical properties of the specimens subjected to different treatments. For example, after SP deformation, one observes a greater uniformity (equality) of the yield stress of the alloy than the degree of uniformity observed after HTMT. The time durations required for the failure of specimens of the VT9 alloy subjected to SP deformation or to HTMT are virtually identical. However, it is important to note that after SP deformation, the strength as well as the ductility are significantly superior to those obtained after the conventional deformationless heat treatment.

Based on the structural studies, one can describe the reasons for the observed differences in the properties of the alloy after different treatments. The excellent combination of mechanical properties obtained after SP deformation

Table 6.10. Mechanical properties of the VT9 alloy after different treatments. τ is the time before fracture (t = 500°C, σ = 600 MPa). The numerator and denominator designate properties in the longitudinal and transverse directions, respectively

Treatment	σ_u [MPa]	$\sigma_{0.2}$ [MPa]	δ [%]	ψ [%]	KCU [MJ/m^2]	τ [h]
HTMT	1490/1450	1380/1310	6/8	14/16	0.27/0.28	140
SPD + quenching[a]: $\dot{\varepsilon} = 6.2 \times 10^{-2}\,s^{-1}$	1490/1500	1390/1310	7/7	18/19	0.26/0.29	120
SPD + quenching[a]: $\dot{\varepsilon} = 2.6 \times 10^{-3}\,s^{-1}$	1460/1480	1310/1270	7/7	17/16	0.27/0.26	160
CT + quenching heating at 950°C for 1 h	1360	1170	5.0	12.0	–	–

[a]Tempering at 570°C for 2 h

is due to the structural uniformity of the alloy [6.13]. As was shown earlier, after deforming the alloy with a dispersed lamellar structure at high strain rates, the structural nonuniformity is not only retained but increases. By comparing the results obtained after conventional treatment and HTMT, the effect of microstructural heterogeneity (existing in the form of the zones of intensive flow) on the mechanical properties of the alloy can be identified [6.13]. Table 6.11 indicates that the combination of the mechanical properties obtained after SP deformation is substantially superior to that obtained after conventional treatment. In particular, the characteristics of strength, impact strength (KCU and KCT), and fatigue limit are improved. In this case, the increase in mechanical properties is due to the distinct difference in the structural states of the alloy produced by conventional treatment and SP deformation. The alloy subjected to CT is characterized by a distinct macro- and micro-structural nonuniformity because of which the resulting mechanical properties are deteriorated.

Table 6.11. Mechanical properties of the VT9 alloy after conventional treatment and SP deformation. τ is the time before fracture (t = 500°C, σ = 600 MPa). The numerator designates properties obtained after aging at 530°C for 6 h, and the denominator gives properties obtained after additional holding at 500°C for 200 h. σ_{-1} is given for $N = 2 \times 10^7$ cycles

Mode of processing	σ_u [MPa]	$\sigma_{0.2}$ [MPa]	δ [%]	ψ [%]	KCU [MJ/m^2]	τ [h]	σ_{-1} [MPa]	KCT [MJ/m^2]
CT	1150/ 1180	1070/ 1090	14.0/ 9.0	42.0/ 36.0	0.64/ 0.36	150	490	0.18
SPD	1230/ 1260	1160/ 1170	13.0/ 9.0	44.0/ 36.0	0.72/ 0.69	134/ 110	540 –	0.21 –

From a practical standpoint, an important fact emerges from the obtained results. By using the optimum regimes of SP deformation, post-deformation annealing is not essential, and it is possible to reduce the duration of exposing the preforms prior to working [6.13].

It is known that the alloys having an initially coarse-grained plate-like microstructure exhibit low strength and poor ductility. Working under superplastic conditions can be recommended as a method for rectifying such a microstructure, as confirmed by data in Table 6.12. After forging the alloy, the preforms were subjected only to aging. It can be seen that the greater the degree of deformation, the higher the ultimate tensile strength and yield stress, and the ductility parameters (δ and ψ) increase significantly as compared to those observed in the heat treated state. However, high values of ductility (comparable with the typical data obtained for the microcrystalline alloy) can be achieved only at relatively high strain levels which provide for the required refinement of the microstructure of the alloy during hot working.

Table 6.12. Mechanical properties of the VT9 alloy having an initial lamellar microstructure

Treatment	Degree of deformation [%]	σ_u [MPa]	$\sigma_{0.2}$ [MPa]	δ [%]	ψ [%]
Deformation at 960°C with $\dot{\varepsilon} = 1.7 \times 10^{-3}\,s^{-1}$	28	1120	1010	8.0	15.0
	50	1210	1040	11.0	23.0
Air cooling, exposing for 6 h at 530°C	68	1190	1090	14.0	42.0
Heat treatment at 960°C for 1 h air cooling, exposing for 6 h at 530°C	–	1110	980	7.0	12.0

Finally, we note that with reference to titanium alloys, SP deformation must be treated not only as a method of product shaping (forming), but also as a method of thermomechanical treatment that makes it possible to improve the required mechanical properties. The effectiveness of processing titanium alloys in the superplastic state is comparable to that of HTMT.

7. Structural Superplasticity of Steels

The ductility of steels can be increased dramatically by transforming them into the superplastic state. Due to polymorphic transformations in iron and a strong dependence of the microstructure and phase constitution of these alloys on the quantity of the alloying elements, additional possibilities for the refinement and stabilization of the microstructure arise. In the temperature range 0.4–0.7 T_m (which is characteristic of SP-deformation) in steels it is possible to obtain: (i) a matrix-type microstructure based on the α- or γ-phases and carbides (intermetallics); (ii) a microduplex-type structure based on the α- and γ-phases; (iii) a mixture of the α- and γ-phases and carbides (intermetallics). A second phase stabilizes the microstructure and promotes SP flow. To analyze the conditions required for SP flow in various phase fields, it is most convenient to start with the iron-carbon alloys. Chemical compositions of the iron-based alloys which will be discussed and parameters of SP deformation are given in Table 7.1.

7.1 Superplasticity of Iron, Plain Carbon, and Alloyed Steels

First of all we examine the dependence of the ductility of pure iron on the deformation temperature and establish its relationship with phase transformations and microstructural changes during heating. The ferrite grains in iron are unstable in the course of heating; the grain growth is rather intensive.

Nevertheless, superplasticity was observed in the region of the $\alpha \rightarrow \gamma$ phase transformation [7.1,2], but according to other data [7.2], ductility dropped in this region. A systematic study of the abnormal plasticity of iron has also been carried out [7.3]. There, the mechanical behaviour of armco-iron (Table 7.1) in two structural states, i.e., cold-worked (ε = 50%) and annealed, was studied.

The dependence of the relative elongation on the deformation temperature of armco-iron is nonmonotonic (Fig. 7.1). In the annealed state a continuous increase of the relative elongation with the rising temperature of deformation is observed. The elongation abruptly decreases in the phase transformation region and increases again in the γ-region. A more perceptible increase in the value of the relative elongation in the temperature range of 600–800°C is characteristic for the cold-worked state, which agrees with the results reported in [7.2]. Plasticity is almost independent of the structural state at temperature above 850°C.

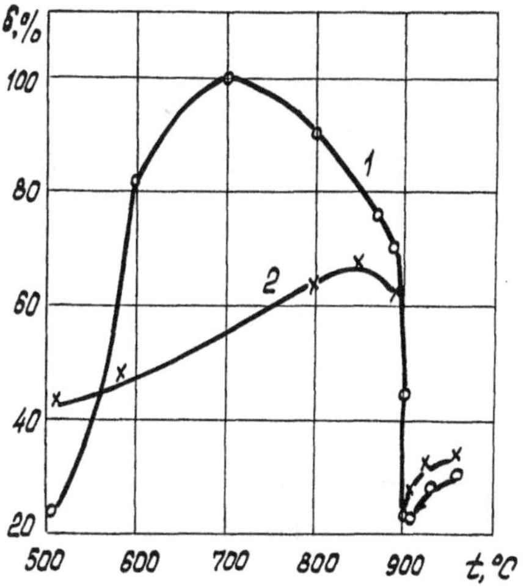

Fig. 7.1. Dependence of elongation δ of armco-iron on the deformation temperature. Curve (*1*) fine-grained state; (2) as-received state (cold worked)

The dependence of flow stress on the deformation temperature is nonmonotonic, too. The flow stresses of a cold-worked material are lower than those of an annealed alloy at temperatures above 500°C. The difference in the flow stresses is the largest in the temperature range of 600–700°C. The value of the iron flow stress decreases with a further rise of temperature, being almost independent of the original treatment at $t > 800$°C. At test temperatures exceeding 880°C the flow stresses first increase and then decrease. The microstructure of armco-iron is also significantly dependent on the temperature. In the annealed state at 600°C, the grain size is 32 μm and becomes 60 μm when the temperature is increased to 970°C. During heating in the cold-rolled iron state recrystallization develops, and the grain size changes from 10 to 45 μm at the same temperatures.

These data may explain the difference in the behaviour of annealed and cold-worked armco-iron. Faint symptoms of superplasticity have been observed in the cold-worked armco-iron at $t = 700$°C. The coefficient m reached the value of 0.27 and the relative elongation was 100% at $\dot{\varepsilon} = 6.6 \times 10^{-4}\,\mathrm{s}^{-1}$ due to the formation of a fine-grained structure. Hence, a relatively small grain growth in this state at $t > 700$°C causes an abrupt drop in the iron ductility, and its mechanical properties become similar to those of the annealed iron.

What causes the abrupt decrease of the iron ductility when the phase-transformation temperature is reached?

A high-temperature metallographic study has indicated that the $\alpha \rightarrow \gamma$ shear transformation developed at the phase transformation temperature during heating, and the γ-phase plates appeared in the ferrite matrix. The development of phase-transformation affect the nature of the deformation relief of the armco-iron samples deformed in the region of the phase-transformation (Fig. 7.2). In contrast to the deformation relief obtained after the tensile deformation of the samples in the α (870°C) and γ (940°C) regions, micro-cracks located in the points of the intersections of γ-phase plates with the boundaries of α-grains are clearly visible on the surface of the samples deformed in the $(\alpha + \gamma)$-region.

Table 7.1. Parameters of SP deformation of steels

Class of steels	Steel grade	Chemical composition [wt %]	Initial grain size [μm]	Optimal conditions of deformation			Parameters		References
				Phase	t [°]C	Strain rate [s^{-1}	δ [%]	m	
α-iron	armco-iron	0.04 C; 0.35 Si; 0.3 Mn 0.28 Cu; 0.02 S	10	α	700	6.6×10^{-4}	100	0.27	[7.3]
Pre-eutectoid steels	A4OG	0.04 C; 1.3 Mn; 0.2 Si 0.23 S	5.2	$\alpha + Fe_3C$	725	1.1×10^{-4}	260	0.36	[7.6]
	1340	0.4 C; 1.9 Mn	1.5	$\gamma + \alpha$	727	3×10^{-4}	460	0.5	[7.14]
	1 Cr-1 Si--1 Ni- 1 Cu--0.15 C	0.14 C; 0.82 Cr; 0.5 Si; 0.45 Ni; 0.32 Cu; 0.03 S	3	$\alpha + \gamma$	800	3.3×10^{-4}	275	0.4	[7.12]
	KNb	0.15 C; 1.16 Mn	< 5	$\alpha + \gamma$	790	5×10^{-3}	738	0.62	[7.36]
	SUI-2	0.75 C; 1.29 Cr; 0.32 Mn; 0.02 Mo; 0.29 Si	α-0.8	$\alpha + Fe_3C$	710	1.4×10^{-4}	745	0.33	[7.37]
			Fe_3C-0.3	$\alpha + \gamma$	730	2.8×10^{-4}	840	0.33	
		0.16 C; 1.54 Mn; 1.98 P; 0.05 Al; 0.3 V	2	$\alpha + \gamma$	900	1.7×10^{-4}	376	0.55	[7.13]
		0.34 C; 0.47 Mn; 2 Al	2	$\gamma + \alpha$	900	1.7×10^{-4}	372	0.48	[7.13]
		4 Ni; 3 Mo; 1.6 Ti	3–4	$\alpha + \gamma$	900	3.3×10^{-4}	800	0.7	[7.13]
	13 Cr-0.2 C	0.2 C; 13 Cr	5	$\alpha + \gamma$	800	6.6×10^{-4}	190	0.32	[7.12]
Eutectoid steels	U8	0.8 C	5	$\alpha + Fe_3C$	710	1.1×10^{-4}	320	0.45	[7.6]
		0.8 C	< 1	$\alpha + Fe_3C$	716	7×10^{-5}	133	0.42	[7.21]
	R6M5	0.8 C; 4 Cr; 6 W; 5 Mo; 1.8 V		$\gamma + Fe_3C$	820	5.8×10^{-4}	100	0.31	

Table 7.1. cont.

Pro-eutectoid steels	AISIO1	0.88 C; 1.15 Mn; 0.5 W; 0.22 V; 0.31 Si	0.5	$\alpha + Fe_3C$	650	1.3×10^{-4}	1200	0.5	[7.38]
	AISI52160	1.6 C; 1.5 Cr	< 3	$\alpha + Fe_3C$	650	1.7×10^{-4}	1220	0.6	[7.7]
	LINS	1.75 C	0	$\alpha + Fe_3C$	650	1.7×10^{-4}	1500		[7.39]
				$\gamma + Fe_3C$	770	1.7×10^{-4}	800		
	U12	1.2 C; 0.23 Mn; 0.25 Si; 0.25 Cu; 0.25 Ni; 0.25 Cr							[7.12]
	U12A	1.2 C; 0.23 Mn; 0.2 Si; 0.2 Cu							[7.12]
Cast irons		2.13 C; 1.5 Mn; 0.2 Si		$\alpha + Fe_3C$	650	1.7×10^{-4}	526	0.5	[7.8]
		3.5 C; 0.6–1.2 Si; 0.4–1.1 Mn; P 0.08	2–3	$\alpha + Fe_3C$	700	8×10^{-5}	150	0.35	[7.9]
		5.25 C; 1.5 Cr	3–4	$\alpha + Fe_3C$	725	1×10^{-4}	100	0.62	[7.10]
				$\gamma + Fe_3C$	1000	1×10^{-4}	610	0.7	
Austenitic stainless steels	SUS310	25 Cr; 20 Ni		$\gamma + \sigma$	800	1.7×10^{-3}	305	0.5	[7.16]
				γ	850	2×10^{-4}	200	0.5–1.0	
	18 Cr-10 Ni-1 Ti-0.12 C	0.12 C; 18 Cr; 10 Ni; 0.5 Ti	2	$\gamma + \sigma$	800	3.3×10^{-3}	240	0.5	[7.12]
Austenitic ferritic steels	D1	0.017 C; 24.7 Cr; 6.8 Ni; 0.5 Si		$\gamma + \sigma + \delta$	900	4×10^{-3}	2300		[7.20]
		0.9 Mn; 9.8 Mo		$\gamma + \delta$	1050		2300		
		25 Cr; 6.5 Ni;	10	$\gamma + \sigma + \delta$	900	4×10^{-3}	> 2500	0.45	[7.40]
		3 Mo; 0.14 N		$\gamma + \delta$	1000	10^{-3}	> 800	0.76	
	12 Cr-14 Mn-4 Ni-3 V	C⩽0.01: 12 Cr: 14 Mn; 4 Ni; 3 V	1–2	$\gamma + \alpha$	900	5.5×10^{-4}	400	0.43	[7.12]

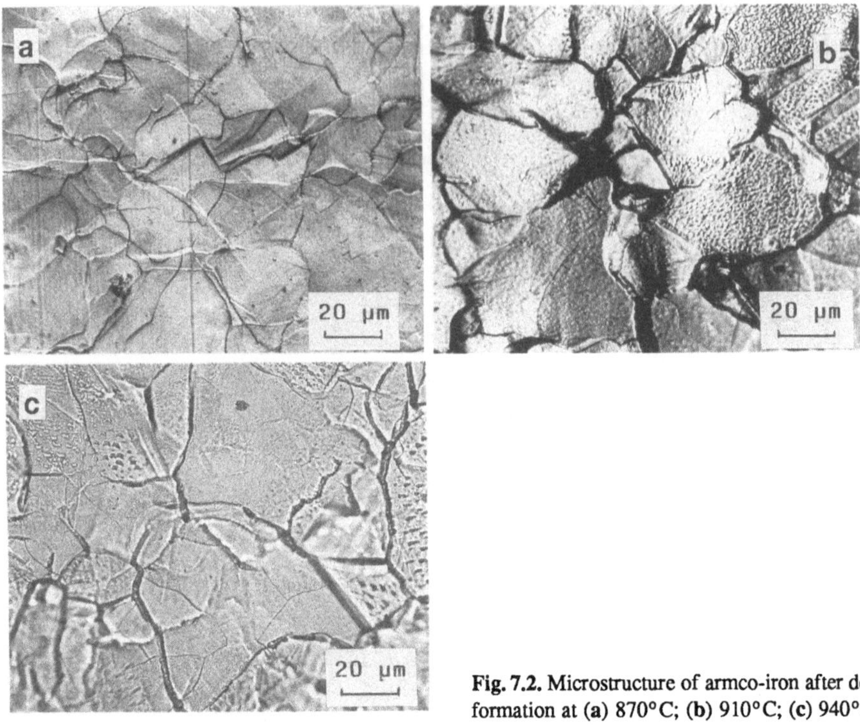

Fig. 7.2. Microstructure of armco-iron after deformation at (**a**) 870°C; (**b**) 910°C; (**c**) 940°C

Thus, the nature of the temperature dependence of the relative elongation of iron in the region of phase transformation is the same as that of the ductility of the commercial-grade titanium. The abrupt drop of the ductility of armco-iron at the moment of phase transformation is due to the localization of deformation in the near-boundary zones. Moreover, conditions required for the stress concentration are created via the formation of the plate-shaped γ-phase by means of the shear mechanism. The obtained results allow us to conclude that anomalies of ductility at the moment of phase transformation are determined not by the development of the transformation proper, but by a change in the structure prompted by the phase transformation. If during heating the fine-grained structure is retained in the two-phase region, superplasticity is possible. For example, high values of m (0.6) and δ ($>$ 150%) in iron were reported in [7.4] in the temperature range of polymorphic transformation during rapid heating of the samples up to $(\alpha+\gamma)$-region. Apparently, in this case, the $\alpha \rightarrow \gamma$ transformation takes place under a conventional diffusional mechanism. Thus, difficulties encounted during the formation of a stable fine-grained microstructure diminish the parameters of superplasticity in iron in the ferrite region.

The situation is quite similar in the austenite region. Steels containing precipitates of carbides, oxides, and intermetallics hindering the grain growth were investigated in [7.5], but the relative elongation never exceeded 200%. Fracture was caused by the formation of voids and micro-cracks at the boundaries of grains grown during SP flow. Let us now examine the peculiarities of superplasticity in the ferrite-cementite, austenite-cementite, and austenite-ferrite regions of the iron-carbon alloy system.

Ferrite-Cementite Region

The quantity of cementite or other structure-stabilizing phases in steel is of great significance in developing superplasticity in this region. In the fine-grained steels having a carbon content of 0.15–0.2% (3% cementite), high elongation levels have not been observed in this phase-field. However, in A40G steel (7% cementite) (Table 7.1) having a grain size of 5.2 μm, the optimum conditions of superplasticity (δ = 260%, m = 0.36 at 725°C and $\dot{\varepsilon} = 1.1 \times 10^{-4}\,s^{-1}$) correspond to the ferrite-cementite region [7.6]. The authors assume that the sulfur-bearing phases present in the steel can additionally facilitate stabilization of the microstructure at these temperatures.

The microstructural stability increases significantly in the eutoctoid steel containing 12% cementite. Based on a comparison of the results of the studies conducted on the ductility of the U8 and A40G steels, *Pohl* et al. [7.6] have noted that at a virtually identical grain size, the U8 steel exhibits higher values of relative elongation. Under the optimum conditions ($\dot{\varepsilon} = 1.1 \times 10^{-4}\,s^{-1}$ and t = 710°C), its elongation is 320% and m = 0.45.

The favourable effect of cementite on the stability of the microstructure makes it possible to realize superplasticity in the proeutectoid steels. The steels U12 and U12A having a granular pearlitic structure also exhibit superplasticity at a temperature below Ac_1, i.e., in the ferrite-cementite region.

Increasing the carbon content of a proeutectoid steel increases the SP effect. Superplasticity manifests itself most distinctly in steels containing 1.3–1.9% C [7.7]. By subjecting these steels to a prior treatment, it was possible to obtain a microstructure having ferrite (grain size of 0.5–1.5 mm) and globular cementite particles (0.1–0.5 μm). The volume fraction of cementite in the steel containing 1.9% C was approximately 30%. In these steels, high values of relative elongation (up to 1500%) are obtained at 650°C, a temperature rather low for this class of steels corresponding to the ferrite-cementite regions. The values of the coefficient m are 0.35–0.50 at this temperature.

Superplasticity was even observed in cast irons [7.8]. Fine-grained cast irons containing 2.12 and 2.36% carbon exhibit an m value of 0.5 at 650°C, and the relative elongations are 560 and 291%, respectively, at $\dot{\varepsilon} = 1.6 \times 10^{-4}\,s^{-1}$. White cast iron containing a large quantity of carbon ($\sim$ 3.5%) was investigated in [7.9]. A microstructure with a ferrite grain size from 2 to 3 μm and a ferrite-to-cementite volume of 1:1 was produced. At t = 700°C and $\dot{\varepsilon} = 8 \times 10^{-4}\,s^{-1}$ its elongation was 100% with the strain rate sensitivity coefficient m = 0.32. At the same test temperature and a tensile strain rate of $8 \times 10^{-5}\,s^{-1}$, it was established that δ = 150% and m = 0.35.

Cast iron containing 5.25% C and 1.5% Cr was studied in [7.10]. The ferrite-to-cementite ratio was 1:4, and the corresponding grain sizes were 3–4 and 2–

3 μm. At $t = 725°C$ and $\dot{\varepsilon} = 1. \times 10^{-4}\, s^{-1}$ the relative elongation is 100% with $m = 0.62$, and $\sigma = 40\,MPa$. It can be seen that the superplastic effect is less pronounced in cast irons than in high-carbon steels. This is not only because the grains are coarser in the studied cast irons, but because of the normal reduction of diffusional parameters with the increasing volume fraction of cementite.

As shown in [7.7] the parameters of superplasticity are improved if a steel is alloyed by chromium up to 1.5%. In this case the microstructure becomes more stable, the carbide growth diminishes during SP flow, and the coarsening of ferrite grain reduces. The graphitization leads to the deterioration of the parameters of superplasticity. To prevent graphitization, the silicon content must be reduced to 0.2%.

Thus, when steels possess a microcrystalline structure in the initial state, superplasticity is observed in the ferrite-cementite region. However, the analyzed data show that superplasticity is attained only in steels containing a significant amount of cementite and because of its stabilizing effect on the microstructure of steels.

Austenite-Ferrite Region

In this region, the stability of the ultrafine-grained microstructure is ensured by the austenite or the ferrite phase, depending on the carbon content and the holding temperature. Regarding the results of investigations on the elongation of iron in the austenite-ferrite region, a low ductility of the steels in this region can be expected. Indeed, a sharp drop of the ductility of low-carbon steels in the $(\alpha + \gamma)$-region was observed [7.2, 11]. However, it must be noted that in these investigations the steels had an initially coarse-grained structure, and the pearlite-to-austenite transformation could occur by a conventional diffusional mechanism. Therefore, a more detailed study of the variation of the relative elongation in a low-carbon steel with a fine-grained initial microstructure in the austenite-ferrite region was made.

A low-carbon low-alloy 1Cr-1Si-1Ni-1Cu-0.15C steel ($A_{c1} = 720°C$, $A_{c3} = 880°C$) with a grain size of 3 μm [7.12] was used. At ambient temperature the microstructure consisted of ferrite and pearlite. The data obtained are given in Fig. 7.3. In the ferrite-cementite region, the values of the relative elongation of the steel are not high. At a test temperature exceeding A_{c1}, the ductility of the steel increases and attains its maximum at 800°C ($\delta = 170\%$). At a temperature above 800°C, ductility decreases. At 900°C (above A_{c3}), the relative elongation drops to 60%. The study of the strain rate dependence of the mechanical properties of the steel at 800°C showed that the maximum relative elongation (215%) was attained at $m = 0.4$ and $\sigma = 50\,MPa$ at $\dot{\varepsilon} = 3.3 \times 10^{-4}\, s^{-1}$.

Thus, in the fine-grained state, the low-carbon 1Cr-1Si-1Ni-1Cu-0.15C steel exhibits all the symptoms of superplasticity at a temperature of 800°C that corresponds to the ferrite-austenite region. Superplasticity is not observed in the temperature range below 800°C.

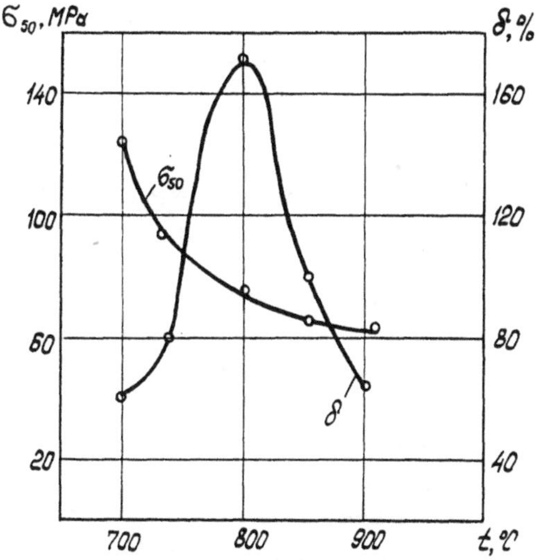

Fig. 7.3. Dependence of flow stress σ and relative elongation δ of 0.15C-Cr-Ni-Si-Cu on the deformation temperature ($\dot{\varepsilon} = 6.6 \times 10^{-4}\,s^{-1}$)

The variation of the mechanical properties is directly dependent on the variation of the phase composition and microstructure of the steel during heating. At a temperature of SP deformation (800°C), an average grain size is equal to 3 µm and is stable due to the presence of a large quantity of a second phase (40% austenite). With a subsequent rise of temperature, the amount of the austenite-phase and the grain size of the steel increase. These data agree with the results obtained on other low-carbon steels [7.13, 14].

The growth of the carbon content in steels (as well as the rise of temperature) leads to an increase in the content of the austenite-phase within the austenite-ferrite region. Apparently, this is why maximum elongations in medium-carbon steels are observed at lower temperatures than in low-carbon steels. In type 1340 steel with a grain size of 1.5 µm maximum elongation of 460% ($m = 0.5$) was observed at $t = 727$°C [$(\alpha + \gamma)$-region] and $\dot{\varepsilon} = 3.3 \times 10^{-4}\,s^{-1}$ [7.14]. However, these results contradict those reported in [7.6]. In the fine-grained (6 µm) A4OG steel, ductility abruptly decreases and flow stresses increase in the austenite-ferrite region as compared with the ferrite-cementite region under strain.

The discrepancy of these data can be explained, most likely, by a difference in the compositions of the steels. As reported in [7.6], during heating a noticable coarsening of microstructure takes place in the ferrite-austenite region due to the dissolution of phases containing sulfur which stabilize the microstructure. In [7.13], low- and medium-carbon steels containing up to 2% Si, 2% Al, and 2% P were investigated. In spite of a significant influence of cavitation on the maximum elongation, at t = 900°C and $\dot{\varepsilon} = 1.7 \times 10^{-4} - 3.3 \times 10^{-4}\,s^{-1}$ the strain rate sensitivity coefficient was 0.5 and relative elongation was 400% in steels containing 2% Al and 2% P. The favourable effect of additives on SP flow was most probably due to precipitations which stabilized the microstructure. Apparently, the stability of the microstructure was also achieved by the iron alloying with nickel, molibdenum and titanium [7.13]. AT t = 900–960°C and

$\dot{\varepsilon} = 3.3 \times 10^{-4} - 8.3 \times 10^{-4}\,s^{-1}$ the alloy with a grain size of 3–4 μm had the following parameters: $m = 0.50$–0.7, $\delta = 800\%$, $\sigma = 10$–40 MPa.

Austenite-Cementite Region

In the presence of a fine-grained microstructure and at temperatures ensuring its stability, it is possible to observe superplasticity in the austenite-cementite region. A study of alloys containing 1.3–2.1%C established that at temperatures ranging from A_{c1} up to 850°C, the relative elongation is 500–800% [7.15]. Within the austenite-ferrite region, the ductility decreases with increasing temperature [7.2]; this is due to the decreased cementite content because of the increased dissolution of carbon in austenite and, thereby, to a loss of microstructural stability. Apparently, in a steel containing 0.91% C, austenite stability is not ensured because of the decrease in the amount of cementite (up to 1–2%); consequently, the ductility is adversely affected during the transition into the austenite-cementite region as compared to the ferrite-cementite condition [7.6], where the cementic content amounts to 13%.

High parameters of superplasticity can be obtained in fine-grained cast iron (5.25% C and 1.5% Cr) in the (γ+Fe_3C) region. At 1000°C and $\dot{\varepsilon} = 1 \times 10^{-4}\,s^{-1}$, δ approaches 610% and $m = 0.7$ [7.10].

Thus, superplasticity of plain-carbon steels can be realized in different structural states with an ultrafine-grained microstructure. The temperature range within which superplasticity of any given steel is realized depends, essentially, on the stabilizing effect of a second phase on the microstructure of the steel. In low-carbon steels, at a temperature above A_{c1}, the SP effect is observed if the growth of the ferrite phase is stabilized by austenite, and in high-carbon steels it is realized if the growth of the austenite phase is stabilized by cementite. Apparently, the deformation characteristics of the phases and their quantitative proportion also influence the attainment of superplasticity in any given condition.

The aforementioned specific features of development of superplasticity in plain carbon steels whose composition corresponds to different structural regions of the Fe-C phase diagram can form guidelines for evaluating the possibility of converting alloy steels into the superplastic state. At temperatures exceeding the recrystallization temperature, the microstructure of alloy steels mainly consists of the solid solutions, ferrite or austenite; in addition to these phases, secondary phases (carbides and intermetallic compounds) are usually found.

It is practical to begin an analysis of the development of superplasticity in alloy steels with steels having a matrix microstructure. The austenitic 18Cr-10Ni-1Ti-0.12C steel forms a typical example showing a matrix structure [7.12]. The microstructure of such a steel consists of austenite grains and carbides that are distributed uniformly along the grain boundaries and within the grains. In a coarse-grained state (grain size of 20 μm) the relative elongation of steel is 50% at ambient temperature, and at temperatures exceeding 700°C, δ is less than 20% (Fig. 7.4). Since these steels have an unsatisfactory high-temperature ductility, they cannot be deformed at temperatures from 700 to 900°C, but deformation

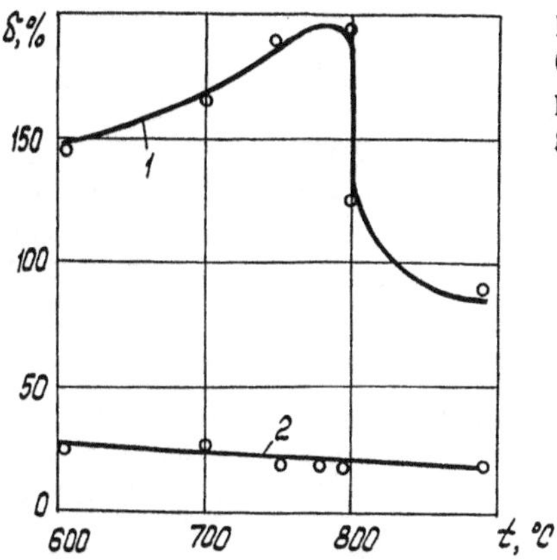

Fig. 7.4. Dependence of relative elongation δ of the 0.12C-18Cr-10Ni-1Ti steel on the deformation temperature. Curve (*1*) fine-grained state; (2) coarse-grained state

in the single-phase region, i.e., at $t = 1200°C$ is recommended. Let us examine specific features of the development of superplasticity in the 18Cr-10Ni-1Ti-0.12C steel with an initial grain size of 2 μm. The maximum ductility ($\delta = 240\%$) and strain rate sensitivity coefficient ($m = 0.5$) are reached at $t = 800°C$ and $\dot{\varepsilon} = 3.3 \times 10^{-3}\ s^{-1}$. The characteristic dependence of ductility on the deformation temperature is determined by the microstructural changes occurring during the processes of heating and cooling. Measurement of the grain size of the specimens deformed by 100% at different temperatures showed that at 740°C, it increases up to 1.5 times of the original grain size, whereas at 800°C it increased 2.2 times, and at 900°C, 6 times. The decreased ductility of the steel at the temperatures above 780°C is due to the loss of microstructural stability that is probably caused by the dissolution and coarsening of carbides. Thus, superplasticity of austenitic steels is due to the stability of the matrix-type fine-grained microstructure. A similar conclusion was made in [7.15]. In the austenitic 25Cr-20Ni steel at $t = 850°C$, relative elongation was 200% and m was more than 0.5, whereas the precipitation of the δ-phase stabilizing the fine-grained structure of austenite provided for an increase in ductility up to 305% at 800°C.

It should be noted that in the austenitic steel, superplasticity is realized in that temperature range where the steel, in its usual non-superplastic condition, loses its ductility. Thus, the available data indicate that the transformation of the 18Cr-10Ni-1Ti-0.12C steel into the superplastic state makes it possible not only to eliminate the ductility loss, but also to improve the ductility up to the levels unattainable even at 1200°C. This example is particularly indicative of the difference in realizing the ductility reserve by deforming under isothermal conditions and under the conditions of superplastic flow. It is possible to obtain the maximum ductility only by controlling the microstructure, the temperature, and the strain rate. Increasing the temperature alone has less effect on the ductility and, furthermore, in a number of cases, e.g., in austenitic steels, the ductility can decrease with increasing temperature.

Thus the superplastic behaviour of the austenite-ferrite steels that exhibit deterioration of the ductility characteristics during hot working compared to the austenitic steels [7.17–19] was examined by *Bannykh* and *Pedan* [7.17] who studied the characteristic features of superplastic deformation of the two-phase austenite-ferrite steels. After preliminary processing of the 12Cr-14Mn-4Ni-3V steel, they obtained a microstructure having austenite and ferrite grains measuring 1–2 μm. The tests conducted at the temperatures corresponding to the two-phase region at a strain rate of $5.5 \times 10^{-4}\,s^{-1}$ revealed all the symptoms of superplastic flow. At 950°C, the maximum relative elongation was 400% with $m = 0.43$. It is precisely at this temperature that a microduplex-type structure with an almost equal volume ratio of ferrite and austenite was obtained.

Upon increasing the test temperature, a significant grain coarsening of both phases was observed and this was accompanied by a decrease of the indices of superplasticity. Similar results were obtained in the studies on SP deformation of the ferrite-austenite chromium-nickel steel [7.18, 19].

In a number of chromium-nickel steels, complex phase transformations associated with the polymorphic $\delta \rightarrow \gamma$ transformation and eutectoid reaction $\delta \rightarrow \gamma + \sigma$ take place during heat treatment. As a result, two types of microduplex structures (δ/γ and γ/σ in the corresponding temperature ranges) can be created. The study of the mechanical behaviour of such steels [7.20] showed that superplasticity is observed in both phase regions. It was established that peaks of ductility correspond to these regions. It is of interest that SP deformation stimulates the eutectoid transformation which induces the formation of an equiaxial micro-duplex γ/σ structure. Cold-worked (ε = 50%) D1 steel at $\dot{\varepsilon} = 10^{-3}\,s^{-1}$ exhibits elongation exceeding 2000% at the temperatures of the $(\delta + \gamma)$ and $(\gamma + \sigma)$-regions (1050 and 900°C, respectively).

Thus, the formation of a micro-duplex structure comprising an approximately equal volume fraction of phases in the two-phase austenite-ferrite steels facilitates their transformation into the superplastic state. The aforementioned data concerning the austenitic and austenitic-ferritic steels indicate that, in the superplastic condition, a sharp increase in the ductility characteristics occurs in the temperatures of hot processing which, in the case of conventional steels, is usually characterized by low ductility and is, consequently, not recommended for hot deformation.

Let us evaluate the specific features of superplasticity in the 13Cr-0.2C pearlitic steel (A_{c1} = 820°C). The structural changes occurring during heating are similar to those observed in a plain-carbon pro-eutectoid steel. The carbide content of the steel is 10–13%. The microstructure is matrix, i.e., carbide precipitates are distributed within the ferrite grains (grain size is 5 μm) and in the grain boundaries.

The study of temperature dependence of the relative elongation and flow stress of the 13Cr-0.2C steel showed that ductility strongly depends on the deformation temperature (Fig. 7.5) and attains its maximum values (160%) in the ferrite-carbide region at t = 800°C and $\dot{\varepsilon} = 3.3 \times 10^{-3}\,s^{-1}$. Increasing the test temperature to above A_{c1} decreases the ductility of the steel. The flow stress decreases on increasing the test temperature, but, upon entering the ferrite-austenite region, the flow stress increases and only at 900°C does it attain the value

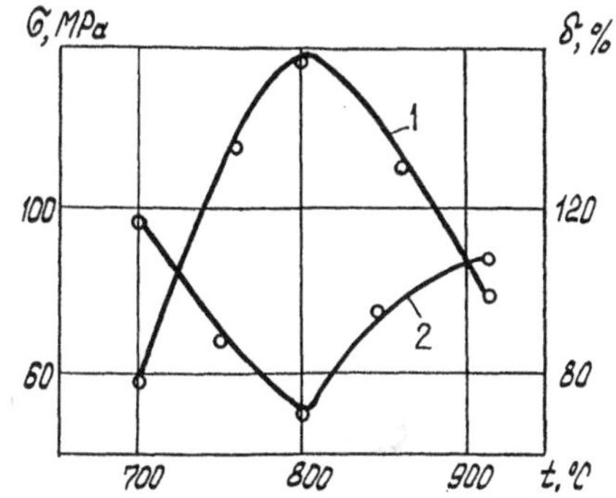

Fig. 7.5. Dependence of flow stress σ (*1*) and relative elongation δ (2) on the deformation temperature of the 0.2C-13Cr steel

equal to that obtained at 700°C. A study of the strain rate dependence showed that it exhibits all the symptoms of superplasticity at the optimum strain range ($6.6 \times 10^{-4}\,s^{-1}$), $\delta = 190\%$ and $m = 0.32$. Thus, the fine-grained 13Cr-0.2C steel of the pearlitic class exhibits superplasticity in the ferrite-carbide region. The stability of the ferrite grains determine the presence of carbide phases in the microstructure. Similar results were obtained in the steels of the ledeburite class, where SP deformation is observed in the ferrite-carbide region at a temperature below (by 10–20°C) A_{c1} [7.12]. These data indicate that in the alloy steels one can identify regions of superplasticity where heating and deformation of the steels ensure retention of a stable ultrafine grained microstructure.

The optimal parameters of SP deformation of steels are given in Table 7.1. According to *Pohl* et al. [7.6], not only the grain size is important for obtaining superplasticity of steel, but also the subgrain structure developed in the steel during the process of prior treatment. However, these data require further confirmation.

Yoder and *Weiss* [7.21] observed SP deformation in a plain-carbon eutectoid steel having a plate-like microstructure. This is due to the transformation of pearlite into a microstructure consisting of equiaxial crystallites of cementite and ferrite during the process of heating and subsequent hot working. However, in general, the conversion of the steels into a superplastic condition actually involves searching for the technological routes for obtaining a fine-grained microstructure. Let us consider this problem in more detail.

7.2 Preparation of Fine-Grained Microstructure in Steels

All types of treatments (Chap. 3) conducted for microstructure refinement in metals and alloys are applicable to steels. The process is facilitated by the occurrence of polymorphic transformations. Heat treatment is the most suitable method for obtaining an ultrafine-grained microstructure in steels.

Quenching followed by high-temperature tempering is employed in medium and high-carbon steels. This treatment leads to the formation of a microstructure having equiaxial cementite precipitates in a recrystallized ferrite matrix. The effectiveness of refinement can be increased by subjecting the steels to fast heating up to the temperature corresponding to the beginning of austenite formation. If, after improving the structure, the steel is subjected to rapid induction or contact heating and second quenching, then additiona grain refinement is achieved. After subjecting the A4OG and U8 steels to such a treatment, the obtained grain size was 5.2 and 6 μm, respectively [7.6].

In structural steels, fine-grained (5–6 μm) austenite can be obtained only under optimum temperature-time conditions during prequenching heating [7.22]. For example, the optimum regime for the 1Cr-1Si-0.38C steel consists of furnace heating at a rate not less than 0.5 K/s up to Ac_3 and subsequent quenching. Holding for more than 1–2 min leads to a greater degree of coarsening of austenite grains than when heated to a temperature exceeding Ac_3 by 30–40°C. However, such a method is applicable only for small-sized preforms.

In a number of cases, spheroidizing annealing below Ac_1 (which leads to the formation of granular pearlite) is effective for refining the high-carbon steel structure. The process of spheroidization can be accelerated significantly if the original structure is dispesed. For example, in a steel containing 1% carbon, granular pearlite can be obtained from bainite by annealing at 700°C for 1–2 h (it takes 10–15 h to achieve this from an initially pearlitic structure.

The microstructure of steels can be successfully refined by thermal cycling (Chap. 3 and [7.23]). However, it should be noted that during thermal cycling, the finest structure can be obtained if fast heating ist additionally employed [7.24]. In the 1Cr-0.15V-0.65Mn-0.4C steel, austenite grains are refined from 45 to 2 μm when it is subjected to multiple inducting heating at a rate of 215 K/s and subsequent quenching [7.24].

Microstructural refinement can be achieved by thermal cycling even in the austenitic steels like 18Cr-8Ni [7.25]. Its range of martensitic transformation is below room temperature. Thermal cycling at 25°C $\rightleftarrows$ 196°C with intermediate heatings up to the crystallization temperature decreases the grain size from 15 to 4 μm.

According to [7.26], a thermal cycling treatment can be better than hot deformation for obtaining a fine-grained structure. The grain sizes of steels containing 0.4% C and 4% Ni were compared following the two kinds of treatments. After four-stage thermal cycling (860-0°C) and high-temperature tempering, equiaxial grains having a size of 3 μm were obtained. On the other hand, after hot rolling at 860°C, quenching from this temperature, and high-temperature tempering, the ferrite grains with $d = 4\,\mu m$ were nonequiaxial. Deforming at 695°C and $\dot{\varepsilon} = 10^{-3}\,s^{-1}$ showed that the maximum values of the coefficient m of the steels in the heat treated and the rolled conditions are equal to 0.55 and 0.35, respectively. However, this result most probably shows the importance of selecting optimal thermomechanical parameters of rolling than to the advantage in the refining of microstructure by thermocycling.

Fine-grained structure can be obtained by recrystallization annealing in the steels that are characterized by the absence of polymorphic transformation. In

12Cr-10Ni-1Ti-0.12C stainless steel, a combination of cold working and subsequent rapid heating up to the temperature of recrystallization annealing makes it possible to obtain a grain size smaller than 2 μm [7.12].

If a metastable phase can be obtained at room temperature, grain refinement becomes possible by inducing its decomposition by cold working and subsequent heating up to the recrystallization temperature. For instance, the austenite-ferrite steels 12Cr-14Mn-4Ni-3V and 12Cr-14Mn-4Ni-2Al, on quenching from 1250°C, showed the presence of δ-ferrite [7.27]. After cold working by 50% and annealing at 800°C, a microduplex structure with a grain size of 1–2 μm was obtained.

It is hardly possible to completely eliminate micro-structural inhomogeneity during the process of recrystallization annealing. *Treheux* et al. [7.19] investigated an austenite-ferrite steel containing (%) 21.12 Cr, 6.75 Ni, 2.39 Mo, 1.62 Cu, and 1.84 Mn and having an initial microstructure represented by alternate bands of ferrite and austenite grains that are elongated in the rolling direction. In order to eliminate microstructural nonuniformity, they carried out cold working followed by recrystallization annealing. Since the embrittling σ-phase can be formed in a certain temperature range, the recrystallization temperature was chosen so as to exceed the upper boundary of its formation.

To eliminate banding, cold rolling (ε = 50%) was carried out in two stages such that the material was deformed in the direction perpendicular to the original rolling direction. After each pass, annealing was performed at 950°C. As a result, the steel had relative elongation of 500% at 1000°C and $\dot{\varepsilon} = 3 \times 10^{-4}\,s^{-1}$. At this temperature, the grain size was 2–5 μm.

Powder metallurgy is employed for microstructure refinement in low-ductility steels and improving the ease of technological processing. Cast iron and steel powders are obtained by atomizing their metals in an inert atmosphere [7.9], a technique that ensures the evolution of a fine-grained structure within each spherical particle due to the high cooling rates (10^5 K/s). The particle size was 60–350 μm. The powders were evacuated, heated to 680–700°C, and compacted. Thermomechanical treatment of the compacts was used for obtaining dense preforms. With this method a steel containing 0.81% C, a microstructure comprising ferrite having a grain size of 0.5–1.0 μm and cementite grains of size 0.1–0.5 μm could be obtained. Similarly, a cast iron containing 3.5% C, 0.6-1.2% Si, and 0.4-1.1% Mn was processed to obtain ferrite grains having a size of 2–3 μm.

A fine-grained Fe-20% Ni-15% Co steel was obtained by granulation [7.12]. After sintering and cold working by 50% in the 700–730°C range, the steel exhibited superplastic flow. At 710°C and $\dot{\varepsilon} = 1.1 \times 10^{-3}\,s^{-1}$, the steel had m = 0.37 and δ = 320%. Hot deformation is a more universal method for producing a fine-grained structure, in which case the fine-grained structure is stipulated by dynamic recrystallization.

To obtain the finest possible grains, plain-carbon steels are subjected to deformation in the two-phase region. For example, a microstructure with a grain size from 3 to 5 μm in low-carbon low-alloy steels can be created by their deformation in the austenite-ferrite region. In the 1Cr-1Si-1Ni-1Cu-0.15C steel such a microstructure is formed, e.g., after deformation at 750–800°C and $\dot{\varepsilon} = 10^{-3} - 10^{-1}\,s^{-1}$ with a degree of compression exceeding 50% [7.12].

In high-carbon steels as well as in tool steels, it is necessary to ensure a combination of hot deformation with the precipitation of second phases, which yields a uniform microduplex structure. For this objective, prior homogenization in the austenite region is usually employed to transfer carbon into a solid solution. The fine-grained structure is formed during subsequent cooling to a temperature below A_1. After a total true strain of 3.0, the ferrite grain size reduces to 1.0 μm, and the cementite particles become as small as 0.1 μm.

Deformation in isothermal conditions is rather efficient. For example, in the R6M5 steel with hot deformaiton at 950°C and a strain rate of $10^{-3} - 10^{-2}\,s^{-1}$ with a strain level 80% or more, makes it possible not only to completely destroy the carbide network, but to obtain an equiaxial fine-grained structure[1]. A steel with such a structure exhibits superplastic properties at 820°C and $\dot{\varepsilon} = 5.8 \times 10^{-4}\,s^{-1}$ ($\delta = 100\%$).

Dynamic recrystallization as a method of producing micro- and submicron-crystalline structure is widely used in stainless steels without phase transformations [7.28]. A systematic study of the formation of a fine-grained structure during hot deformation was carried out on the austenitic 18Cr-10Ni-1Ti-0.12C steel[2]. Hot deformation at t = 800–1000°C results in the refining of the microstructure to a grain size from 2 to 6 μm. The size of recrystallized grains is independent of the initial grain size and is determined by the temperature-strain rate conditions of deformation, i.e., grain size increases with the increasing temperature and decreasing strain rate. The higher the strain level, the larger the recrystallization volume. Similar results were obtained on the 18Cr-11Ni and 18Cr-8Ni steels [7.28, 29].

Dynamic recrystallization was also observed in ferritic stainless steels, in which, with high values of stacking fault energy, the development of dynamic recrystallization is hindered by active recovery. However, due to the development of dynamic recovery many subgrain boundaries are formed, and during further deformation these boundaries are transformed into high-angle boundaries [7.30–32]. The size of recrystallized ferrite grain exceeds the size of austenitic steel grains several times, the temperature and strain rate being the same. When steels are alloyed with Nb and Ti, second phase particles are formed, which stabilize the microstructure and facilitate the formation of a microcrystalline structure [7.33].

The possibility of the formation of a microduplex-type structure due to dynamic recrystallization was demonstrated in an austenitic-ferritic steel containing 0.04C-18.68Cr-5.23Mn-1.13Ni-1.25Mo-0.52Si-0.18Al [7.34]. At temperatures ranging from 680 to 900°C in the strain rate interval from 10^{-4} to $10^{-1}\,s^{-1}$, a microduplex structure with equal volume fractions of γ- and α-phases having a grain size of less than 10 μm is formed in the steel. After such a treatment the steel acquires superplastic properties.

---

[1] Results obtained by Yu.B. Timoshenko and M.V. Karavaeva

[2] Results obtained in collaboration with R.G. Zaripova

Thus, it is possible to refine the microstructure of steels by means of numerous techniques. However, the most universal method is hot deformation under the conditions ensuring the development of dynamic recrystallization. In this case it is important to coagulate a second phase in order to improve the structure stability and enhance the parameters of superplasticity.

7.3 Effect of Superplastic Deformation on Mechanical Properties

Microstructural refinement of steels makes it possible to achieve significant improvement in their mechanical properties such as the ultimate tensile strength, ductility, impact strength, and fatigue limit, and also leads to a decreased threshold of cold shortness of the steels and almost eliminates their tempered martensitic embrittlement. These effects are related to the sharp increase in the total area of the grain boundaries which are known to be preferential sites for impurity segregation. During grain refinement, impurity concentration per unit grain boundary area decreases, and the carbides are more uniformly distributed, which causes the mechanical properties of the steels to be improved. Furthermore, it follows that the temperatures at which superplastic deformation and subsequent heat treatment are carried out do not always coincide (in contrast to other alloys, such as titanium-based ones). In particular, such behaviour is characteristic of the steels in which polymorphic transformations occur. Thus, it is of interest to discuss the intrinsic effect of superplastic deformation on the properties of steels and the effect of SPD on these properties during subsequent heat treatment that leads to a significant change in their microstructure.

Let us examine the effect of microstructural refinement and superplastic deformation on the properties of the steels not having polymorphic transformations. The 18Cr-10Ni-1Ti-0.12C steel was studied for this purpose [7.12]. After grain refinement up to 2 μm, the properties significantly exceeded those of a coarse-grained steel (20 μm), i.e., in the coarse-grained state $\sigma = 730$ MPa and $\delta = 55\%$; and in the fine-grained condition $\sigma = 890$ MPa and $\delta = 50\%$. Deformation under superplastic conditions leads to a slight decrease of steel strength, apparently due to the grain growth during deformation; however, the strength is retained at a sufficiently high level.

In addition to mechanical properties, the tendency of the steels in the coarse-grained and fine-grained states toward intergranular corrosion was evaluated. After holding in a boiling acid for 15 h, the symptoms of intergranular corrosion were observed at the surface of the coarse-grained specimens. However, no signs of corrosion were observed in the superplastically deformed samples even after holding in the boiling acid for 200 h. Apparently, this result is due not only to the grain refinement, but to a decrease in the chemical inhomogeneity.

At temperatures below 0°C, the $\gamma \rightarrow \alpha$ martensite transformation takes place in austenitic stainless steels, this transformation being stimulated by deformation.

Table 7.2. Mechanical properties of 12Cr-18Ni-1Ti-0.12C steel after cryogenic deformation

Initial State	Treatment	Mechanical Properties					HRC
		σ_u [MPa]	$\sigma_{0.2}$ [MPa]	δ [%]	ψ [%]	KCV [J/cm^2]	
As-received	Deformation at $-196°$C $\varepsilon = 50\%$	750	530	41.4	63.7	–	32
SP Deformed		1415	1360	38.0	48.9	21.4	44
As-received	Deformation at $-196°$C, tempering at 200°C, aging at 400°C	835	705	32.8	72.9	–	35
SP Deformed		1815	1800	9.5	44.1	16.1	50.5

An investigation of the effect of microstructure produced during SP deformation on the phase transformations and mechanical properties of steels was carried out on the 18Cr-10Ni-1Ti-0.12C steel[3]. Two states were chosen: I – as received state with a grain size of 30 µm; II – after 80% SP deformation, the grain size being 6 µm. Both states were subjected to cooling down to $-196°$C and subsequent deformation by 50%.

Mechanical properties of the steel after different treatments are given in Table 7.2. It can be seen that after SP deformation and subsequent cryogenic deformation the steel has improved strength and satisfactory ductility parameters. Differences in mechanical properties are retained after annealing at 200 and 400°C.

Microstructural analysis indicated that (Fig. 7.6) treatment at $-196°$C led to the formation of strain-induced martensite, its fraction being 45% in state I and 80% in state II. The martensite plate size is quite different, too. In the state I

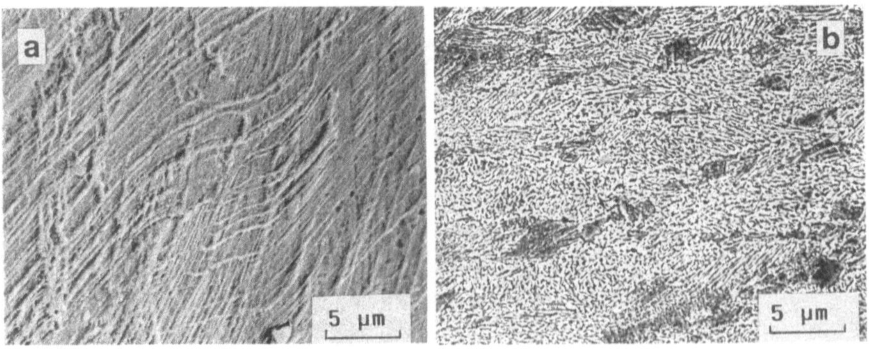

Fig. 7.6a,b. Microstructure of 0.12C-18Cr-10Ni-1Ti steel after deformation at temperature $-196°$C (a) as received state; (b) after superplastic deformation

[3] Results obtained by R.G. Zaripova and K.G. Farkhutdinov

samples isolated martensite plates have a common orientation within one grain, an average plate size being 0.2–2.0 μm. In state II under cryogenic deformation, massive martensite with various orientations within the initial austenite grain is formed, the plate size being 0.05–0.2 μm.

The observed differences in the structure and mechanical properties of steel is the result of SP deformation which retains the fine-grained structure with chemical homogeneity. With subsequent cryogenic deformation, the fine-grained structure and dispersed carbides promote an increase in the number of deformational martensite nuclei. Thereafter, the morphology and dispersivity of martensite change. Thus, it is possible to depict the effect exerted by SP treatment on the phase transformation that follows.

Let us now examine some other examples showing the influence of SP deformation on subsequent transformation in steels. In [7.30] a stainless chrome-nickel steel was used to study the effect of SP deformation on the development of eutectoid transformation. Superplastic deformation provided for a more uniform distribution of σ-phase within material, thus changing the properties of steel. A considerable increase of the yield point and ultimate strength was detected, with a certain decrease in ductility as compared with the as-annealed state.

A study of the mechanical properties of the steels that undergo polymorphic transformations was carried out in the isothermally deformed and heat treated condition. However, the deformation conditions did not correspond to those of the superplastic state due to which the effect of SPD on the properties of the steels could not be revealed.

Nevertheless, the necessity of conducting such a study is evident since it is not clear how the properties of steels vary during subsequent heat treatment as a result of phase transformations. In this case, a comparison with one of the effective methods of strengthening the steels, e.g., with high-temperature thermomechanical treatment (HTMT) is advisable. Such a study was performed on two steels, viz., the low-carbon 1Cr-1Si-1Ni-1Cu-0.15C and the corrosion-resistant 13Cr-0.2C steels [7.12].

Superplastic deformation of the 1Cr-1Si-1Ni-1Cu-0.15C steel having an initially fine-grained microstructure was carried out using the compression method (by 50%) 800°C and $\dot{\varepsilon} = 6.6 \times 10^{-3}\,s^{-1}$, which form the optimum conditions for SPD. The specimens were then subjected to heat treatment at 930°C for 1 h and water quenched, and subsequently tempered at 500°C for 1 h. Some of the preforms were processed according to this shedule without carrying out SP deformation (deformationless treatment DLT). Superplastic deformation of the 13Cr-0.2C steel having an initially fine-grained microstructure was performed at 800°C and $\dot{\varepsilon} = 6.6 \times 10^{-3}\,s^{-1}$, the optimum conditions of SPD, and by upsetting at 50%. Thereafter, heat treatment was performed by holding for 1 h at 1050°C and oil quenching and tempering at 700°C for 2 h. Here again, some of the as-received preforms were treated according to this schedule but without conducting deformation (DLT).

To compare the mechanical properties, high-temperature thermomechanical treatment (HTMT) was used. The 13Cr-0.2C steel was subjected to HTMT

Table 7.3. Mechanical properties of 1Cr-1Si-1Ni-1Cu-0.15C and 13Cr-0.2C steels after different treatments

Steel Grade	Treatment	σ_u [MPa]	$\sigma_{0.2}$ [MPa]	δ [%]	[%]
1Cr-1Si-1Ni-1Cu-0.15C	DLT, quenching from 930°C, tempering at 500°C	880	815	15	61
	SP, quenching from 930°C, tempering at 500°C	970	905	18	58
13Cr-0.2C	DLT, quenching from 1030°C, tempering at 700°	705	450	23	64
	SP, quenching from 1030°C, tempering at 700°C	840	620	17	59
	HTMT	815	660	20	68

according to the following schedule: preform-holding temperature of 1050°C, $\dot{\varepsilon} = 5\,s^{-1}$, and $\delta = 50\%$. The time lag between preform withdrawal from the furnace up to oil quenching of the forging was 5 s. Tempering was carried out at 700°C for 2 h.

The experimentally determined mechanical properties of the 1Cr-1Si-1Ni-1Cu-0.15C and 13Cr-0.2C steels following different treatments are given in Table 7.3, which shows that the 1Cr-1Si-1Ni-1Cu-0.15C and 13Cr-0.2C steels possess higher levels of mechanical properties after SPD with subsequent heat treatment than after DLT. In the 1Cr-1Si-1Ni-1Cu-0.15C steel, the ultimate tensile strength and yield stress obtained after SPD are greater by 90 MPa as compared to the values obtained after DLT; and in the 13Cr-0.2C steel, they exceed by 130 and 170 MPa, respectively. By comparing the mechanical properties of the 13Cr-0.2C steel after HTMT and SPD, it can be seen that they are close to each other. There is no significant difference in either strength properties or in ductility.

Thus, superplastic deformation has a favourable effect on the mechanical properties of the steels investigated and significantly improves them over those obtained after deformationless treatment. Mechanical properties of the 13Cr-0.2C steel after SP deformation are just as good as those after HTMT.

The retention of fine-grained structure after SP deformation must affect the viscosity and temperature of the cold shortness threshold. In order to verify this assumption, the 0.45%C (type 45) steel was studied[4].

The tensile and impact strength tests were conducted at temperatures ranging from ambient to −110°C. The following 3 states were chosen:

[4] Results obtained by K.G. Farkhutdinov and F.F. Fairushin

I. As-received state (hot rolling followed by normalizing); the grain size of the ferrite-pearlite mixture was 40 µm.

II. Upsetting by 60% in the $(\alpha + \gamma)$ region ($t = 760°C$) at $\dot{\varepsilon} = 1 \times 10^{-3}\,s^{-1}$ for obtaining a fine-grained (10 µm) structure.

III. The same as II, followed by upsetting at 40% under SP conditions at 740°C and $\dot{\varepsilon} = 3 \times 10^{-4}\,s^{-1}$.

After such a treatment the microstructure changes. On the one hand, refinement of grains in the ferrite fraction continues; on the other, the pearlite fraction changes remarkably and acquires the features of a mixture of both plate-like and granular pearlite. In the latter the cementite grains are located at the ferrite grain boundaries. The average grain size of both fractions is $< 1\,\mu m$. The pearlite grains have an average cementite plate thickness of 0.01 µm, and the spacing between them is 0.1 µm; they are more dispersed and more uniformly distributed than in state I. Properties obtained at ambient temperature are given in Table 7.4.

SP deformation significantly enhances the impact strength of the steel in the temperature range from ambient to $-110°C$ as compared with the initial state (Fig. 7.7). For example, the impact strength at $t = -110°C$ increases from 20 J/cm^2 to 40 J/cm^2.

The fractographic analysis indicated that, after SP deformation, the fracture at 20°C was completely ductile, whereas in state I there was about 20% of a brittle constituent. At $-100°C$, the fracture surface of state III contained up to 30% of a ductile constituent, whereas in state I the fracture was completely brittle.

Table 7.4. Mechanical properties of 0.45%C steel

State	Flow stress [MPa]	Tensile strength [MPa]	δ [%]	KCV [J/cm^2]
Original	380	630	18	36
Fine-grained	390	670	24	–
SP deformation (ε = 40%)	490	800	26	64

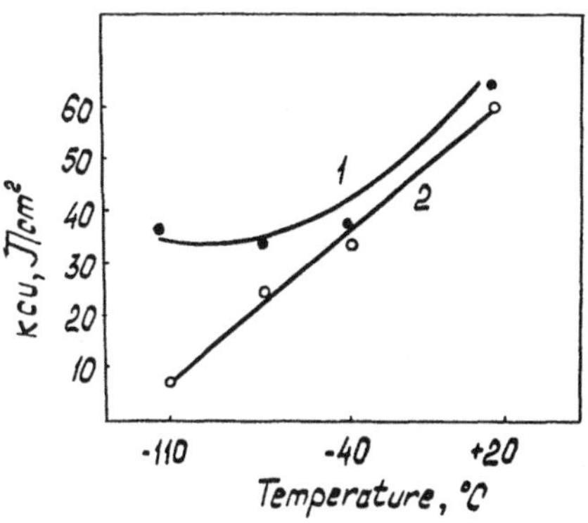

Fig. 7.7. Change of characteristic of viscosity steel 45 in the temperature range 20°C – (–110°C): Curve (*1*) after SP deformation; (2) initial state

The increase of yield stress and ultimate strength as well as the ductility of type 45 steel after SP deformation correlates with structural variations during such treatment. It is known that the grain refinement as well as a decrease of spacings between the plates in pearlite leads to the increase of σ_y and σ_u [7.35]. However, the decrease of the inter-plate spacings must lead to a decrease of ductility. The increase of ductility after SP deformation can be attributed to the increase of the ferrite fraction in the course of the structure preparation and hot deformation under SP conditions. Indeed, quantitative metallographic analysis showed that ferrite fractions are 31, 46, and 56% in states I, II, and III, respectively.

Thus, the deformation of construction steel under SP conditions not only enhances the strength parameters, but also improves ductility and lowers the temperature of the steel shortness threshold.

The conducted microstructural studies revealed that the general improvement of microstructure observed during SPD, i.e., evolution of a structural state with a high degree of structural and chemical homogeneity, is inherited during the subsequent heat treatment. This fact is particularly noticeable in the 13Cr-0.2C steel in which the nonuniformity of grain size distribution coexists with the nonuniform distribution of carbides; under the conditions of superplastic deformation, both these undesirable features are eliminated.

Thus, working under superplastic conditions can be an effective method of strenghtening steels. The specific aspects of the effect of superplastic deformation on the structure and properties of steels require further investigation.

8. Superplasticity of Nickel-based Superalloys

Nickel-based alloys are one of the most extensively used groups of heat resistent materials (superalloys) and are frequently found in engines operating at elevated temperatures up to $0.8\,T_m$. The practical significance of nickel-based alloys motivated intense studies on their structure and properties [8.1, 2]. These alloys consist of several structural constituents. The matrix of the alloy (γ-phase) is a nickel-based fcc solid solution containing a large fraction of an element that is soluble in nickel (cobalt, chromium, molybdenum or tungsten). The intermetallic γ'-phase, $Ni_3(Al,Ti)$, having an ordered fcc structure, is usually the strengthening constituent. The coherence between the lattices of the γ and γ' phases and the closeness of the lattice parameters (the mismatch between the parameters is less than 1%) create the possibility of developing phase a interface with a low surface energy. This situation imparts a high degree of stability to the γ'-phase.

Usually, these superalloys contain 0.05–0.2% C which forms MC type carbides with the alloying elements. During heat treatment and also when in place as an engine component, they decompose into the $M_{23}C_6$ and M_6C type carbindes which often precipitate along the grain boundaries.

Regardless of the fact that many aspects concerning the relationship between the stability and the morphology of the nickel-based superalloys require further investigation, a number of basic factors ensuring their strength at high temperatures can be identified on the basis of general experience and experimental data [8.3, 4]. One factor is the high degree of "alloyability" of the solid solution (γ-phase) which causes lattice distortion and slows down diffusion processes. The main factor, however, is the precipitation hardening due to the precipitation of γ' particles during the postquenching aging process. In this case, the coherency of the γ'-phase lattice and the matrix is retained up to high temperatures, which leads to the appearance of interphase boundaries that hinder the movement of dislocations and to the increase of the microstructure stability. A characteristic feature of the high-temperature strength of the nickel-based superalloys is the dependence on the grain size, i.e., it has been established reliably that increasing the grain size improves the service life and creep resistance. Grain boundary structure also exerts a significant influence on the heat resistance of these alloys [8.3].

Establishing these factors determining the high-temperature strength, i.e., creep resistance, allowed redirected efforts toward improving the strength properties of these superalloys. At the present time, success has been achieved in increasing the content and the degree of dispersion of the γ'-phase, alloying the matrix and the γ'-phase for increasing the elastic stresses at the coherent boundaries of the two phases, and decreasing the rate of coalesence of the precipitate particles. Complex alloying is also being tested to find the most suitable processing regimes [8.1, 3]. This has allowed an increase in the operating temperature of the new alloys up to 1000–1050°C.

However, attempts to increase the high-temperature strength of the alloys required labourious preparation procedures and decreased the yield. This is because increased high-temperature strength is associated with a decreased ductility. Therefore, with superalloys one finds a conflict between the aim of ensuring high reliability and lifetime and the reality of the costs in time and material of obtaining such products. When in service, superalloys must be highly resistant to deformation, but during processing they must have a high degree of ductility. A solution of this problem is obtained by a complex approach involving provision of a sufficient degree of engineering ductility and subsequent restoration of the desired operational characteristics. As a rule, the high-temperature creep properties of the alloys are determined by the thermal treatment. It is more difficult to control ductility. Among the newer methods is a way to the alloys into a superplastic state [8.5–7].

Superplastic deformation yields a high degree of structural uniformity and facilitates elimination of the differences in the grain size distribution which otherwise adversely affects the high-temperature strength. The latter is one of the main causes of sample rejection in industrial production of superalloy components.

The nature of superplastic deformation (Chap. 2) requires that in order to convert the superalloys to their SP state, it is important not only to obtain microstructural refinement of the matrix of the γ-phase, but also to change the structure of the grain boundaries betweem thr γ- and γ'-phases. When these phases are coherent, the γ'-phase acts as a strengthening agent and the grain boundary cannot slide easily. Therefore a microduplex structure with nearly identical grain size in the γ- and the γ'-phases (Fig. 8.1) is constructed. The main consideration is that the interfaces between the γ- and the γ'-phases must be high-angle boundaries which facilitate grain boundary sliding.

This chapter discusses the morphology of the γ'-phase in the development of superplastic deformation, examines the parameters determining the ductility and the superplasticity of the superalloys, gives an analysis of the methods of coverting them into the superplastic state, and deals with the principles of restoring and improving their high-temperature properties. Since the strength and ductility characteristics of nickel-based superalloys significantly depend on the structural features and the conditions of deformation, the effect of these factors will also be examined.

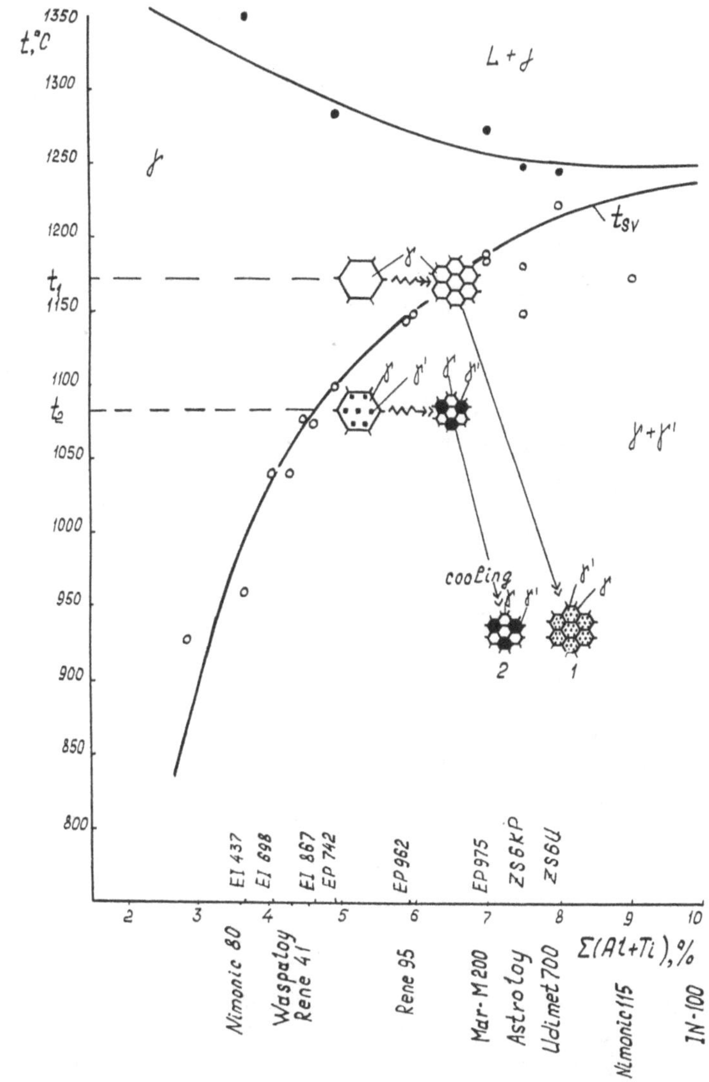

Fig. 8.1. Scheme illustrating the phase and structural transformation of nickel-based superalloys in conversion to the superplastic state. t_{sv} is the temperature of complete dissolution of the γ'-phase

8.1 Effect of Structure, Deformation Temperature and Strain Rate on the Ductility of Alloys

A complete set of systematic data on the influence of microstructure, phase constitution and temperature-strain rate conditions on the plasticity and superplasticity of superalloys are not available at present. In this section we review that which is already known for nickel-based alloys of increasingly complex composition, from commercial-purity nickel and the single-phase 80% Ni-20% Cr alloy up to superalloys.

The compositions of the alloys studied are given in Table 8.1. All of these alloys are industrial materials and their properties have been studied in detail. We have included typical low-alloyed (EI437), medium-alloyed (EI867, EP742) and high-alloyed (EP962, EP975, ZS6U) species that correspond to the foreign alloys from Nimonic 80 to Udimet 700 (Fig. 8.1).

The diverse chemical compositions and phase constitutions make it possible to evaluate the effect of alloying and amount of γ'-phase on the SP properties of superalloys. Systematic studies of this set of alloys allows us to establish general regularities and to predict SP properties of practically any nickel-based superalloy.

A detailed study on the ductility of commercial purity NP-2 nickel and nichrome 80% Ni-20% Cr has been carried out [8.5]. It was shown that these alloys show qualitatively different behaviour in their coarse-grained (80–100 μm) and fine-grained (2–4 μm) states. In the fine-grained state, both materials exhibit superplasticity. In pure nickel the superplastic condition appears at 800°C and $\dot{\varepsilon} = 5 \times 10^{-3}\,s^{-1}$. Low flow stresses, a high value of m, and a significant relative elongation prior to fracture (curve 2) are observed (Fig. 8.2). The magnitude of the maximum elongation ($\delta \sim 180\%$) is less than that of the typical superplastic materials because of the abrupt grain growth during deformation.

The effect of microstructure, temperature and strain rate conditions on the ductility of nichrome is qualitatively similar to that observed in nickel. Its maximum SP properties are realized at 1000°C, i.e., at a temperature exceeding that of nickel by 200°C; the maximum strain rate of ductility is an order of magnitude higher. Fine-grained nichrome has a high strain rate sensitivity of the flow stress at the threshold temperature for superplasticity.

The results of these studies indicate the decisive effect of the grain size on the ductility of these materials at high temperatures. However, the instability of the microstructure of nickel and nichrome does not make it possible to realize their superplastic properties completely. In spite of the qualitative similarity in the development of the effect, there is a significant difference in the behaviour of the two due to the difference in their chemical composition. An addition of chromium slows down the diffusional processes. At 800°C, the dislocation density in deformed nichrome is significantly higher than that observed in nickel. Apparently the decelerated diffusion in nichrome considerably hinders absorption of the dislocations by the grain boundaries which leads to a significant deterioration of the ductility. Thus in order to improve the ductility, it is necessary to increase the deformation temperature. Chromium addition must also have a positive effect, due to some degree of microstructure stabilization. It is only because of this that superplasticity is attainable in nichrome at 1000°C.

The second-phase particles efficiently reduce the tendency of the grains to grow. This can be seen in the behaviour of the regular features of hot deformation of a fine-grained EI437 alloy. With respect to chemical composition, this alloy is close to nichrome but also contains titanium and aluminium that go into the

Table 8.1. Chemical composition and amount of γ'-phase in nickel-based superalloys

Alloy	Volume fraction of γ'-phase [%]	Composition [wt. %]											
		C	Mn	Si	Cr	Mo	Co	W	AL	Ti	Fe	B	Other alloying elements
Commercial purity nickel NP-2	–	0.02	0.1	–	–	–	–	–	–	–	0.04	–	–
80%Ni-20%Cr Nichrome	–	0.08	0.36	–	20	–	–	–	0.02	0.22	0.35	–	–
EI437	14	0.06	0.019	0.032	19.78	–	–	–	0.8	2.82	0.63	0.01	0.02 Ce
EI698	24	0.06	0.22	0.25	14.43	3.02	–	0.08	1.51	2.52	2.56	0.005	2.11 Nb
EI867	34	0.02	–	0.12	9.76	9.62	5.08	4.93	4.62	–	0.27	0.02	–
EI742	35	0.04	0.03	0.13	14.14	4.99	10.08	–	2.39	2.50	0.36	0.01	2.62 Nb
EI929	38	0.05	0.13	0.18	10.3	5.04	13.4	5.5	4.04	1.65	1.01	0.02	0.37 V
EP962	45	0.10	0.02	0.13	13.09	4.63	10.74	2.82	3.22	2.64	0.58	0.01	3.44 Nb
EP975	55	0.08	0.02	0.10	8.43	1.04	15.80	10.3	4.72	2.24	0.50	0.02	1.55 Nb
ZS6KP	55	0.13	0.10	0.10	10.5	5.4	8.7	3.7	4.5	3.0	0.2	0.01	–
ZS6U	60	0.14	–	–	8.8	1.6	10.3	9.7	5.5	2.6	2.0	0.035	1.1 V, 0.04 Zr, 0.01 Y,

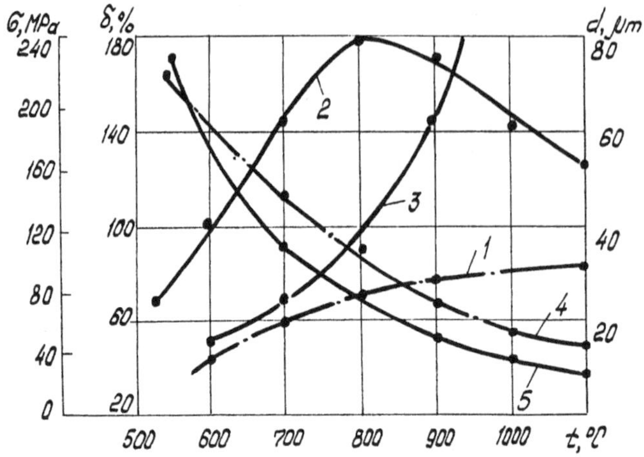

Fig. 8.2. Dependence of the relative elongation (curves *1,2*) and the flow stress (curves *4,5*) of NP-2 nickel at $\dot{\varepsilon} = 10^{-1}\,s^{-1}$ and the grain size (curve *3*) after a tensile deformation of 50% on the test temperature

formation of γ'-phase. Its maximum content in the alloy after aging at 700°C amounts to 10–14% and the temperature of complete redissolution is 960–970°C.

In order to obtain a fine-grained structure, the alloy is subjected to hammer-upsetting at 700°C and a 30-min annealing at 1000°C. The grain size is then equal to approximately 7 μm and the γ'-phase exists mainly in the form of dispersed particles. Subsequent annealing at 1000°C does not lead to a significant grain growth: after 1 h the grain size increases only to 8–9 μm. It is only at a higher temperature (1040–1050°C) that one observes an abrupt coarsening of the microstructure which is similar to that observed in single-phase nichrome after annealing at 1000°C. However, the ductility properties of the thus processed fine-grained EI437 alloy are not high and do not differ significantly in the fine-grained and the coarse-grained states (Fig. 8.3). In this case, $m \simeq 0.25$.

The ductility of the alloy at high temperature is closely related to its structure. The low values of relative elongation observed at 800–900°C are easily understandable, because compared to nichrome, the alloy not only has a slightly larger grain size, but also contains strengthening particles of the γ'-phase. It is more difficult to explain its low ductility at 1000°C, i.e., above the temperature of complete dissolution of γ'-phase although the grain growth is less than that observed in nichrome. Apparently, short-range ordering of the dissolved atoms constituting the γ'-phase is still retained because hardening of the alloy can occur.

Plasticity of the fine-grained ($d = 3\,\mu m$) EI929 alloy containing dispersed γ'-precipitates has been studied. It was established [8.5] that superplasticity appears only at a temperature higher than 1150°C when dissolution of precipitates occurs.

Based on these results of ductility of superalloys and a comparison of these data with that concerning single-phase nickel and nichrome, it can be concluded that the γ'-phase plays an important role in the superplastic flow, and in realizing the superplastic effect in nickel-based superalloys which requires special

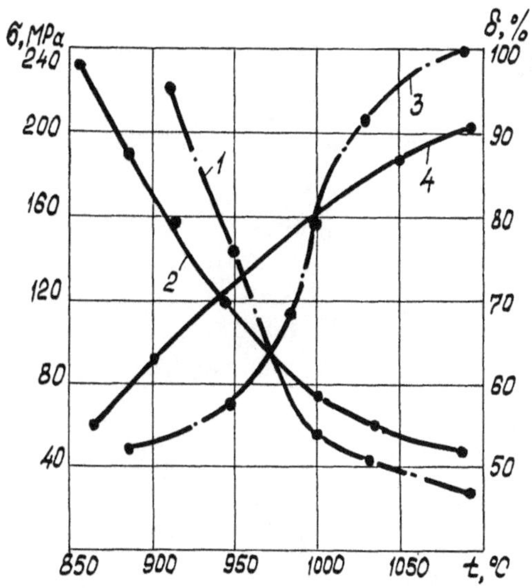

Fig. 8.3. Dependence of the flow stress curves (*1*, *2*) and the relative elongation curves (*3*, *4*) of the EI437 alloy at $\dot{\varepsilon} = 10^{-3}\,s^{-1}$ on the test temperature: Curves (*1*, *3*) $d_0 = 7\,\mu m$; curves (*2*, *4*) $d_0 = 25\,\mu m$

experiments for determining the relationship between the structural state and the properties of the alloys. There are, however, numerous methodological features which are not always given their due structure of superalloys. First of all, the structure is usually studied at room temperature, which may not correspond to the high-temperature state. When heating superalloys, one observes significant phase and structural transformations where coagulation and dissolution of the γ'-phase and carbides occur. These changes are significant in the fine-grained materials. In addition, precipitation of the γ'-phase during the cooling of alloys from high temperature must be considered (Fig. 8.1). In many superalloys, this precipitation is suppressed only when quenching extremely thin specimens. The behaviour of the γ'-phase must be considered when studying its role in controlling superplastic deformation.

In this context, data concerning the structural changes taking place during the process of heating the fine-grained ZS6U alloy [8.8] are of considerable interest. This alloy, having a grain size of $2\,\mu m$ after hot extrusion and cooling, contains a large quantity of dispersed cubic particles of the γ'-phase (Fig. 8.4a). However, after heating at a temperature exceeding 1050°C, dispersed particles of γ'-phase dissolve completely, and during subsequent furnace cooling at a rate of less than 50 K/h a fine-grained structure is retained. In this structure the γ'-phase exists mainly in the form of incoherent grain boundary precipitates (Fig. 8.4b). The characteristic γ-fringe contrast is seen at the boundaries between the dispersed γ'-precipitates and the matrix (Fig. 8.4c) which indicates coherency of these boundaries. At the same time, in the annealed specimens, the typical fringe contrast of the common high-angle boundaries is observed at the γ- and γ'-phase interface (Fig. 8.4d).

The quantity and the morphology of the precipitates of the γ'-phase on superplastic behaviour were studied on the ZS6KP alloy [8.9]. Three states of the

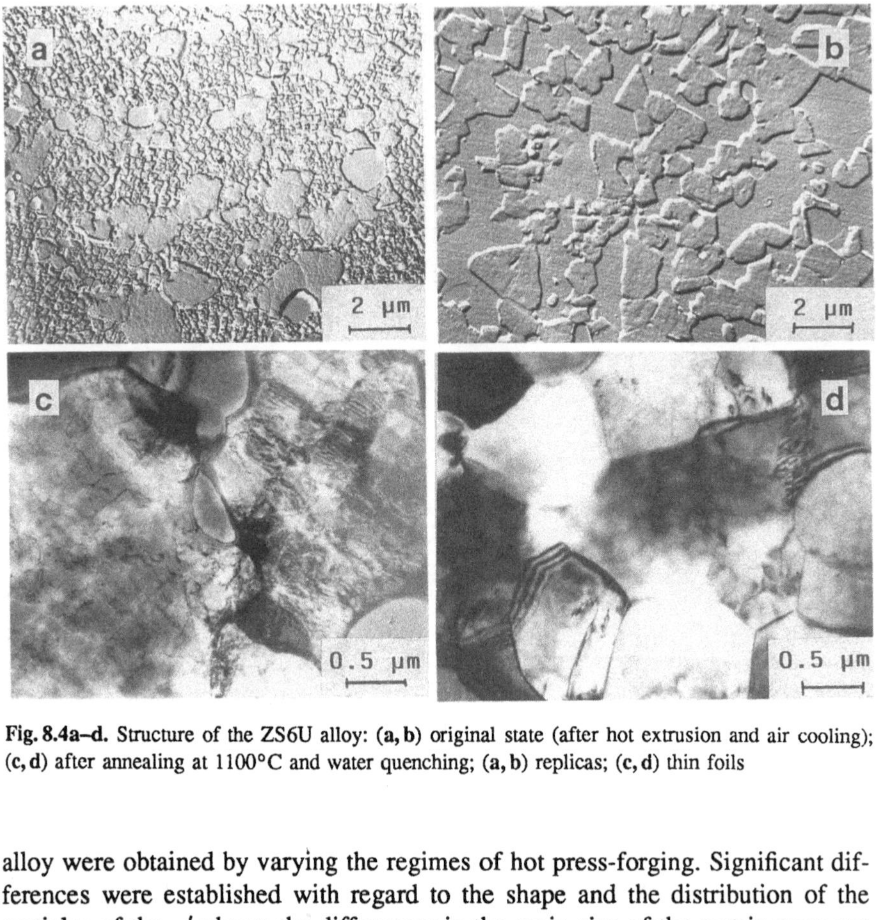

Fig. 8.4a–d. Structure of the ZS6U alloy: **(a, b)** original state (after hot extrusion and air cooling); **(c, d)** after annealing at 1100°C and water quenching; **(a, b)** replicas; **(c, d)** thin foils

alloy were obtained by varying the regimes of hot press-forging. Significant differences were established with regard to the shape and the distribution of the particles of the γ'-phase; the differences in the grain size of the matrix were not as significant as shown by the following data:

State of the alloy	I	II	II
Degree of drawing	9	15	9 and 5.7
t [°C]	1140	1140	1140 and 1060
Speed of press-forging [mm/s]	100	100	100 and 16
d [μm]	6	11	4

In structural state I, the γ'-phase is represented by coarse round-shaped particles of 1–3 μm size (grains of the γ'-phase) at the grain boundaries of the γ-phase and by dispersed cubic precipitates, having a size of 0.2–0.3 μm and distributed within the matrix grains. In state II, the γ'-phase exists mainly in cubic precipitates measuring 0.3–0.4 μm and individual grains with a size of 1–2 μm are also present; in this case, the density of the individual grains is significantly less than that observed in state I. Cubic γ'-precipitates are absent in state III; all the particles are essentially rounded and have a size of 0.5–2 μm. There is also a large quantity of finely dispersed spherical precipitates, measuring 0.04–0.08 μm (the so-called 'rash').

Upon heating the alloy to the temperature of superplastic deformation, significant structural changes that depend on its original condition are observed. When heated to 1150°C, the grain size of the matrix changes insignificantly in all states, while the degree of dispersion and the distribution of the γ'-precipitates change significantly. Heating and holding state I specimens for a period of 20 min leads to coagulation of the γ'-phase from 1050°C. The density of the cubic particles decreases and the coarse γ'-particles grow and the zones free from the dispersed γ'-particles from around them. By increasing the temperature to 1100°C, the change becomes more distinct, and a significant portion of the cubic particles dissolve. Increasing the temperature to 1150°C leads to a total disappearance of the dispersed γ'-precipitates and to partial dissolution of the grains of the γ'-phase.

Heating and holding the specimens leads to coagulation of the γ'-phase at 1100°C in state II, also. These changes are similar to those occurring in state I at this temperature; however, the density of the cubic particles remains approximately 1.5 times higher than in state I. In state III under the same annealing conditions, complete dissolution of the finely dispersed γ'-precipitates and a rapid growth of the coarser precipitates is observed; in this case, the size of the γ'-particles increases to 0.8–3.0 μm. It should be noted that at identical temperatures and times of annealing, the volume fraction of the γ'-phase varies in the same way in all the three states.

Depending on the duration of holding at the deformation temperature prior to deformation, different microstructures evolve in the alloy corresponding to different states. In states I and III, there is formation of a stable fine-grained microstructure consisting of grains of the γ- and the γ'-phases which have approximately equal size. In state I, dispersed particles of the γ'-phase are present right up to 1100°C. At the same time, in state II, the γ'-phase is mainly distributed within the grains in the form of dispersed cubic precipitates.

Alyoshin et al. [8.9] established that after press-forging, the structure of the alloy is completely recrystallized. A study of the diffraction patterns obtained from the neighbouring grains of the γ- and the γ'-phases showed that the misorientations between them exceed 15°, i.e., the phase interfaces are high-angle boundaries. The fact that the delta-fringe contrast is not observed at the grain boundaries of the γ and γ'-phase indicates an absence of coherency. It is important that the cubic precipitates of the γ'-phase, existing in the aged and unrecrystallized alloy, retain their coherency with the γ'-phase during the process of heating to the temperature of superplastic deformation.

The mechanical properties of the ZS6KP alloy for different states are given below (t_{def} = 1100°C):

State of the alloy	I	II	III
σ_{40} [MPa]	45	62	24
δ [%]	750	110	> 800
m	0.42	0.32	0.51

The material in state II exhibits the highest flow stress and low ductility ($m \approx 0.3$ and $\delta \approx 100\%$). State III ($m = 0.5$) is characterized by high ductility ($\delta > 800\%$) and low values of σ. The properties of the alloy also depend on the temperature-strain rate conditions of deformation. In the entire range of experimental temperatures (1000–1150°C), the maximum ductility in state III is obtained at 1100°C and $\dot{\varepsilon} = 10^{-3}\,s^{-1}$. Under these conditions high ductility ($\delta = 800\%$ and $m = 0.4$) and low flow stress levels are also observed in state I. However, in contrast to state III, the ductility characteristics in state I decrease rapidly when decreasing or increasing the temperature and the strain rate. The behaviour of the material during compression is similar to that during tensile deformation except that the range of the strain rates required for the development of superplastic characteristics is shifted towards the high strain rates.

The observed distinct differences in the mechanical properties is closely related to the difference in the structure of the material in different states. The slight difference in the grain sizes cannot be the reason since the alloy exhibits superplastic flow even at a grain size of 10 μm [8.10]. Since the volume fraction of the γ'-phase is virtually identical in all states, it can be concluded that differences in the behaviour of the alloy are mainly due to the differences in the degree of dispersion and the morphology of the γ'-precipitates at the temperature of deformation. The ductility properties increase sharply with decreasing density of the dispersed coherent precipitates of the γ'-phase having incoherent high-angle boundaries.

Similar results were obtained in a study of the effect of the morphology of the γ'-phase on the ductility properties of the EI867 alloy [8.11]. The fine-grained alloy containing dispersed γ'-precipitates showed low ductility up to the temperature of dissolution of the γ'-phase (1075°C) (Fig. 8.5). The alloy having a microduplex structure was superplastic in the 950–1075°C range. Furthermore,

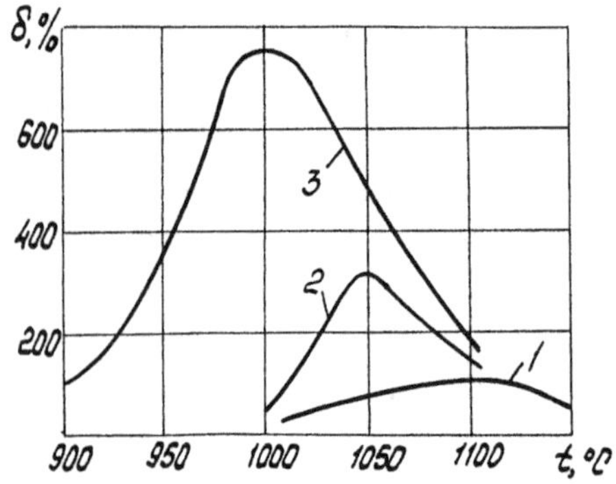

Fig. 8.5. Dependence of the ductility of the E1867 alloy on the temperature and the quantity of the γ'-phase existing in the form of grains: curve (*1*) 0% (matrix structure – all the γ'-phase exists in the form of dispersed coherent precipitates); curve (2) 65%; curve (*3*) 100% (microduplex structure)

in this case with all other conditions being the same, the ductility of the alloy increases significantly with an increasing number of grains of the γ'-phase.

The results of the experimental studies clearly indicate that ductility enhancement of nickel-based superalloys is obtained not only by controlling small grain size of the matrix γ-phase, but also by the shape and the distribution of the particles of the γ'-phase which, in turn, depend on the prior treatment and on the condition of heating before initiating deformation. From the standpoint of increasing ductility, it is most desirable to convert the coherent precipitates into grains of the γ'-phase, i.e., to obtain general interfaces between phases during the treatment.

The SP behaviour of superalloys with a wide variety of alloying elements was investigated and confirmed the above assumption [8.12]. To obtain a microduplex microstructure in all alloys, pre-annealing for heterogenization was performed, a treatment that involves heating up to the temperature of the γ-field and subsequent furnace-cooling at a rate of 20–30 degrees/h to a temperature where the γ'-phase starts to dissolve. Then a two or three stage isothermal compression ($\varepsilon = 80\%$) in the $\gamma + \gamma'$-field was performed, followed by annealing for 4–16 hours at a temperature equal to or 25–50°C below the deformation temperature. Microstructures with specific characteristics were produced.

Table 8.2 shows that the matrix grain (γ-phase) size of all alloys are similar. Following this treatment the γ/γ' interphases become noncoherent general boundaries. Microduplex microstructures with diverse volume fractions of the γ'-phase were also obtained.

During tensile tests (Fig. 8.6), even in the EI437 alloy, the maximum elongation at 900°C was 180%, which was considerably higher than ductility of the alloy with coherent γ'-precipitates (Fig. 8.3).

The relative elongation in the EI698 alloy at 900°C and $\dot{\varepsilon} = 5 \times 10^{-4}\,\mathrm{s}^{-1}$ was 450%. Samples of medium- and high-alloyed specimens (EP742, EP962, EP975) did not fail even under a strain of 500–700% and at a strain rate of $1.3 \times 10^{-3}\,\mathrm{s}^{-1}$. The maximum values of elongation ($\delta > 550\%$) and strain rate sensitivity coefficient ($m = 0.5$) at a strain rate of $\dot{\varepsilon} = 1.3 \times 10^{-2}\,\mathrm{s}^{-1}$ were reached in the EP975 alloy, but in low-alloyed species the coefficient m did not exceed 0.2, and elongation was less than 190%.

Table 8.2. Structural characteristics of some nickel-based superalloys after different thermomechanical pretreating

Alloy	Grain size [μm]		Volume fraction of the γ'-phase
	γ-phase	γ'-phase	[%]
EI437	3.5	0.5	9
EI698	3.5	1.0	14
EP742	4	1.2	20
EP962	5	1.7	29
EP975	5	1.9	30

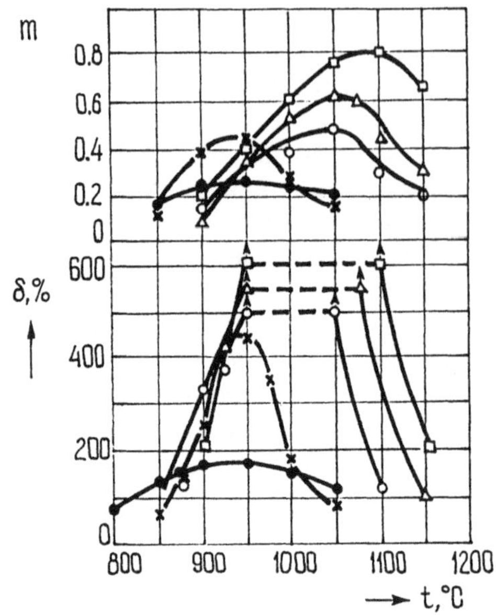

Fig. 8.6. Dependence of the strain rate sensitivity coefficient m and relative elongation δ on the deformation temperature of nickel-based superalloys with ultrafine-grained structure. • – EI437, x – EI698, o – EP742, △ – EP692, □ – EP975

Thus, a microduplex structure is optimal for SP deformation. An increase in the amount of the γ'-phase stabilizes the microstructure. As a result, the SP characteristics are enhanced, and the temperature and strain rate ranges of SP deformation widen. For alloys with more complex compositions the SP temperature is shifted towards higher homologous temperatures, i.e., from $0.67\,T_m$ for low-alloyed species to $0.88\,T_m$ for high-alloyed ones.

The obtained results agree with theoretical concepts concerning the structure of superalloys in the SP state (Fig. 8.1). Let us now consider these aspects in more detail.

The phenomenology of the development of superplasticity in superalloys fits into the characteristic of structural superplasticity (Chap. 1). It can be assumed that, in general, the behaviour of the superalloys and the mechanisms of their superplastic deformation should be similar to those observed in other materials. This is supported by results on the structural changes taking place during superplastic flow. For example, our studies [8.13] show that during superplastic deformation of the ZS6U alloy, there is no significant change in the shape of the grains, but a characteristic grain growth occurs with decreasing strain rate and an increasing degree of deformation. The grain size in the deformed portion of the specimens is greater than that observed in the heads (gripped portions), which is indicative of deformation-induced grain growth. These data agree with results obtained on other materials (Chap. 1).

By studying the mechanism of SP deformation in superalloys, we are able to understand various deformation mechanisms. For example, *Kaibyshev* et al. [8.13] succeeded in observing pair dislocations in the γ'-phase in thin foils prepared from samples after SP deformation; these dislocations are characteristic of an ordered structure.

Alyoshin [8.14] obtained direct evidence for the intragranular slip taking place in the γ-phase during SP deformation. In the ZS6KP alloy with dispersed γ'-precipitates the lattice dislocations were observed directly, because they were inhibited by the precipitates. The SP state was retained, but a certain increase in the flow stress, as compared to an alloy without dispersed γ'-precipitates, took place. Diffusion creep during SP deformation of the superalloy was also observed directly by the appearance of precipitation-free zones near the transverse grain boundaries [8.15].

Assessments based on these experimental results showed that when the structure is microduplex, it is GBS that makes the largest contribution (89%) to deformation, and the total contribution of DC and IDS is 11%. *Alyoshin* et al. [8.14] also analyzed the role played by grain and interphase boundaries in the SP deformation of the superalloy. Optimum SP characteristics are observed when the specific length and fraction of random (general) boundaries are maximal, because such boundaries have the greatest ability for absorption of lattice dislocations and high diffusionl properties.

The mechanism of SPD in superalloys is, in fact, the same as in other alloys. The newest information is that these alloys clearly indicate that it is not the microstructure refinement proper that determines their conversion into an SP state, but an increase in the area of the general boundaries. SP deformation is impeded when the γ/γ' interfaces are coherent and, vice versa, the conditions for the development of SP deformation are favourable when the boundaries are of a general type.

8.2 Methods for Converting Alloys into a Superplastic State

Preparation of the alloys for achieving superplasticity (Chap. 3) consists of a thermomechanical treatment to refine the microstructure by means of fixing the initial stage of recrystallization. When applied to superalloys, such a preparation procedure must take into account several aspects. First, it is necessary to obtain an optimal microstructure which must not only be fine-grained, but have incoherent precipitates of the γ'-phase that do not strengthen the alloy. Second, the thermomechanical treatment is difficult to carry out since the superalloys are heat resistant, are usually difficult to work, and have low ductility. Third, the complex and diverse chemical compositions of the various superalloys require different processing treatments for each species.

The importance of the third factor is illustrated by the scheme of dependence of the temperature ranges during melting and the existence of the γ'-phase on the total content of aluminium and titanium that constitute the γ'-phase (Fig. 8.1), which shows that in alloys containing a small quantity of the alloying elements, mechanical working can be used either in the single-phase γ-region or in the γ+γ' region. On the other hand, alloys containing a large quantity of alloying elements cannot safely be worked in the γ-region because of its narrow temperature range; they can be deformed only in the $\gamma + \gamma'$ region.

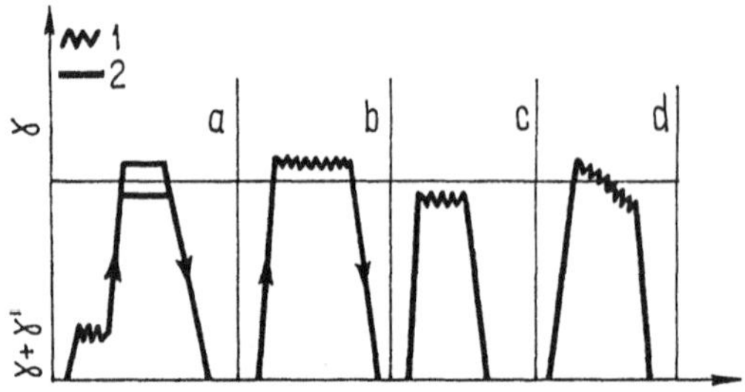

Fig. 8.7. Schemes of the main routes of processing superalloys to obtain a microcrystalline structure: (*1*) deformation; (2) heating

Depending on the temperature range of deformation and the regime of working (treatment), the methods of microstructure preparation of superalloys can be conveniently divided into four basic schemes (Fig. 8.7) that involve deformation in the γ and the $\gamma + \gamma'$ regions. We shall now examine the specific features of each of these schemes.

Figure 8.7a shows the simplest method for developing a fine-grained structure. It involves severe strain hardening of the alloys by cold or warm working and subsequent recrystallization annealing. However, due to the extremely high strength and low ductility, this treatment is possible only in the case where the alloys have a low content of alloying elements. In this case they attain a considerable degree of strain hardening after quenching in order to retain the solid solution (solution treatment) and after simple deformations, such as rolling or upsetting. In the case of medium-alloyed superalloys, the required cold or warm working can be accomplished only by hydraulic pressing (hydroextrusion) [8.16]. The effectiveness of this method was verified on the EI929 alloy [8.5]; hydroextrusion of this alloy was carried out at 250°C, maintaining a reduction factor of 43%. The alloy was then heated rapidly to 1150–1165°C which led to the formation of a matrix-type ultrafine-grained structure. One can obtain significantly different structures depending on the heating regime of the superalloys deformed at low temperatures. Rapid heating of the cold worked material is the most effective method for obtaining a matrix-type fine-grained structure [8.5].

Formation of a microduplex-type structure is possible during slow heating of the cold worked EI437 alloy. The disadvantages of the scheme, presented in Fig. 8.7a, are caused by poor workability of the alloys. Thus, the method has not yet found wide applications.

Another regime of processing (Fig. 8.7b) involves working in the γ-region, which is easily done for the low-alloyed specimens containing 3–4% of Al and Ti (alloys EI437 and EI698), but is very difficult for the medium-alloyed superalloys containing 4.6–5.7% of Al and Ti (alloys EI867 and EI929). During this treatment, grain refinement can be achieved only after deforming by more than 40% in the $\dot{\varepsilon} = 10^{-2} - 10^{-2}\,s^{-1}$ range [8.11]. Note that it is necessary to take into account the high rate of occurrence of recrystallization of the superalloys in the γ-region and, also, the instability of the evolved microstructure. Investigations carried out on the EI867 alloy showed [8.11] that, during isothermal upsetting at 1100°C, a microstructure having a grain size of approximately 10 µm can be obtained only after exceeding $\varepsilon = 70$–80% and at $\dot{\varepsilon} = 10^{-1}\,s^{-1}$ with subsequent rapid cooling of the alloy. Increasing the temperature to 1125–1150°C or decreasing the strain rate or holding in the press-tool (die) for a long time leads to an abrupt increase of the grain size as well as the appearance of a nonuniform size distribution of the grains. Increasing the strain rate to 10–$10^2\,s^{-1}$ also led to some grain growth due to adiabatic heating of the specimens during deformation.

When cooling the specimens that were deformed in the γ-region, deposition of the dispersed precipitates of the γ'-phase occurs. Therefore, as shown in Sect. 8.1, the subsequent SP treatment can be carried out only near the range of complete dissolution of the γ'-phase and only when a stable microstructure is retained. All these facts indicate why the second scheme is not widely used.

Structure preparation in the $\gamma + \gamma'$ two-phase region (Fig. 8.7c) is a more widely employed method. The 'Gatorizing' process described by *Moore* and *Athey* [8.17] is a modified form of this method. This technique is used to process the highly alloyed superalloys such as IN-100, Astroloy, and Waspaloy that contain 5–9% of (Al+Ti). Processing is carried out in two stages. During the first stage, structure preparation is accomplished by heating the alloy at a temperature slightly below the recrystallization temperature and subjecting it to several deformations – forging, rolling, and pressing – maintaining a reduction of not less than 4:1. During working, the temperature is increased to the recrystallization temperature due to adiabatic heating of the alloy; consequently, a microduplex-type structure evolves. During the second stage, components are shaped from the superplastic semifinished products under the conditions of superplasticity; then heat treatment is used in order to restore the high-temperature (creep) properties. In this case cast materials as well as the powder metallurgy products can be used as the initial preforms.

Processing in the two-phase region for obtaining fine grains has also been described in [8.10, 18, 19]. In order to decrease the force required for deformation, one uses preannealing treatments which facilitate a reduction of the strengthening effect of the γ'-precipitates (due to their coarsening) [8.19].

Processing according to the scheme presented in Fig. 8.7d may be effective for preparing superalloys containing low and medium quantities of alloying elements. This treatment ranks between the two other schemes, since it is carried out initially in the γ-region and subsequently in the $\gamma + \gamma'$-region.

The main advantage of the treatments presented in Figs. 8.7d is that they offer the possibility of obtaining a microduplex two-phase ($\gamma + \gamma'$) structure that is optimal for realizing superplastic deformation. The small amount of precipitates formed during the process of cooling after completing the treatment do not have a significant effect on the subsequent deformation under conditions of SP, since the particles of the γ'-phase coarsen and dissolve during heating.

To convert superalloys into the SP state, it is necessary to obtain a microduplex microstructure. The formation of a microduplex structure (Chap. 3) is a complex process stipulated by a correlation of recrystallization and decay of a solid solution. Therefore, this process is governed not only by the conditions of structure heterogenization, but by microstructural changes taking place during

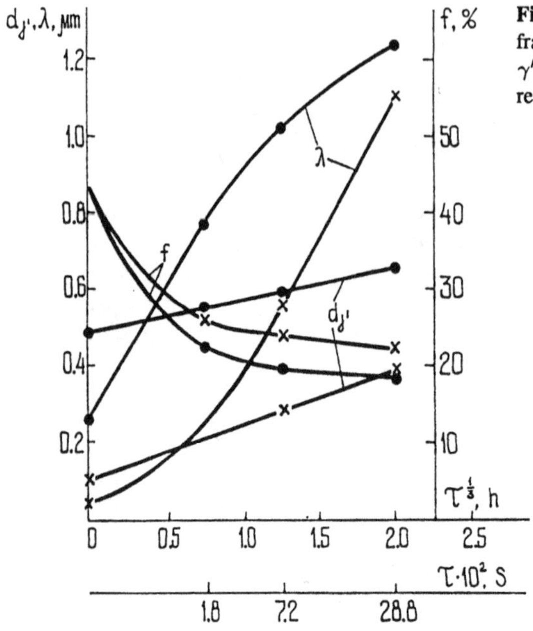

Fig. 8.8. Changes of the size, volume fraction and interparticle spacing of the γ'-phase of the EP962 superalloy as a result of heating

heating and hot deformation. Let us now consider the role of each process, on the example of the EP962 alloy.

Heterogenization. It was necessary to discover the influence of the γ'-phase morphology and its coherency with the matrix (the γ-phase) on the formation of a microduplex structure. The alloy was heated to 1180°C and cooled in two ways: in air (state 1) and in a furnace at a cooling rate of 25 K/h (state 2). Following this treatment, a disperse (0.1 μm) γ'-phase precipitated homogeneously within the grains in state 1, and in state 2 the γ'-phase particles grew to 0.5 μm. The changes of the γ'-phase morphology in the process of heating to 1100°C were studied.

Figure 8.8 shows that the volume fraction of the γ'-phase decreases during annealing to an equilibrium value, and the size of the γ'-particles as well as spacings between them increase. Along with the coarsening of γ'-particles, changes in the structure of the interphase boundaries occur. In state 2 (with a pre-coarsened γ'-phase), at the beginning of heating most of the γ/γ' boundaries are coherent, but following annealing of more than 30 minutes, misfit dislocations appear in the γ/γ' boundaries, indicating a partial loss of coherency (Fig. 8.9a). At the same time, in an alloy with an originally disperse γ'-phase, coherency is retained even after annealing for 8 hours, although a coarsening of particles of up to 0.4 μm is observed (Fig. 8.9b).

Thus, after heating and holding at 1100°C the heterogenization of the alloy structure (state 2) takes place, which strongly affects the alloy properties under hot deformation. For example, the values of ductility and the coefficient m are

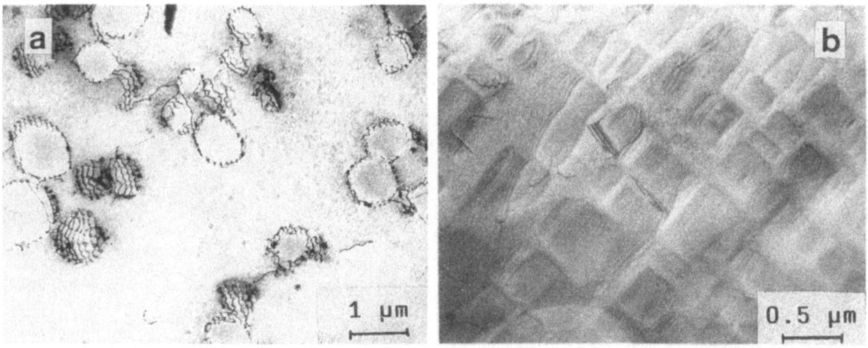

Fig. 8.9a,b. Structure of the EP962 alloy after isothermal annealing at 1100°C and 8 h exposure: (a) state 2, (b) state 1

1.5–2 times higher and the flow stress is 15–30% lower in state 2 as compared with state 1, when the samples are compressed at 1100°C and $\dot{\varepsilon} = 10^{-3}\,s^{-1}$.

Thus, heterogenization significantly enhances ductility of the EP962 alloy by coarsening the γ'-particles, increasing their spacings, and promoting the transformation of the coherent γ/γ' interfaces into partially coherent.

Recrystallization. Heterogenization has significant influence on the structure formation during recrystallization. The alloy was deformed at 20°C to different strain levels and subsequently heated up to 1100°C. In state 1 recrystallization is extremely uneven and occurs predominantly near initial grain boundaries.

During this process, the γ'-phase precipitates within recrystallized grains in the form of lamellae which are coherent with the matrix (Fig. 8.10). Following 50% prestraining and annealing for one hour, the γ'-precipitates had lamellar shape, and dispersed γ'-phase was retained within the nonrecrystallized areas. At the same time, in state 2, recrystallization annealing after the 50% prestraining leads to the formation of a uniform microduplex structure throughout the specimen. In this structure, nearly the entire γ'-phase precipitates at the bound-

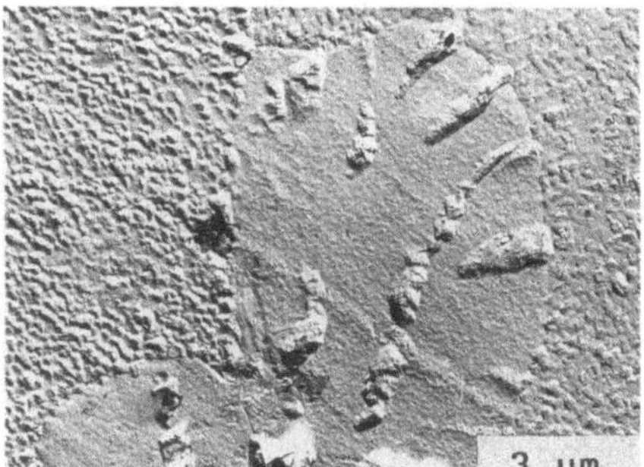

Fig. 8.10. Structure of the EP962 alloy in state 1 after deformation at 20°C to a 10% strain and 1 h annealing at 1100°C (replica)

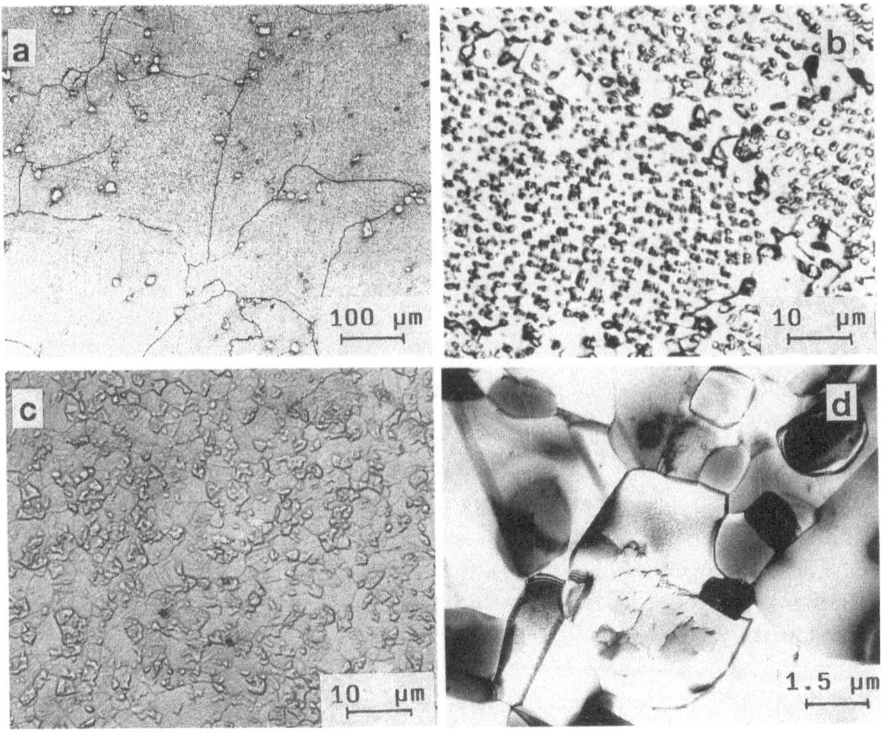

Fig. 8.11a–d. Structure of the EP962 alloy after hot deformation at 1100°C with a strain rate of $\dot{\varepsilon} = 10^{-3}\,s^{-1}$: **(a, b)** state 1; **(c, d)** state 2; **(a)** $\varepsilon = 25\%$ and postdeformation annealing at 1100°C for 2 h; **(b)** $\varepsilon = 70\%$, without annealing; **(c, d)** $\varepsilon = 70\%$ and postdeformation annealing at 1100°C for 8 h

aries of recrystallized γ-grains ($d = 4.2\,\mu m$) in the form of incoherent globular grains with an average size of 1.3 μm.

Hot Deformation. Structure heterogenization also significantly affects the process of structure formation during hot deformation. The 20–30% deformation of state 1 samples with a strain rate of $10^{-4} - 10^{-3}\,s^{-1}$ followed by annealing for 1 hour at the deformation temperature (1100°C) yields no recrystallized grains. At the same time initial grain boundaries become meandering (Fig. 8.11a), probably due to the subgrain formation within zones free from γ'-precipitates and local migration of high-angle boundaries.

In the case of 70% strain at the same temperature and strain rate, recrystallization develops only at the boundaries of original grains (Fig. 8.11b), i.e., it leads to a 'necklace'-like structure. The peculiarity of state 1 is that after deformation with low strain rates, e.g., $10^{-5}\,s^{-1}$, the γ'-particles within the γ-grains grow predominantly in one crystallographic direction. Whereas in state 2 at any studied strain rate ($10^{-5} - 10^{-1}\,s^{-1}$), a microduplex-type structure is formed, whose characteristics and recrystallized volume are determined by the deformation temperature-strain rate conditions. For example, the volume fraction of microduplex constituent is 50% after 70% strain at 1100°C and $\dot{\varepsilon} = 10^{-3}\,s^{-1}$. Subsequent annealing for 8 hours at the deformation temperature causes the recrystallized volume to increase up to 85% (Fig. 8.11c,d). This increase of a hot-worked material can be due to an additional generation of recrystallization nuclei through a mechanism related to the dissociation of a high-angle γ/γ' interphase boundary into a semi-coherent γ/γ' interface and a high-angle γ/γ' grain boundary [8.20].

Heterogenization and static and dynamic recrystallization have a significant effect on the superalloy microstructure, which offers several methods for obtaining a microduplex structure in superalloys. Specific treatments depend on the degree of alloying and the temperature of the complete dissolution of γ'-phase (Fig. 8.1). At the hot deformation temperatures approaching or exceeding the temperature (T_{sv}) of the complete dissolution of γ'-phase (i.e., $T_1 \geqslant T_{sv}$), a dynamically recrystallized structure of the matrix-type is formed, and as mentioned above, is unstable.

Favourable conditions for the formation of the microduplex structure are created when $T_2 < T_{sv}$. With increasing alloying content of Al and Ti, the volume fraction of γ'-phase increases and the quantitative ratio of the phases changes. For example, in the EI437 alloy the quantitative ratio of γ and γ'-phase is $d_\gamma/d_{\gamma'} = 8$, but in the EP975 alloy it is only 2.5. The volume ratio of γ and γ'-phases is 10:1 in the EI437 alloy, but in the high alloyed EP975 alloy, it approaches 1:1.

When the deformation temperature decreases in the $\gamma + \gamma'$ region, a more complete precipitation of the γ'-phase occurs, causing production of a more fine-grained microduplex structure. However, this leads to a decrease in the recrystallization volume fraction, which is why a fractional deformation followed by prolonged annealing in the two-phase region is necessary. The temperature mode providing for the microduplex structure formation depends on the chemical composition. For example, the optimal temperature range of hot deformation is 0.65–0.71 T_m for the EI437 and EI698 alloys, and 0.80–0.85 T_m for the EP742, EP692, and EP975 alloys.

Following a specific mode of processing nickel-based superalloys in order to convert them into the SP state is a rather difficult task, but at the present time there are several examples of its success, Gatorizing being one of them.

Pressing, hydro-pressing and rolling have been used most frequently for obtaining semifinished products with an SP structure [8.10, 18, 19]. The first two modes of structure preparation require significant expenditures of power. Short service life of equipment and limited possibilities of producing large-dimension blanks are further disadvantages. Therefore, using another method, namely, the upsetting of superalloys in isothermal or near-isothermal conditions [8.11] is of more interest. This scheme of deformation in the SP temperature-strain rate modes allows us to reduce significantly power required for straining and widens the variety of SP blanks produced, which, in turn, can be used for manufacturing various parts under SP conditions. The basic modes of prestraining for a number of alloys and parameters defined in the course of this treatment are presented in Table 8.3.

Table 8.3. Modes of preparing a micro-grained structure and SP properties of nickel-based superalloys

Alloy	Total content of Al and Ti [wt,%]	Microstructure preparation: Treatment	Microstructure preparation: t [°C]	Microstructure parameters: grain size [μm] γ-phase	Microstructure parameters: grain size [μm] γ′-phase	Superplastic characteristics: t [°C]	$\dot{\varepsilon}$ [s^{-1}]	σ [MPa]	δ [%]	m	Ref.
Waspaloy	4.4	PM[a] + forging	960			982	1.1×10^{-2}	112	295		
		PM + pressing or forging	840– 965			895– 995					[8.21]
EI-867	4.6	Isothermal upsetting	1025	5–6	1.8	1050	10^{-3}	30	320	0.5	[8.11]
EI-929	5.7	Hydroextrusion Annealing	200 1165	3	0.1	1165	10^{-2}	60	200	0.3	[8.5]
NASATAZ-8A	6.0	PM				1038	4.3×10^{-4}	6.9	600		[8.22]
MAR-M-247LC	6.3			1–1.5	0.1–1.5	1050	3.5×10^{-4}	1.5	800	0.63	[8.22]
IN-738	6.8	PM		0.5	0.1	900			500	0.4	[8.22]
MAR-M200	7.3	PM		2.8	0.4–1.5					0.4– 0.6	[8.22]
Astroloy	7.5	PM + rolling PM + forging	1034 1034			1034 1034	1.1×10^{-2} 1.2×10^{-3}	77 17	1335 1275	0.68	[8.21]
ZS6KP	7.9	Rolling Pressing Hydropressing	1160 1100 1100	2.3		1100 1100 1100	1.6×10^{-3} 1.6×10^{-3} $10^{-3} \times 10^{-4}$	40	650 390 500	0.5 0.38 0.78	[8.10]
ZS6K Udimet 700	8.0	PM + extrusion PM		1.0 1–2	1.0	1100 1000	8×10^{-3} 7×10^{-3}	 37.5	1100 700	0.5 0.4	[8.23] [8.22]
ZS6U	8.1	PM + pressing	1150	2–3	1–2	1120	8.3×10^{-3}	17.5	1200	0.55	[8.13]
IN-100	10.2	PM + pressing or forging	1034– 1145	0.5		920– 1050			850	0.5– 0.65	[8.21]

[a]powder metallurgy

8.3 High-Temperature Properties of Alloys After Processing in the Superplastic Condition

As finished products in daily service, superalloys are subjected to a complex action of varying temperatures and loads. In addition to the common parameters, such as σ_t, $\sigma_{0.2}$, δ, ψ, and impact strength, which are determined for all structural materials, specific parameters such as creep limit (limiting creep stress) and rupture stress must be evaluated. The first parameter is the magnitude of the stress which causes a given value of deformation or a given strain rate within a certain specified time and at a given temperature; the second parameter is the maximum stress withstood by the material without failing at a given temperature and test duration under standard working conditions. Securing the high-temperature properties represented by these characteristics allows development of a particularly stable structure that guarantees reliable long-term service. This is related to the realization of the optimum values of the main factors affecting heat resistance, and in particular, the strengthening coherent γ'-precipitates and the formation of a relatively coarse-grained structure. In practice, this is achieved by employing a standard head treatment involving high-temperature annealing in the single-phase γ-region, quenching, and subsequent aging. As a result, a coarse-grained structure with a γ-grain size of 100–300 μm containing disperse γ'-particles is formed.

Employing heat treatment is most effective for alloys working at a temperature exceeding 800°C. For moderate temperatures (600–700°C), it is useful to use high and low temperature thermomechanical treatments in combination with plastic deformation [8.24]. The aim of these methods is to obtain additional strength of the alloys by developing the substructure within the grains. By employing warm working of the superalloys in the two-phase $\gamma + \gamma'$ region, a 'necklace' type of microstructure forms [8.25]. This is characterized by the formation of a network of fine grains of the $\gamma + \gamma'$ phases ('necklace') around the deformed coarse grains. An alloy having this structure exhibits high levels of mechanical properties at moderate temperatures, because the properties at lower temperatures are close to those of a fine-grained alloy having a microduplex structure. However, the thermomechanical treatment for obtaining such a microstructure has not yet found wide industrial application because of its technological complexity.

When employing the conventional processing methods involving deformation of the superalloys in the γ-region, it is difficult to ensure development of a uniform structure because of the marked growth tendency of their grains; the subsequent heat treatment also has no favourable effect. Products having a coarse structure with differences in grain sizes forms the majority of rejections in the superalloy industry. There are a number of reasons for the development of nonuniform grains size during the production of the components from nickel-based superalloys [8.26]. These include the effect of heterogeneous chemical and phase composition and inhomogeneous deformation which can arise due to the action of external forces and thermal stresses. Therefore, an examination of the properties of the superalloys after subjecting the material deformed in the superplastic condition to the standard heat treatment is in order.

The structure of the fine-grained superalloys becomes highly uniform after superplastic deformation and, in contrast to the treatments in the nonsuperplastic condition, it shows little dependence on the degree of deformation. Furthermore, due to the active development of the mechanisms leading to the redistribution of the grains and accelerated diffusion (Chap. 2), superplastic deformation eliminates the nonuniformity of chemical and phase compositions. Evidently, superplastic deformation with subsequent heat treatment can also be used for improving high-temperature properties. In recent years, numerous investigations have been carried out in this direction.

It is clear that structures of the microduplex type cannot ensure high temperature strength because superplasticity is actually an antipode of high-temperature strength. Therefore, additional heat treatment is required for the recovery of high-temperature properties. The easiest solution is conventional treatment including heating of the material in a one-phase temperature range and subsequent aging. According to [8.5], the EI929 alloy high temperature properties completely recover after this treatment.

Rakhmanov and *Livanov* [8.27] showed that the semifinished products of the ZS6KP alloy that were pressed in the superplastic condition have a more uniform structure than those subjected to hot rolling. After conducting the standard heat treatment, the strength characteristics of the pressed and hot rolled products were at the same level; however, the average level of the rupture stress of the pressed bars was somewhat higher. The maximum improvement in the rupture stress (25–30%) of the pressed semifinished products was obtained after employing a special heat treatment regime involving annealing for 6 h at 1240 ± 10°C followed by air cooling and subsequent aging at 950 ± 10°C for a period of 6 h. This cycle differs from the standard regime by virtue of the higher annealing temperature (30–50°C) and longer holding time. Similar results were reported for the EP220, ZS6UD, and EP791 alloys, also.

Kolpashnikov et al. [8.18] compared the mechanical properties of the ZS6KP alloy having fine-grained and coarse-grained structures obtained after hydropressing and hot rolling, respectively. They showed that the standard heat treatment (1220°C, 4 h; 950°C, 2 h) improve the high-temperature short-term properties of the originally fine-grained alloy to a significantly higher degree than those of the coarse-grained alloy. At the same time, the rupture stress was found to be independent of the condition of the alloy prior to heat treatment but, in this case, the hydro-pressed bars showed a smaller (narrow) time interval for fracture due to the greater uiformity of the structure as compared to the rolled ones. A similar effect of the initial structure on the properties was observed in the EI867 alloy.

Table 8.4 shows the results of mechanical testing of the EI867 alloy in different conditions, including the alloy obtained after a heat treatment involving quenching from the γ-region and aging at 950°C for 8 h. It can be seen that the alloy in the superplastic state with a microduplex structure shows a higher level of strength properties after heat treatment as compared to the previously hot

Table 8.4. Mechanical Properties of the EI867 alloy after different treatments

Type of treatment	σ [MPa] at t [°C]			$\sigma_{0.2}$ [MPa] at t [°C]			δ [%] at t [°C]			τ [h] (t = 900°C)]	
	800	900	950	800	900	950	800	900	950	σ = 400 [MPa]	σ = 220 [MPa]
Microduplex	–	280	80	–	–	–	–	90	400	–	–
Microduplex + heat treatment Annealing at 1200°C, 5 h, air-cooling at 950°C, 8 h	1060--1120	930--950	720--740	1000--1020	890--900	650--680	2–3	4–6	6–10	2–4	
Microduplex + heat treatment Annealing at 1150°C, 4 h, air-cooling at 950°C, 8 h	1000--1050	700--720	530--570	920--950	600--620	470--500	6–9	8–10	12–14	1.5–3	60–80
As-shipped state (hot-rolled) + heat treatment: annealing at 1200°C, 5 h air-cooling, aging at 950°C, 8 h	920--940	660--680	560--620	820--850	600--650	520--540	4–8	6–8	8–10	1–2	57–70
In accordance with technological requirements to the alloy	900--950	550--700	450--500	600--650	400--500	289--350	3–10	7–18	10–20	1	55

rolled condition. Similarly, these specimens showed longer duration up to fracture during short-term tests carried out at higher loads. In addition, the relative elongation of the specimens having a microduplex structure prior to heat treatment is slightly less than that of the hot rolled alloy; however, by decreasing the quenching temperature, it is possible to increase the level of relative elongation with simultaneous permissible reduction of the strength properties.

Metallographic studies confirmed that, after heating the alloys initially having a microduplex structure up to 1200°C and holding them for 5 h, they possess a more uniform structure than the hot rolled alloy subjected to a similar treatment. The indices of the grain size differences [8.28] of the alloy initially having a microduplex structure (evaluated according to the ratios $d_{\max}/d_b$ and $\Delta d/d$, where $d_{\max}$, d_b, and d are the maximum, the most probable, and the average grain size, respectively) were 1.3 and 1.5 times less than those observed in the heat treated hot-rolled alloy.

Improvement of the service properties after superplastic deformation was also established in the ZS6U alloy obtained from granules. After heat treatment, this alloy shows superior strength properties during the short-term tests carried out in the 900–1100°C range as compared to the cast alloy as well as the alloy deformed according to the conventional scheme. With regard to the rupture stress, after superplastic deformation, the alloy exhibits better properties at medium temperature (750–800°C): $\sigma_{100}^{750} = 700\,\text{MPa}$, but is inferior to the other conditions at the temperatures exceeding 800°C. The fatigue parameters (2×10^{-7} cycles) at 20°C, 750°C, and 800°C also exceeded the properties of the cast material.

Thus, conventional heat treatment leads to the recovery of heat strength in all cases. In addition, an analysis (Sect. 8.2) of the changes in the microstructures of superalloys gives us an opportunity to propose a number of specific ways for processing these alloys ensuring various combinations of low- and high-temperature properties.

The EP962 alloy was used for the investigation, and alloys with coarse-grained, microduplex, and necklace-type structures were compared. The microduplex structure (Fig. 8.11c,d) was produced by 50% upsetting of the samples with an initially fine-grained structure under SP conditions ($t = 1050°\text{C}$, $\dot{\varepsilon} = 10^{-3}\,\text{s}^{-1}$). Coarse-grained structures with grain sizes of 30 and 120 µm were formed by holding the specimens for 4 hours at 1130 and 1150°C, respectively, and their subsequent quenching from these temperatures. The necklace-type structure was formed by straining a coarse-grained alloy containing disperse γ'-precipitates under SP conditions ($t = 1100°\text{C}$, $\dot{\varepsilon} = 10^{-4}\,\text{s}^{-1}$, $\varepsilon = 70\%$). During this treatment, recrystallization occurred in the near-boundary regions only. Under smaller strains (25–30%), followed by quenching from the deformation temperature (1100°C), a coarse-grained structure with meandering grain boundaries is formed (Fig. 8.11a). The alloy in all structural states underwent final aging of 850°C for 6 hours, air-cooling; 800°C for 16 hours, air-cooling.

The mechanical properties of the alloy in different microstructural states are given in Table 8.5. At room temperature the properties of the alloy with the microduplex structure are higher than those of the coarse-grained alloy. How-

Table 8.5. Mechanical properties of the EP962 alloy with different microstructures

Microstructure	At room temperature					At 650°C	
	σ_u [MPa]	$\sigma_{0.2}$ [MPa]	δ [%]	ψ [%]	KCU [J/cm^2]	Stress [MPa]	Time before failure [h]
Microduplex	1650	1176	24.0	22.4	39.2	1000	74
Coarse-grained (d = 120 μm)	1500	1020	22.5	17.0	30.4	1000	284
Coarse-grained (d = 30 μm)	1590	1150	22.8	14.8	31.4	1000	164
'Necklace'	1630	1255	21.9	18.2	33.3	1000	444
						1050	188
Coarse-grained with meandering boundaries	1562	1222	20.7	15.0	31.4	1000	520
						1020	318
						1050	243
						1080	102

ever, when the testing temperature increases, the situation is the opposite. Heat treatment, consisting of sample heating to a complete dissolution of the γ'-phase (1150°C; d = 120 μm) and aging, leads to the complete recovery of the high-temperature properties. However, a slight deterioration of the low-temperature properties is observed.

Heat strength together with enhanced short-term properties were ensured by quenching from a temperature twenty degrees below that of the complete dissolution of the γ'-phase. The γ'-phase grain size was rather small (30 μm) because of the incomplete quenching.

A good combination of low- and high-temperature properties was obtained in specimens with the necklace-type structure. Apparently, this was due to the fact that the matrix in this state was additionally hardened by a substructure, stabilized by coherent γ'-precipitates of 0.6 μm. Coarser incoherent γ'-precipitates (1.5 μm) are distributed in the near-boundary region with recrystallized γ'-phase grains of 5 μm. The best high-temperature properties were observed in specimens with a coarse-grained structure and meandering grain boundaries (Table 8.5), the short-term properties being rather moderate.

Thus, deformation in the SP flow temperature-strain rate range makes it possible to produce various microstructural states in superalloys and, therefore, to achieve the necessary combination of short-term and high-temperature properties. The achieved level of properties is similar to that produced by thermomechanical treatment only. This method allows the creation of a higher thermal stability of hardening as compared with conventional thermomechanical treatments, because, in this case, SP deformation occurs at low strain rates.

In conclusion, we note that by controlling structural transformations in superalloys we can vary their plasticity as well as high-temperature and short-term properties on a broad scale by processing these alloys under conditions required for inducing SP flow.

9. Superplasticity of Intermetallic Compounds

Intermetallic compounds are superior to conventional metallic materials primarily because of their high-temperature strength parameters. The enhancement of these properties in the transition from metals to intermetallics is due to the change of the interatomic bonding as well as to a more complicated crystal structure. For example, in intermetallics, strong covalent bonds between the atoms provide a higher cohesive strength [9.1–5]. The more complicated crystal structure results in the growth of the elementary cell size, reduction of the symmetry and growth of the Burgers vector, which in turn lead to an increase in the Peierls stress and limit the number of active slip systems. These are the reasons why the majority of intermetallics are low-plasticity materials. In ordered intermetallic compounds, deformation is accomplished through the motion of superdislocations consisting of superpartial dislocations separated by a planar defect, i.e., an antiphase boundary [9.6]. Contrary to metals and disordered alloys where the slip plane is determined by the value of the Peierls forces or by the possibility of crystallographic splitting of the slipping dislocations, another criterion determines the slip plane choice. This is the energy of an antiphase boundary created during the motion of superdislocations in this plane. The anisotropy of the antiphase boundary energy can impede the process of cross slip and even slow down the motion of superdislocations [9.1–3, 5]. These features provide for the enhanced high-temperature strength and heat and corrosion resistance of intermetallics and make their application as high-temperature structural materials rather promising.

However, the low plasticity of the majority of intermetallics hinders their commercial application. Various suggestions have been made concerning the causes of low plasticity, viz., the limited number of slip systems, impeded cross-slip and grain-to-grain transfer of shear, creation of dislocation barriers specific for ordered alloys, and segregation of harmful impurities at grain boundaries [9.1–3, 5, 7, 8]. Another fundamental factor causing fracturing at grain boundaries and which is typical only of ordered structures is the low cohesive strength of grain boundaries [9.1, 2, 5]. The strength balance between the grain interior and grain boundaries is broken because of the features of atomic bonds at the grain boundaries of intermetallics. A classical example of grain boundary brittleness due to the violation of this balance is found in Ni_3Al (superlattice $L1_2$).

Most recent investigations indicate that the plasticity of intermetallics can be substantially enhanced by means of alloying, complex loading, and varying the microstructure and temperature-rate conditions of deformation. The most vivid example of the first method is the effect of an abrupt enhancement of the plasticity

Table 9.1. Parameters of superplasticity in intermetallic compounds

Material superlattice	d [μm	t [°C]	$\dot{\varepsilon}$ [s^{-1}]	δ_{max} [%]	m	Ref.
Ni_3Si-based $L1_2$	10–12	1080	6×10^{-3}	650	–	[9.9]
Ni_3Al-based $L1_2$	13	–	–	640	–	[9.10]
Ti_3Al-based DO_{19}	–	–	10^{-4}	500	0.4–0.5	[9.10]
Ni_3Al $L1_2$	1.6	700	$< 1.04 \times 10^{-4}$	160	> 0.3	[9.11]
Ti_3Al-based DO_{19}	5	1000	10^{-4}	550	0.6	authors' data
TiAl $L1_0$	~ 10	1025	8.3×10^{-4}	250	0.43	– " –
TiAl $L1_0$	0.4	800	8.3×10^{-4}	225	> 0.3	– " –
Mn-Al-C based $L1_0$	1	700	10^{-3}	–	0.42	– " –

of polycrystalline Ni_3Al at room temperature after the addition of 0.1% boron [9.5, 8]. Considerable plastic deformation can be attained in intermetallics using hydrostatic extrusion. In [9.5], a magnetic alloy $Fe(Si_{0.65}Al_{0.35})$ ("Sendust") and TiAl, which are typical brittle materials, were extruded.

However, the most efficient way of changing the plasticity of intermetallics is to convert them to an SP state. Recently published papers have shown that SP can be realized in intermetallics (Table 9.1); for example, a relative elongation to rupture of 650% was obtained at 1080°C at an initial strain rate of $6 \times 10^{-3}\,s^{-1}$ in an alloy based on the Ni_3Al intermetallic compound [9.9]. A recently published review [9.10] reported on the SP behaviour of alloys based on Ni_3Al ($\delta = 640\%$) and Ti_3Al ($\delta = 500\%$). These alloys, however, comprised a disordered phase coexisting with the ordered intermetallic one. The most interesting and important study was performed on the ordered intermetallic compound Ni_3Al (24 mol%Al) with an addition of 0.24 mol% of boron [9.11], which displayed SP behaviour at 700°C and initial strain rates lower than $1.04 \times 10^{-4}\,s^{-1}$, the maximum elongation being 160%.

The SP of intermetallics has been thus far insufficiently studied, probably due to the fact that they are "fatally brittle" materials. Many aspects are still obscure, the most important among them being

1) the mechanisms of deformation operating during SP flow in intermetallics (are they the same as in conventional metals?);
2) the methods of microstructure refining applicable to intermetallics;
3) the properties of intermetallics after SP deformation.

Development of SP in intermetallics would be a substantial advancement towards their commercial applicability. However, the low plasticity of intermetallics at room temperature remains the main obstacle for their actual application as structural materials. Meanwhile, SP treatment not only provides for an abnormally high deformability of intermetallics at elevated temperatures, but offers an efficient means for enhancing the plasticity of intermetallics at room temperature and improves other physical properties, because it also increases the structural and chemical homogeneity of materials.

Developing methods of obtaining fine-grained microstructures in intermetallics is an important problem related to their conversion to the SP state. We now present new results in the SP of intermetallics which can be used for solving these problems [9.12–20]. To present the data systematically, we start our discussion with alloys where the intermetallic phase coexists with a metallic phase. We note that such alloys also include nickel-based superalloys (Chap. 8), but because of their chemical composition, structure, and mechanical behaviour, these materials will be discussed separately.

9.1 Superplasticity of Alloys with Both Intermetallic and Metallic Phases

The most detailed studies were performed on X30K23, which refers to a group of hard magnetic materials of the precipitation hardening class. From this alloy, magnets are produced, predominantly by casting, because forming is complicated by its low deformability caused by an embrittling intermetallic σ-phase (superlattice $L1_0$) which forms in the structure in a temperature range from 700 to 1050°C. However, as shown below, the creation of a fine-grained microstructure and deformation under certain temperature-rate conditions not only provides for SP in the alloy, but also, due to the SP treatment, improves its magnetic properties.

The chemical composition of the alloy X30K23 is Cr-30.5 wt.%; Co-23%; Ti-0.4%; V-0.6%; Si-0.4%; balance Fe. In a temperature range from 1250 to 1050°C, the alloy consists of two phases, i.e., α (bcc) and γ (fcc). At temperatures below 1050°C the intermetallic σ-phase with superlattice $L1_0$ arises. In a temperature range from 1050 to 950°C all three phases of the alloy – α, γ and σ – are in equilibrium, the fraction of the α-phase decreases from 30 to 2% and that of the γ-phase from 68 to 45% with a decrease in temperature to 950°C. In a temperature range from 800 to 950°C the alloy also has a two-phase structure, consisting of the γ- and σ-phases.

Let us consider the alloy behaviour in a microduplex state, with a mean grain size of the γ- and σ-phase equiaxial grains being 2 µm. The temperature dependence of the mechanical properties is shown in Fig. 9.1a, which shows that the relative elongation to rupture δ first increases with the rising temperature, reaches the maximum at 950°C, and decreases with a further rise of temperature. With the rising temperature the flow stress monotonously decreases.

The strain rate sensitivity of flow stress at the maximum plasticity temperature t = 950°C is of a sigmoidal shape, typical fo SP behaviour, with a clearly distinguished region of enhanced strain rate sensitivity that corresponds to the maximum values of the relative elongation to rupture δ and of the strain rate sensitivity coefficient m (Fig. 9.1b).

The dependence of the relative elongation on the initial strain rate at 950°C is of an extreme character. The maximum elongation was attained at an initial

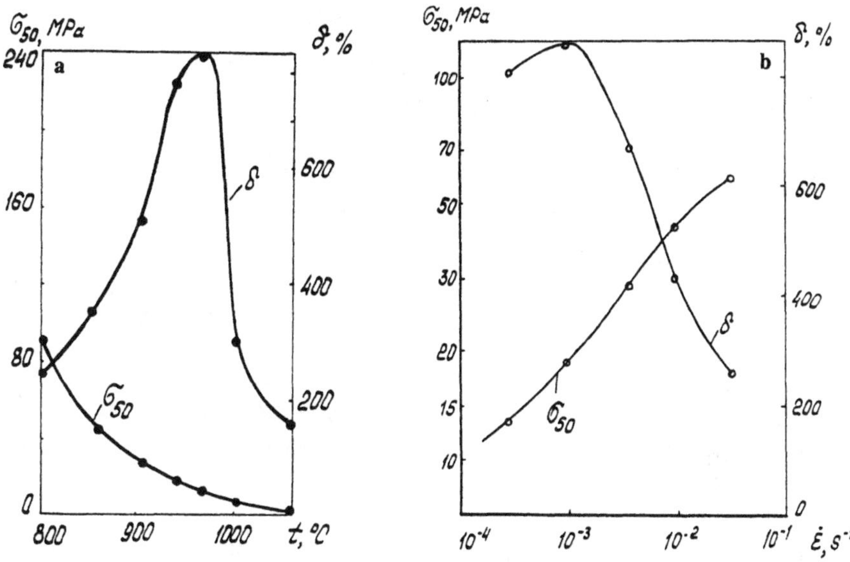

Fig. 9.1. Dependence of flow stress σ_{50} and relative elongation δ on temperature at $\dot{\varepsilon} = 0.8 \times 10^{-3}\,s^{-1}$ (a) and on strain rate at 950°C (b) in the X30K23 alloy in a fine-grained state

strain rate of $0.8 \times 10^{-3}\,s^{-1}$ and amounted to 810%. The strain rate dependence of coefficient m at $\varepsilon = 50\%$ and 950°C also looks like a curve with a maximum at $\dot{\varepsilon} = 0.8 \times 10^{-3}\,s^{-1}$.

When the sample is tensile strained at 950°C with an initial strain rate of $0.8 \times 10^{-3}\,s^{-1}$, the hardening stage is followed by a smooth transition to the steady flow stage at a strain level of 50–60%. Within the strain range from 20 to 150%, the coefficient m remains constant and equals 0.47.

Metallographic analysis of the alloy indicates that grains of γ- and σ-phases grow under SP flow remaining equiaxial up to 810% strain. At $t = 950$°C and $\dot{\varepsilon} = 0.8 \times 10^{-3}\,s^{-1}$, within the first 100% of straining, the mean grain size increases from 2 to 3.8 μm and within the subsequent 700% of straining it increases only to 5 μm. In the heads of tensile strained samples the grain size increases only to 3 μm, which indicates that in the intermetallic phase containing the alloy X30K23, SP deformation induces surplus grain growth as in conventional metallic materials.

Microstructural analysis indicates that after 100% SP deformation the intermetallic σ-phase grains form chains transverse to the tensile axis, thus producing an impression of a transverse banded structure. As a result, at the 200% strain level the majority of the γ/σ type interphase boundaries make an angle of 90° with the tensile axis, and the majority of the γ/γ and σ/σ boundaries are parallel to the tensile axis. When the strain level is increased to 400%, the direction of the transverse structure bands changes, making an angle of 45° with the tensile axis. The number of the σ-phase chains decreases if the strain rate deviates from its optimal value. The formation of the σ-phase chains in the course of SP flow results in the growth of the specific surface of the σ/σ and γ/γ boundaries

Fig. 9.2

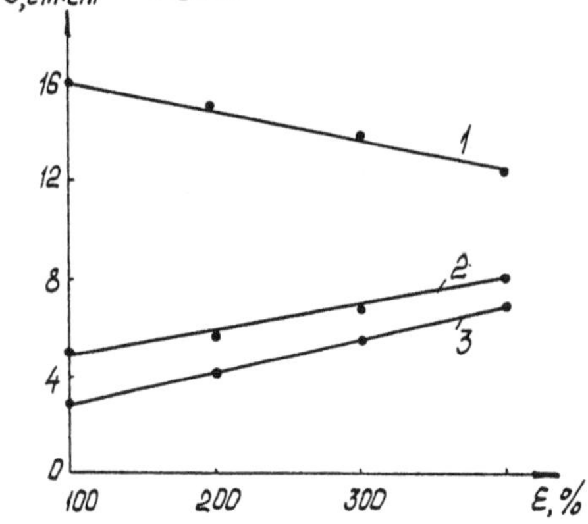

Fig. 9.3

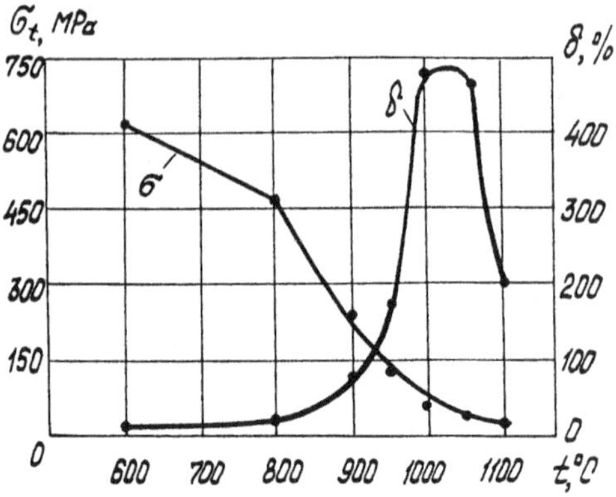

Fig. 9.2. Dependence of the specific surface S of boundaries of different types on strain level in the X30K23 alloy after tensile straining at 950°C and $\dot{\varepsilon} = 0.8 \times 10^{-3}\,s^{-1}$; Curve (*1*) γ/σ; (2) γ/γ, (3) σ/σ

Fig. 9.3. Dependence of flow stress σ_t and relative elongation to rupture δ on the temperature of deformation t in an alloy based on the Ti_3Al intermetallic compound in a fine-grained state at $\dot{\varepsilon} = 3.3 \times 10^{-4}\,s^{-1}$

and, respectively, in the reduction of the specific surface of the γ/σ boundaries (Fig. 9.2). These facts provide convincing evidence that the grain boundary sliding is more enhanced in the γ/σ boundaries than in γ/γ and σ/σ ones during SP deformation. (The effect of the type of grain boundaries on the development of grain boundary sliding under SP deformation was described in Chap. 2).

Electron microscopy analysis indicates that, after SP deformation, the density of dislocations in the alloy is low both in the intermetallic σ-phase and in the γ-phase, its value being $\sim 10^{-12}\,m^{-2}$. Thus, the alloy X30K23, in which the metallic γ-phase coexists with the intermetalic σ-phase with a mean grain size in both phases of 2 µm, displays features of structural SP typical of conventional metals and alloys, when it is deformed at temperatures from 900 to 1000°C and at strain rates from 10^{-2} to $10^{-4}\,s^{-1}$.

An interesting example of an alloy which has a metallic phase along with an ordered intermetallic phase and displays SP behaviour is a titanium alloy based on the intermetallic compound Ti_3Al (superlattice $D0_{19}$) alloyed with niobium with small molybdenum and zirconium additives. In the temperature range from 20 to 1100°C it has two phases: α_2 (Ti_3Al) and β (bcc), their fractions being 60 and 40%, respectively. Figure 9.3 shows the temperature dependence of the mechanical properties of this alloy with an equiaxial fine-grained (d = 5 µm) microstructure produced during tensile tests with an initial strain rate of $3 \times 10^{-4}\,s^{-1}$. With a rising test temperature, the relative elongation to rupture, δ, first gradually grows in a temperature range from 600 to 800°C, then abruptly increases in the range of 900–1050°C, and at 1000-1050°C reaches 480–500%. A further rise in temperature to 1100°C results in a drop of δ to 200%. The flow stress monotonically decreases with rising temperature.

The strain rate dependence of the mechanical properties indicates that throughout the strain rate interval from 3.3×10^{-5} to $3.3 \times 10^{-3}\,s^{-1}$ the alloy displays an enhanced strain rate sensitivity, the coefficient m being higher than 0.4. The functions $m(\dot{\varepsilon})$ and $\delta(\dot{\varepsilon})$ have maxima at $m = 0.6$ and $\delta = 550\%$ at $\dot{\varepsilon} \sim 10^{-4}\,s^{-1}$. With an increasing strain rate, the flow stress grows abruptly from 15 MPa at $3.3 \times 10^{-5}\,s^{-1}$to 145 MPa at $3.3 \times 10^{-3}\,s^{-1}$.

Metallographic analysis indicates that after deformation under optimal conditions of SP ($t = 1000$°C and $\dot{\varepsilon} \sim 10^{-4}\,s^{-1}$), mean grain sizes in the sample neck and head are 12 and 8 μm, respectively. In the sample neck, grains retain their equiaxial shape. Substantial growth of grains during deformation at 1100°C results in the reduction of the alloy plasticity. Thus, at 1000–1050°C and $\dot{\varepsilon} = 3.3 \times 10^{-5}\,3.3 \times 10^{-3}\,s^{-1}$, the Ti_3Al-based compound displays SP features.

These results, as well as data reported in [9.10, 11] lead to the following general conclusion. When a fine-grained microstructure is created in alloys in which the metallic (or disordered [9.10, 11]) phase coexists with the intermetallic phase, they display SP behaviour under certain temperature-rate conditions, which is similar to conventional metallic materials. Moreover, under SP flow, the intermetallic phase, which concentrates stress and induces brittle fracturing at low temperatures, behaves like a metallic phase. This is possible due to the fine-grained microstructure and intensive grain boundary sliding which occurs during SP deformation. Since a substantial portion of deformation is realized via grain boundary sliding, then in both intermetallic and metallic phases accommodation processes occur which provide for the relaxation of local stresses. In addition, in the alloy X30K23, the development of grain boundary sliding is easier in the interphase boundaries than in the intergranular ones. The existence of the intermetallic phase also provides for the stabilization of microstructure curbing the grain growth during SP deformation. This results in the broadening of the temperature-rate interval in which SP is observed on the alloys considered above.

9.2 Superplasticity of Intermetallic Compounds

We now consider the SP behaviour of alloys comprised of an ordered intermetallic phase only using as an example the compound TiAl (superlattice $L1_0$), which possesses a set of properties typical of the majority of intermetallics, viz., low plasticity and low deformability even at high temperatures [9.21, 22]. TiAl is remarkably superior to existing metallic materials in its specific values of high-temperature properties and elastic modulus in a broad range of temperatures. In addition, it is highly heat-resistant in oxygen and does not burn, which makes TiAl one of most promising high-temperature structural materials. This and the following sections are devoted to new methods for increasing the plasticity of this compound based on SP treatment.

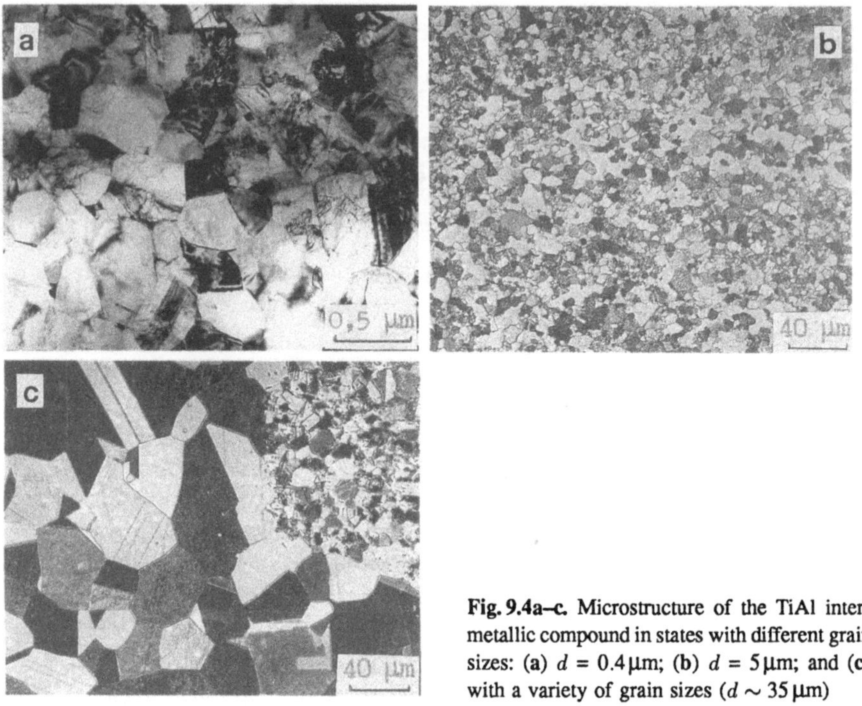

Fig. 9.4a–c. Microstructure of the TiAl intermetallic compound in states with different grain sizes: (**a**) $d = 0.4\,\mu m$; (**b**) $d = 5\,\mu m$; and (**c**) with a variety of grain sizes ($d \sim 35\,\mu m$)

Ti-36 wt.%Al (Ti-50 at.%Al) alloy shaped into ingots and blanks produced by compacting granules was used for the investigation. Although the alloy is of a stoichiometric composition, the ordered γ-phase (TiAl) coexists in it with a small amount ($\sim$ 3% of the volume) of α_2-phase (Ti_3Al). According to high-temperature X-ray structural analysis, this phase composition remains constant in an interval from 20 to 1100°C. Hereafter the alloy Ti-36 wt.%Al will be referred to as TiAl.

Effect of Grain Size. Three states with different grain sizes d were exmained: the first two had relatively homogeneous microstructures with grain sizes of 0.4 and 5 µm, respectively; the third state was characterized by a considerable diversity of grain sizes and a "coarse"-grained fraction (80% of the volume) with a mean grain size of 35 µm coexisting with a "fine"-grained fraction ($d = 3$–$8\,\mu m$) (Fig. 9.4). Comparison of the mechanical properties indicates that the TiAl plasticity grows with diminishing grain size (Table 9.2). The temperature dependence of the TiAl mechanical properties in states 1 and 2 is rather complicated. As the testing temperature increases, the relative elongation to rupture δ first increases monotonically, but upon reaching a maximum (δ = 260% at 850°C for state 1 and 250% at 1025°C for state 2), rapidly decreases. In state 3, the relative elongation to rupture is only 8% even at 1000°C. With rising temperature, the tensile strength σ_t of the states decreases more rapidly the smaller the grain size.

Table 9.2. Effect of grain size on the TiAl mechanical properties at elevated temperature ($\dot{\varepsilon} = 8.3 \times 10^{-4}\,\mathrm{s}^{-1}$)

States	d [μm	Mechanical properties	Temperature of deformation [°C]								
			700	800	850	900	950	975	1000	1025	1050
1	0.4	δ [%]	130	225	260	220	–	–	–	–	–
	σ_t	[MPa]	730	225	120	120	–	–	–	–	–
2	5	δ [%]	–	95	–	155	175	195	215	250	175
	σ_t	[MPa]	–	365	–	195	150	135	120	110	100
3	35	δ [%]	0.1	2	–	2.5	–	–	8	–	–
	σ_t	[MPa]	400	390	–	350	–	–	250	–	–

The analysis of the rate dependence of the mechanical properties indicates that the TiAl intermetallic compound in state 1 displays an enhanced strain rate sensitivity ($m > 0.3$ at $\dot{\varepsilon} < 1.7 \times 10^{-3}\,\mathrm{s}^{-1}$) while in state 2 the coefficient m exceeds 0.3 only at 1000°C (0.74 of the melting point) and at a strain rate of $(0.83\text{–}1.6) \times 10^{-3}\,\mathrm{s}^{-1}$. The coefficient m attains its maximum value at temperatures of deformation of 850°C ($m = 0.47$ at $\dot{\varepsilon} = 3.3 \times 10^{-4}\,\mathrm{s}^{-1}$) and 1025°C ($m = 0.43$ at $\dot{\varepsilon} = 10^{-3}\,\mathrm{s}^{-1}$) in states 1 and 2, respectively.

The $\sigma - \varepsilon$ curves obtained in states 1 and 2 under the temperature-rate conditions providing for maximum plasticity are typical of SP behaviour, i.e., the hardening stage is followed by a steady flow stage.

Thus, the refining of the TiAl microstructure enables the realization of SP in this intermetallic compound. In this case, the smaller the grain size, the lower the temperature at which the intermetallic compound displays SP features.

Effect of Grain Boundary Structure. A specific feature of the TiAl intermetallic compound is its tendency toward creating several special boundaries, i.e., annealing twin boundaries, which allows us to perform a specific experiment comparing the intermetallic compound mechanical properties in states with similar grain sizes, but with different grain boundary structures. Two states, A and B, were studied, and their microstructures are also characterized by a variety of grain sizes (Fig. 9.5a,b). The "fine"-grained and "coarse"-grained volume fractions constitute 15 and 85%, respectively. The sizes of fine grains in both the states are the same, i.e., $d = 2\text{–}8\,\mu\mathrm{m}$. The average sizes of coarse grains are 12 and 15 μm in states A and B, respectively, taking into account the annealing twin boundaries as well. On the whole, these states can be characterized as having similar grain sizes. No difference in the amount, distribution, size (0.1–3 μm), and shape of the α_2-particles in either state was observed. However, these states possess different grain boundary structures. The fraction of twin boundaries in states A and B is 10 and 35% of the total number of boundaries, respectively.

Let us consider the strain rate dependence of the TiAl mechanical properties in states A and B at 1025°C (Fig. 9.6). In state A, in a certain range an enhanced strain rate sensitivity of the σ_{50} flow stress is observed. Thus, in state A, in a strain

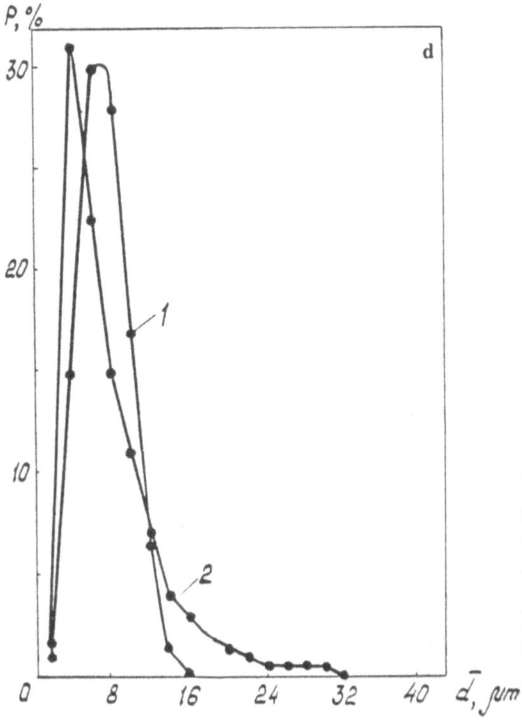

Fig. 9.5a–d. Microstructure of the TiAl intermetallic compound: (**a**) state *A*; (**b**) state *B*; (**c**) after 160% SP deformation of the intermetallic compound in state *A* $t = 1025\,^{\circ}\mathrm{C}$ and $\dot{\varepsilon} = 8.3 \times 10^{-4}\,\mathrm{s}^{-1}$); (**d**) histograms of the grain size distribution in the sample head (*1*) and neck (*2*) at the same strain level

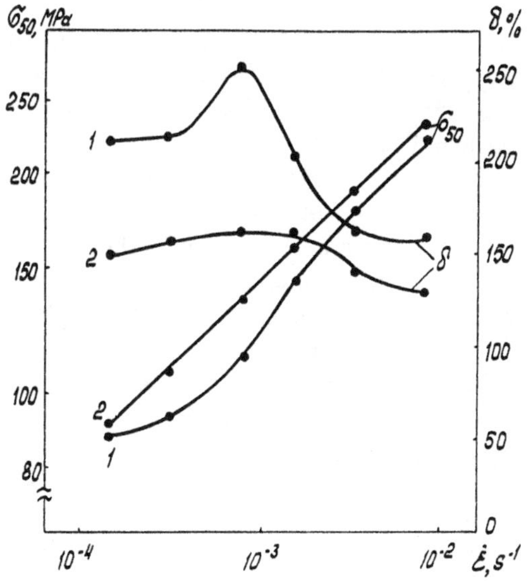

Fig. 9.6. Dependence of flow stress σ_{50} and relative elongation to rupture δ on strain rate at $t = 1025°C$ in TiAl: Curves (*1*) state *A* and (2) state *B*

rate interval of $(0.83–1.6) \times 10^{-3}\ s^{-1}$ the m value varies from 0.33 to 0.43. Higher values of the coefficient m are matched by larger relative elongations to rupture (δ = 200–250%). Under these conditions, the samples are uniformly elongated to rupture. State B is characterized by a constant value of m throughout the considered range of strain rates (m = 0.26), while there is little dependence of the relative elongation to rupture on the strain rate (δ = 130–160%), the samples being strained nonuniformly.

The $\sigma - \varepsilon$ curves obtained during the tensile straining of the TiAl samples in states A and B at an optimal strain rate ($\dot{\varepsilon} = 8.3 \times 10^{-4}\ s^{-1}$) indicated that the steady flow stage in state A intermetallic compound is preceded by a prolonged hardening stage, while in state B the initial stage is marked by a flow stress peak. With further deformation, the flow stress monotonically decreases and approaches the σ level of state A.

Thus, in state A, the TiAl intermetallic compound displays SP features, while in state B the SP behaviour is less pronounced. Since states A and B have similar grain sizes and are identical in amount, distribution, size, and shape of the α_2-phase particles, the decreases of plasticity during the transition from state A to state B is probably associated with the growth of the fraction of twin (special) grain boundaries in the intermetallic compound. The data obtained imply that not only a large grain boundary area, but also the type of grain boundary structure are important in making intermetallics superplastic. The same conclusion was made for metals in Chap. 1.

Effect of Structural Changes During Superplasticity Deformation. An interesting effect was discovered while studying the SP deformation of the TiAl intermetallic compound. Although this material is essentially one-phase, in the course of SP flow its microstructure remains stable. Thus, the intermetallic compound displays relatively high plasticity. Structural investigations helped to establish the origin of the TiAl microstructure stability, which is caused when the rate of SP flow coincides with the rate of the development of dynamic recrystallization in the intermetallic compound. Figure 9.5a,c shows the TiAl microstructure in state *A* in the sample head and neck after 160% SP deformation at an optimal strain rate. It can be seen that SP treatment results in making the grain sizes more uniform and eliminates twins. Comparison of the histograms of the grain size distribution of arbitrarily selected areas of the sample neck and head indicates that the microstructure homogeneity increases due to the simultaneous growth of fine grains and the refining of coarse grains (Fig. 9.5d). The distribution of α_2-phase also becomes more uniform. During SP deformation, grains remain equiaxial until failure. Mechanical twinning and bulging of grain boundaries are observed in coarse grains at the initial stage of deformation (ε = 15%). In the course of further deformation (ε = 30%) along grain boundaries and within the original grains, new grains are created. The mechanisms of this grain formation are similar to those observed during the hot straining of TiAl with a coarse-grained microstructure (Sect. 9.3). However, the process of the "new" grain formation takes place not only at the initial stage of deformation, but also throughout the SP flow until sample failure. In addition, this process develops even at strain rates substantially lower than the optimal rate. The recrystallized grains do not mechanically retwin.

The growth of strain rates over the optimal rate results in an intense refining of the intermetallic compound microstructure; vice versa, strain rates below the optimal rate lead to the growth of the majority of grains and increase the diversity of grain sizes. In these cases, grains also remain equiaxial. At the final stage of flow, voids which are especially numerous in the fracture area, are created.

The evolution of dislocation structure indicates that during SP deformation a high density of dislocations, unusual for superplasticity, is observed in the intermetallic compound. In this case, at different strain levels, primarily individual dislocations are observed in fine grains ($d < 8\,\mu$m), and only some of the grains contain dislocation walls and nets. The density of dislocations ϱ is $10^{12} - 5 \times 10^{12}\,\mathrm{m}^{-2}$. At the initial stage of flow, dislocations are strongly accumulated in coarse grains ($d \geqslant 8$) µm. The value of ϱ at $\varepsilon = 15\%$ is $10^{13}\,\mathrm{m}^{-2}$. By this moment of deformation grains will contain dislocation tangles, nets, walls, subgrains, and mechanical twins (Fig. 9.7a,b).

The higher the strain level, the lower the density of dislocations in coarse-grained areas, but the density of dislocations remains sufficiently high to provide for SP flow and reaches $10^{13} - 5 \times 10^{13}\,\mathrm{m}^{-2}$ at ε = 120%. No mechanical twinning occurs under these conditions. The increase or decrease of the strain rate causes no qualitative change in the dislocation structure, affecting only the value of ϱ. After SP deformation, the α_2-phase particles are found not only in the grain boundaries, but also in the grain interior. It should be noted that, following SP flow, the enhanced density of dislocations was also observed in the Ni_3Al intermetallic compound with boron additives [9.11].

Thus, the TiAl intermetallic compound is unique in that under SP conditions dynamic recrystallization develops with a rate commensurable with the strain

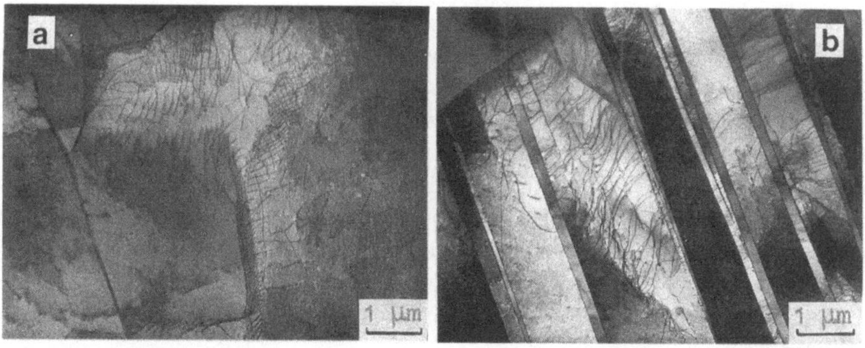

Fig. 9.7a,b. Dislocation structure (**a**) and mechanical twinning (**b**) in coarse grains of the TiAl intermetallic compound after deformation at $t = 1025°C$, $\dot{\varepsilon} = 8.3 \times 10^{-4}\,s^{-1}$, and $\varepsilon = 15\%$

rate, which eliminates the initial diversity of the microstructure grain sizes. This is rather unexpected, because the localization of SP flow in fine-grained areas impedes the development of recrystallization in coarse grains. The facility of dynamic recrystallization in TiAl is probably due to dislocations that are less mobile than in metals. Indeed, the high Peierls barrier and substantial crystallographic splitting of superdislocations $a[101]$, $a[011]$, and $1/2a[112]$ in TiAl impede their slip and transition from one plane to another via cross-slip and climbing [9.22–24]. Therefore, a density of dislocations required for the start of recrystallization is created in coarse grains, while in many fine grains there are dislocations that can be observed in metals only in quenching experiments (Chap. 1). Although the homogeneity of the microstructure in the TiAl intermetallic compound substantially increases by a certain stage of deformation, new grains will be generated continuously until sample failure. The new grains are probably formed in individual grains grown during SP flow. The same pattern is observed in a titanium β-alloy (Chap. 6).

These features of SP behaviour of the TiAl intermetallic compound can be explained in terms of the phenomenological physical model (Chap. 2), according to that model, under SP grain boundaries play a specific role, viz., their surface is a site of the grain boundary dislocation motion providing for SP flow. When the grain size increases, the motion of grain boundary dislocations fails to provide a strain rate specified by the machine, which results in the enhancement of dislocation slip facilitating the relaxation of the arising stresses. This enhancement results in the accumulation of dislocations in grains, followed by the development of dynamic recrystallization in the intermetallic compound. However, in metals, in contrast to TiAl, a coarsening of grains under SP flow does not always result in a substantial accumulation of dislocations, which is due to rapid recovery (Chap. 1), the rate of dynamic recrystallization and the strain rate being considerably different.

During SP deformation, in TiAl there is an approximate dynamic equilibrium between the processes of grain growth and refining. A deviation from the temperature-rate conditions of SP breaks the said equilibrium and results in a decrease in the intermetallic compound plasticity. It should be noted that although under SP flow in metals the growth of grains decreases the metal plasticity, in TiAl the developing dynamic recrystallization limiting the grain growth can contribute to an improvement of plasticity. However, the recrystallization does not occur regularly in the microstructure and probably fails to remove the concentrations of stresses at the grain boundary junctions. The relaxation of stresses due to grain boundary sliding, with the accommodation of grain growth during SP flow being insufficient, usually results in the formation of voids (Chap. 1). The coalescence of voids during deformation leads to the failure of the intermetallic samples.

The simultaneous growth of fine grains and the refining of coarse ones during SP flow improve the homogeneity of the intermetallic compound microstructure as compared to the original state. According to [9.17], the microstructure heterogeneity is one of the factors underlying the low plasticity of the TiAl intermetallic compound. In Sect. 9.4 it is shown that SP treatment lowers the cold shortness threshold.

Similar SP behaviour was also observed in another intermetallic compound with superlattice $L1_0$, namely, magnetic alloy Mn-29.4Al-0.6C with a fine-grained microstructure (d = 1 μm), which was compression strained at a temperature at which a metastable τ-phase exists (t = 700°C), the stain rate being $10^{-3}\,s^{-1}$. In this case the alloy displays an enhanced strain rate sensitivity of flow stress, i.e., $m = 0.42$.

Based on the results, the following conclusions can be made:

1) The effect of the area and structure of grain boundaries on the SP flow on the TiAl intermetallic compound was established. The enlargement of the area of random boundaries substantially lowers the temperature at which SP is attained. At the same time, special boundaries in the intermetallic compound cause a diminishing of the SP parameters.
2) During deformation under SP conditions, dynamic recrystallization develops, which allows improvement of the intermetallic compound plasticity.

9.3 Refining of Microstructure in Intermetallics

At present, the most widely practiced methods of microstructure refining of intermetallics are powder metallurgy and hot straining.

Methods of Powder Metallurgy. Methods of powder metallurgy are used to produce intermetallics as fine as nanocrystalline [9.25], but because of the grain growth during sintering under pressure, it is not always possible to preserve fine grains in bulk blanks. Hence, in actual practice, a compromise must be found between the blank dimensions and the grain size. In addition, these methods often fail to provide the required chemical composition and homogeneity of the blanks.

Hot Deformation. The problem of producing a fine-grained microstructure in intermetallics can be solved by means of hot deformation. In this case, grains are refined by a development of dynamic recrystallization. This technique makes it possible to create fine grains in bulk blanks, both cast and powder compacted. However, irrespective of the original microstructure, the minimum grain size attained during hot deformation is about 0.1 μm.

In this section, TiAl, being one of the most hard to deform compounds, is used to illustrate an approach which applies hot deformation to creating fine grained (d = 1–10 μm) and even submicrocrystalline ($d \sim 0.1$ μm) structures in bulk intermetallic blanks.

Formation of a Fine grained Microstructure (d = 1–10 μm). The compacted state of TiAl was taken as the initial state. Its microstructure was described above and is shown in Fig. 9.4c. It should also be noted that the compact microstructure contains a considerable number of annealing twins predominantly observed in the coarse grained fraction. The twin boundaries constitute approximately 50% of the total number of boundaries in this fraction, while in the fine-grained fraction their content is no more than 10%.

The samples of the intermetallic compound were compression strained from 950 to 1150°C at an initial strain rate of $8.3 \times 10^{-4}\,s^{-1}$, to different strain levels. The 30–40% deformation below 950°C resulted in cracking. The tests were also performed at 1000°C at initial strain rates of 1.6×10^{-4} and $8.3 \times 10^{-4}\,s^{-1}$.

Recrystallized grains of the 80% strained samples are much finer than the original grains (Fig. 9.8), regardless of the temperature-rate condition used. Their average size and specific volume (measured in the sample center) increase with rising temperature and decreasing strain rate (Table 9.3). A temperature of 1000°C and a strain rate of $8.3 \times 10^{-4}\,s^{-1}$ yield the most favourable combination of the intermetallic compound microstructure parameters, viz., a mean grain size of 4.5 μm, with a fraction of nonrecrystallized areas of 13%.

Microstructural analysis indicates that, in TiAl recrystallized grains are formed not only in the original grain boundaries, but in the grain interior as well (Fig. 9.9a,b). The formation of new grains is associated with mechanical twinning

Fig. 9.8. Microstructure of the TiAl intermetallic compound after 80% strain at t = 1000°C and $\dot{\varepsilon} = 8.3 \times 10^{-4}\,s^{-1}$

Fig. 9.9a,b. Microstructure of the TiAl intermetallic compound after deformation at 1000°C and $\dot{\varepsilon} = 8.3 \times 10^{-4}\,s^{-1}$: **(a)** $\varepsilon = 15\%$; **(b)** $\varepsilon = 30\%$

which is in progress in the compound. The high-temperature twinning provides for the development of dynamic recrystallization. On the one hand, twins which divide grains into fragments speed up the accumulation of dislocations, while on the other hand, since their boundaries are high-angle boundaries, they facilitate the formation of new grains while being transformed into random boundaries in the course of deformation. The quantitative relationship between mechanical twinning and recrystallization is shown in Table 9.3, i.e., the larger the fraction of twinned grains (assessed at $\varepsilon = 5\%$), the greater the volume embraced by the dynamic recrystallization (V_{rec}). When the temperature of deformation rises or the strain rate decreases, both microstructural parameters increase.

Thus, the data obtained indicate strong development of dynamic recrystallization in TiAl under hot deformation, which under certain temperature-rate conditions allows it to create a fine-grained microstructure. It should be noted that during hot deformation of an ingot, the processes that take place in the intermetallic compound are the same as that during hot deformation of a compact.

Formation of Submicrocrystalline Structure ($d \sim 0.1\,\mu m$). The initial material was TiAl in a state with a homogeneous fine-grained structure with a mean grain size of 5 µm (Fig. 9.4b). This microstructure makes it possible to do compression straining from 700 to 950°C without cracking.

Table 9.3. Dependence of mean size d, recrystallized volume fraction V_{rec}, and twinned grains η on the temperature of deformation t (at $\dot{\varepsilon} = 8.3 \times 10^{-4}\,s^{-1}$) and strain rate (at $t = 1000°C$)

Microstructure parameters	Temperature [°C]					Strain rate [s^{-1}]		
	950	1000	1050	1100	1150	1.6×10^{-4}	8.3×10^{-4}	8.3×10^{-3}
d [µm]	1–2	4.5	10.5	12.5	13	6	4.5	4
v_{rec} [%]	80	87	90	93	96	90	87	80
η [%]	37	42	47	66	75	55	42	35

Table 9.4. Dependence of mean size of recrystallized grains d and specific recrystallized volume V_{rec} on the temperature of deformation t (at $\dot{\varepsilon} = 8.3 \times 10^{-4}\,s^{-1}$) of the TiAl intermetallic compound

Microstructure parameters	Temperature [°C]					
	700	750	800	850	900	950
	700	750	800	850	900	950
d [μm]	0.13	0.25	0.45	0.75	1.2	1.9
v_{rec} [%]	56	70	78	84	90	96

The analysis of the microstructure of samples subjected to 80% compression straining at 700–950°C indicates that, due to dynamic recrystallization, "new" grains are created in the intermetallic compound which are much smaller than the original grains (Figs. 9.4b and 9.4a). When the temperature rises, the average grain size and specific volume of the recrystallized grains increase (Table 9.4). By comparing Tables 9.3 and 9.4, one can see that, under the same conditions of deformation (t = 950°C), the specific volume of the recrystallized grains is noticeably larger where the microstructure was prerefined.

Thus, a general principle of the formation of submicrocrystalline structure in low-plasticity intermetallics can be formulated. The strain processing is divided into several stages. Because of the limited plasticity of intermetallics, each of the stages must be provided with schemes of deformation and temperature-rate conditions of deformation which ensure the development of deformation and dynamic recrystallization. Refining accomplished at the previous stage of treatment increases the intermetallic compound plasticity, thus enabling deformation at lower temperatures at the subsequent stage, which results in more refining.

Dynamic recrystallization was also observed in the magnesium alloy Mn-29.4%Al-0.6%C during plastic flow under the temperature conditions required for the existence of a metastable intermetallic τ-phase. When this alloy is compression strained at 700°C and $\dot{\varepsilon} = 10^{-3}\,s^{-1}$, the microstructural changes that occur are similar to those described above. As in TiAl, the high-temperature mechanical twinning in Mn-29.4%Al-0.6%C stimulates the development of dynamic recrystallization.

Thus, dynamic recrystallization developing in the considered intermetallics is an efficient and universal means of refining their microstructure and, having much in common with metals, possesses its own specific features associated with high-temperature mechanical twinning. The approach described above could most likely be applied for the refining of microstructures in other hard to deform intermetallic materials.

9.4 Effect of Superplastic Treatment on the Physico-mechanical Properties of Intermetallics

9.4.1 Mechanical Properties Under Ductile-Brittle Transition

The Role of Grain Boundaries. Four states of the TiAl intermetallic compound, distinguished by their mean grain sizes, were studied. The first three states have relatively uniform microstructures, because they are obtained by means of SP treatment at different temperatures; their mean grain sizes are 0.4; 1.2; and 2.4 μm, respectively. The fourth state has a diversity of grain sizes. The average grain size of the coarse-grained fraction ($V = 80\%$) is 30 μm; the sizes of fine grains vary from 3 to 8 μm.

The mechanical properties of the intermetallic compound are shown in Fig. 9.10. With rising temperature, the tensile strength of the submicrocrystalline state monotonically decreases, and the relative elongation increases. The temperature dependence of the tensile strength of the other states is abnormal, i.e., the tensile strength increases with the temperature to its maximum, but on a further rise of temperature, it diminishes. The growth of the tensile strength is accompa-

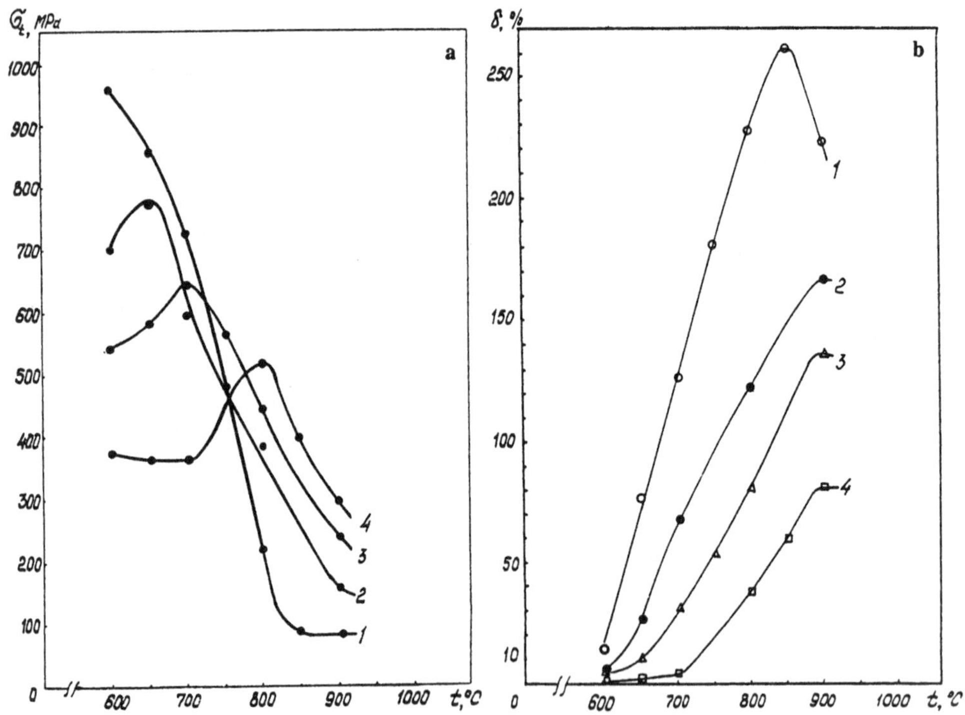

Fig. 9.10. Dependence of the tensile strength σ_t (a) and relative elongation to rupture δ (b) on the temperature of deformation t of the TiAl intermetallic compound with different grain sizes d ($\dot{\varepsilon} = 8.3 \times 10^{-4}\,s^{-1}$): Curve (*1*) 0.4; (2) 1.2; (*3*) 2.4; and (*4*) ~ 30 μm

nied by a growth of the relative elongation in the same temperature range, i.e., the ductile-brittle transition is observed.

The comparison of the mechanical properties indicates that when the grain size is reduced, the ductile-brittle transition interval shifts towards lower temperatures, this shift being most pronounced when the grain sizes vary from 0.4 to 2.4 μm.

Influence of Grain Boundary Structure. TiAl was selected because it allows us to compare its mechanical properties in states with similar grain sizes, but substantially different grain boundary structures.

The microstructures of the three states considered are shown in Figs. 9.4b, 11. State 1 was obtained by means of SP treatment, and has a homogeneous fine-grained microstructure with a mean grain size of 5 μm. The microstructures of states 2 and 3 are characterized by a diversity of grain sizes. In these states the fine- and coarse-grained volume fractions are 15 and 85%, respectively. In both states fine grains are primarily in the range of 2–6 μm. The average grain size of the coarse-grained fraction is 20 μm (twins included in the latter case: see below). On the whole, they can be characterized as states with similar grain sizes. No differences in the amount (about 3% of the volume), distribution, size (0.1–3 μm), and shape of the α_2-phase particles were observed.

Fig. 9.11a,b. Microstructure of the TiAl intermetallic compound: **(a)** state 2; **(b)** state 3

At the same time, the microstructures of these states contain special boundaries, i.e., annealing twin boundaries. They are most numerous in state 3, about 50% of all the boundaries, where they concentrate predominantly in coarse-grained areas; in states 1 and 2 the fraction of twin boundaries is no more than 5%. No α_2-phase particles were observed in the twin boundaries.

A comparison of the mechanical properties of TiAl in each of the considered states is interesting (Fig. 9.12). States 1 and 2 of the intermetallic compound indicate that they are similar in their properties (a slight difference appears only at test temperatures of 800 and 900°C), although they are remarkably different in

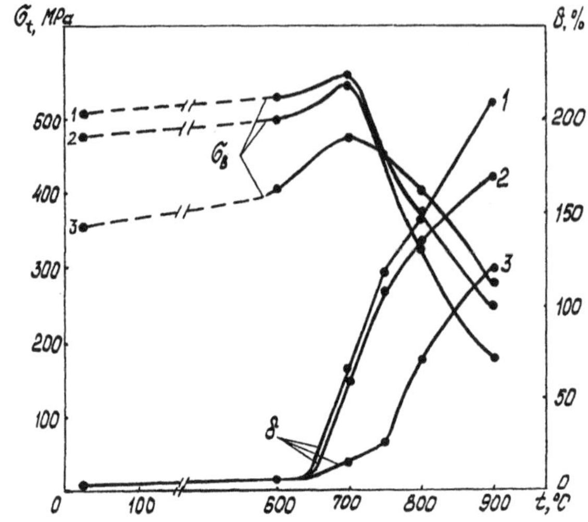

Fig. 9.12. Dependence of the tensile strength σ_t and relative elongation to rupture δ of the TiAl intermetallic compound in states 1–3

their grain sizes and microstructure homogeneity. At the same time, during the transition from state 1 to 2, the properties radically change. Taking into account the similarity of their grain sizes, this change is probably due to the difference in the fraction of twin boundaries. States 1 and 2 have higher values of tensile strength and relative elongation than state 3. Only when the temperature is above 750°C, is the tensile strength in state 3 higher than in states 1 and 2. As shown in Fig. 9.12, the ductile-brittle transition interval in state 3 shifts against that in states 1 and 2 to higher temperatures by about 100°C.

Thus, the growth of the twin boundary fraction in the intermetallic compound reduces its ductility at temperatures above 600°C. The intermetallic compound displays the highest plasticity after SP treatment, becuase the latter provides a homogeneous fine-grained microstructure and eliminates unfavourable grain boundaries.

Influence of the Nonuniformity of Grain Size. Two states of the TiAl intermetallic compound were compared, viz., a uniform fine-grained state with a mean grain size of 5 μm obtained after SP treatment, and a nonuniform (incompletely recrystallized) with fine- and coarse-grained volume fractions of 87 and 13%, respectively. The average size of the fine grains is also 5 μm. Coarse grains are located in patches and have an oblong shape, their longitudinal and transverse sizes are 30–50 and 15–30 μm, respectively.

Analysis of mechanical properties indicates that the presence of individual coarse grains in the intermetallic compound results in a substantial shift of the ductile-brittle transition interval towards higher temperatures. At the same time, SP treatment improves the microstructure homogeneity and leads to a shift at lower temperatures.

Hence, SP treatment is an efficient means of decreasing the cold shortness threshold of intermetallics, because it does not only eliminates the unfavourable

Table 9.5. Influence of grain size on the mechanical properties of the TiAl intermetallic compound at room temperature (at $\dot{\varepsilon} = 8.3 \times 10^{-3}\,s^{-1}$)

d [μm]	δ [%]	Tensile strength [MPa]	Treatment
0.4	0.2	960	SP deformation
1.2	0.5	620	SP deformation
2.4	1.1	510	SP deformation
5	2–3.2	500	SP deformation
20	2–3.1	475	conventional deformation
30	0.6	340	SP deformation + annealing

twin boundaries, but also contributes to the improvement of the microstructure homogeneity with a large area of random boundaries.

9.4.2 Mechanical Properties at Room Temperature

The Role of Grain Size. The effect of grain size and SP treatment on the TiAl plasticity at room temperature is not as clear as at elevated temperatures (Table 9.5). When the grain size decreases, the relative elongation to rupture first increases and reaches its maximum value of 3.2% for a grain size of 5 μm and then decreases. At the same time, the tensile strength increases noticeably with the reduction of grain size.

Influence of Grain Boundary Structure. As mentioned above, SP deformation eliminates twin boundaries, thus decreasing the cold shortness threshold. However, at room temperature this effect is not observed. Analysis of the mechanical properties in the states 1–3 discussed above indicates that at room temperature its plasticity is practically independent of the structure of grain boundaries (Table 9.6). However, the growing number of twin grain boundaries in the alloy results in a considerable decrease of the tensile strength.

Influence of Nonuniformity of Grain Sizes. A homogeneous fine-grained state obtained by means of SP deformation and a state with different grain sizes (both

Table 9.6. Influence of grain boundary structure on the mechanical properties of the TiAl intermetallic compound at room temperature (at $\dot{\varepsilon} = 8.3 \times 10^{-4}\,s^{-1}$)

State	d [μm]	Fraction of twin boundaries [%]	δ [%]	Tensile strength [MPa]	Treatment
1	5	5	2–3.2	500	SP deformation
2	20	5	2–3.1	475	Conventional deformation
3	20	50	2–3.1	360	SP deformation + annealing

Table 9.7. Influence of the nonuniformity of grain sizes on the mechanical properties of the TiAl intermetallic compound at room temperature (at $\dot{\varepsilon} = 8.3 \times 10^{-4}\,s^{-1}$)

State	δ [%]	Tensile strength [MPa]	Treatment
Uniform	2.0–3.2	500	SP deformation
Nonuniform	0.2	480	Conventional deformation

Table 9.8. Mechanical properties of the Mn-29.4%Al-0.6%C alloy subjected to compression straining (at $\dot{\varepsilon} = 1.6 \times 10^{-3}\,s^{-1}$)

Treatment	Mechanical parameters	
	Compression strength [MPa]	Relative elongation to rupture [%]
Conventional deformation	1450	0.2
SP deformation	1800	3.0

described above) were studied. Comparison of the mechanical properties of the intermetallic compound in these states at room temperature indicated that the nonuniformity of grain sizes makes it brittle (Table 9.7).

Another interesting example of the improvement of the mechanical properties at room temperature after SP treatment is seen in the magnetic alloy Mn-29.4%Al-0.6%C. Two states were examined, viz., after conventional deformation resulting in the refining of the original coarse-grained microstructure and after SP deformation. Table 9.8 presents the results obtained during compression straining at room temperature. During the transition from the state after conventional treatment to the state after SP treatment, the compression strength and relative elongation to rupture substantially increase.

Microstructural and X-ray structural analyses indicate that, in conventionally strained samples, a partial decay of the metastable intermetallic τ-phase into equilibrium η- and β-phases takes place, the η-phase having coarse precipitates in the direction of straining. In the microstructure, the fine-grained fraction coexists with individual coarse grains with mechanical twins. After SP deformation the microstructure is homogeneous, the average grain size is 1 μm, and there are no signs of decay of the τ-phase into η- and β-phases. Hence, SP treatment allows the formation of a homogeneous fine-grained microstructure of the τ-phase and delays its decay into equilibrium phases, which, in turn, results in the improvement of the mechanical properties at room temperature.

The improvement of mechanical properties after SP treatment is also observed in another magnetic alloy containing an intermetallic phase, i.e., the X30K23 alloy. Figure 9.13 shows the dependence of the fracture stress (σ_f) during the elongation of an alloy with a microduplex ($\gamma + \sigma$) structure at room temperature

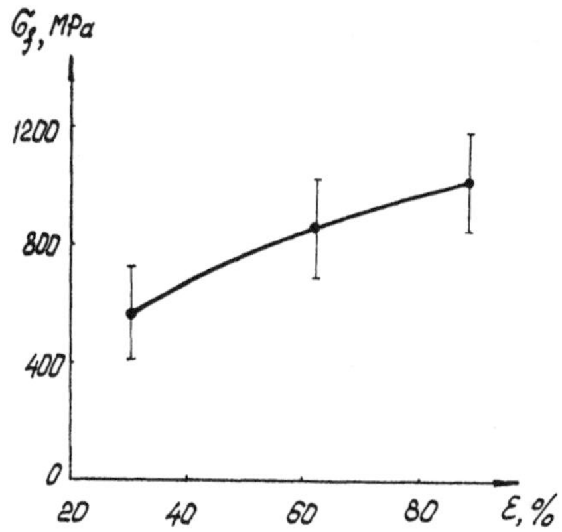

Fig. 9.13. Dependence of the fracture stress σ_f during the alloy tensile straining at 20°C on the level of SP prestraining

on the level of SP prestraining. This dependence shows that an alloy with such a microstructure is inclined to brittle fracture. This figure also shows that the increase of the SP strain level from 30 to 90% results in the growth of the alloy tensile strength from 600 to 1100 MPa. Thus, the SP treatment of the alloy in the two-phase ($\gamma + \beta$) state increases its fracture stress.

The X30K23 alloy strength increases not only immediately after SP deformation, but also after its subsequent processing to a high coercivity state, where tensile tests at room temperature indicate that the tensile strength increases from 270 (as-cast) to 580 MPa. According to the metallographic analysis, the size of the α-phase grains after SP deformation and subsequent processing to a high coercivity state decreases from 900 (as-cast) to 180 μm.

Results discussed in this section prove that the application of SP treatment can improve the mechanical properties of intermetallics at room temperature.

9.4.3 Magnetic Properties

As shown above, SP treatment substantially improves the mechanical parameters of magnetic alloys Mn-29.4%Al-0.6%C and X30K23 containing an intermetallic phase. However, SP treatment is also an efficient means of improvement of the magnetic properties of these materials.

Table 9.9 presents the magnetic properties of the alloy Mn-29.4%Al-0.6%C after conventional and SP deformation. Transition from conventional to SP treatment increases the stability of the τ-phase and results in the enhancement of the saturation magnetization (B_s), residual magnetization (B_r), and, consequently, maximum magnetic energy $(BH)_{max}$. The coercivity also increases both in magnetization ${}_iHc$ and induction ${}_bHc$. The latter is caused by a change of the morphology from plastic to globular fine-grained.

The analysis of the X30K23 alloy indicates that SP pretreatment induces a decrease in the azimuthal heterogeneity of magnetic flow as compared to the as-cast state (Table 9.10), without deteriorating other magnetic parameters.

Table 9.9. Influence of method of treatment on the magnetic properties of the Mn-29.4%Al-0.6%C alloy

Magnetic parameters	Conventional deformation	Superplastic deformation
$_bHc$ [kA/m]	128	200
$_iHc$ [kA/m]	184	280
B_r [T]	0.35	0.50
B_s [T]	0.52	0.69
$(BH)_{max}$ [kJ/m^3]	17.6	37.6

Table 9.10. Magnetic properties of the X30K23 as-cast alloy after superplasticity treatment

Magnetic properties	As-cast	SP treatment
Residual magnetization B_r [T]	1.03	1.03
Coercivity $_bHc$ [kA/m]	55	57
Magnetic energy BH_{max} [kJ/m^3]	30	31
Azimuthal heterogeneity of magnetic flow [%]	7	3

Hence, SP treatment ensures an improvement of both mechanical and magnetic properties of materials which makes it rather promising for manufacturing intricately shaped magnets, because in this case, casting worsens the magnetic properties and performance because of casting defects and other structural heterogeneities. The development of the technologies for producing magnets by means of SP deformation allows not only the enhancements of their magnetic properties, but also an improvement of their mechanical properties.

10. Superplasticity of Ceramics

Ceramics possess a number of unique properties, such as high-temperature strength, wear resistance and hardness, enhanced resistance to corrosion, and other species-specific physical and mechanical properties. These are determined by their electronic, atomic, micro- and macrostructure. In the electronic structure two types of bonds dominate – covalent and ionic. Such compounds can be semiconductors and dielectrics, have enhanced hardness and high elasticity moduli, stable mechanical properties in a broad range of temperatures, and low coefficient of thermal expansion. However, it is the nature of the bonding that also determines the principal limitation of ceramics, i.e., their absolute brittleness. When a load is applied, the bonding force hinders the motion of dislocations because of the high value of the potential Peierls-Nabarro barriers. Hence, under conventional conditions, SP deformation in ceramics is practically impossible. The brittleness of ceramics is caused by microscopic defects such as voids and inclusions that, being strong concentrators of stresses, become sources of cracking. Since the plasticity of ceramics is zero, no relaxation of stresses occurs in such places and under a load exceeding a certain threshold value the material immediately fails. Although ceramic materials possess many useful properties, the lack of plasticity considerably limits their application as construction materials.

Serious technological problems arise when products are manufactured from ceramics. A high melting point ($\sim$ 2000–3000°C) and zero ductility during deformation prevent the use of casting and forming. The method of powder metallurgy is a conventional and virtually the only way of manufacturing ceramic products, which puts some limitations on the application of ceramic materials.

Ductile ceramics would overcome these difficulties and substantially improve the reliability of ceramic parts by preventing the disastrous failures which develop. A relaxation of stresses would also help the construction strength of ceramics.

Finding ways to improve the ductility is a major scientific problem. There are data indicating that the ductility of ceramics can be strongly increased by converting them to an SP state [10.1–5]. However, to date no systematic studies in this regard have been carried out. However, there are a number of questions of both theoretical and practical interest:

1) What is the dependence of the SP of ceramics on their microstructure and what are the phenomenological peculiarities of ceramics compared to other materials?

2) What are the mechanisms of the SP deformation of ceramics, taking into account the variety of structures in these materials?
3) What is the effect of SP deformation on the construction and physico-chemical properties of ceramics?

Answers to these questions would allow us to control the structure and properties of ceramics and would probably allow new application of ceramics as construction materials.

Superplasticity in ceramics can be attained by means of temperature cycling near the temperature of phase transition under load and via deformation at a constant temperature when a microcrystalline and submicrocrystalline microstructure are available.

Superplasticity displayed during the phase transformation in ceramics is described in detail in [10.6, 7]. We shall discuss the structural SP in ceramics associated with the microcrystalline microstructure. It is important to find out how the SP in ceramics is affected by grain size, chemical and phase composition, phase morphology and stress diagrams.

10.1 Influence of Structure and Strain Conditions on the Superplasticity of Ceramics

Various structural states can be obtained in ceramics depending on the type of treatment. The structure can undergo qualitative changes with minimal variations of the composition, crystals, orientation, temperature conditions, etc. We now give an analysis of the influence of the structure parameters and strain conditions on the high-temperature ductility of ceramics and the SP parameters of a broad class of ceramics (Table 10.1).

Grain Size. As mentioned above, structural SP in metals and alloys requires a stable microcrystalline structure. Regarding ceramics, more substantial refining is required to ensure SP. For example, in [10.8], the authors used submicrocrystalline MgO with an initial grain size of 0.1 µm. Compression tests at a constant strain rate indicated that in a range of 950–1150°C large plastic strains can be obtained without failures. At a temperature of 1050°C and at very small strain rates of $6.7 \times 10^{-6} - 10^{-5}\,s^{-1}$, true deformation reached 80%. In that case, grains grew to 1 µm though retaining their original equiaxial shape. At this temperature the strain-stress curves display no noticeable strengthening, and the strain rate sensitivity coefficient m is high, 0.77–0.83. Hence, in MgO the characteristic features of SP flow are the same as in metallic materials.

Superplasticity is retained when the grain size coarsens within certain limits. In [10.9], a dense polycrystalline ceramic material based on Al_2O_3 with a mean grain size of 1.6 µm was deformed under a continuous compressing load to a true strain level of 60% at 1420°C. At a strain rate of $2 \times 10^{-4}\,s^{-1}$ the flow stress was only 30 MPa. By studying the microstructure, the authors detected a change in

Table 10.1. Superplastic ceramic materials and parameters of SP

Material	Load scheme	Grain size [μm]	Temperature [°C]	m	$\dot{\varepsilon}$ [s^{-1}]	ε (δ_r) [%]	Ref.
			Oxide ceramics				
Y-TZP	compression	0.3–0.4	1450	~0.5	8.3×10^{-5}	80%	[10.10]
		0.7	1450	~0.5	8.3×10^{-5}	75%	[10.10]
		1.5	1450	~0.5	8.3×10^{-5}	60%	[10.10]
		2.3	1450	~0.5	8.3×10^{-5}	45%	[10.10]
	tensile	0.3–0.4	1450	~0.5	2.8×10^{-4}	160%	[10.28]
		0.3–0.4	1450	0.4–0.5	4.8×10^{-5}	246%	[10.32]
		0.3	1550	~0.3	$\sim 10^{-4}$	350%	[10.17]
Al_2O_3 (with MgO additives)	compression	0.75	1450	~0.75	$\sim 10^{-6}$	40%	[10.6]
		1.0	1450	0.5–1	$\sim 10^{-4}$	40%	[10.6]
		1.6	1420	–	$\sim 10^{-4}$	45%	[10.9]
Y-TZP 20%Al_2O_3	compression	0.9	1500	0.5–0.8	$\sim 10^{-4}$	85%	[10.6]
	tensile	0.5	1450	~0.5	$\sim 10^{-4}$	200%	[10.16]
		0.5	1650	~0.5	$\sim 10^{-3}$	500%	[10.17]
MgO	compression	0.1	1050	~0.5	10^{-5}–10^{-3}	80%	[10.8]
UO_2	compression	2	1350	~0.1	$\sim 10^{-5}$	70%	[10.6]
$BaTiO_3$	compression	0.45	1150	~0.5	$\sim 10^{-4}$	40%	[10.6]
β-spodumene glass ceramics	tensile	~1	1150	~1	$\sim 10^{-4}$	135%	[10.23]
		~1	1020	~1	$\sim 10^{-4}$	400%	[10.23]
$MgO{\cdot}2Al_2O_3$	compression	3.0	1050	~0.47	3×10^{-4}	60%	[10.13]
$YBa_2Cu_3O_{7-x}$	compression	lamellar microstructure, $d_{\parallel} = 25\,\mu m$ $d_{\perp} = 5\,\mu m$	930	–	2×10^{-4}	80%	[10.14]
Bi_2O_3	compression	10	700	–	10^{-4}	80%	authors[a] data
			Oxygenless ceramics				
α-SiC	compression	1.5–2.8	1900	~1	10^{-4}	40%	[10.6]
β-SiC	compression	0.3–1.4	1600	~0.5	10^{-6}–10^{-7}	20%	[10.29]
$TiC_{0.5}$/Ti	compression	10	950	~0.3	10^{-3}–10^{-4}	80%	[10.38]
$MOSi_2/MO_2B_5$	bending	1.6–2.4	1400	–	10^{-3}–10^{-4}	60–80%	[10.25]
Fe_3C	tensile	4	1025	0.5–0.7	10^{-4}	610%	[10.16]

[a]O.A. Kaibyshev, N.G. Zaripov; unpublished work.

the original grains and their rapid growth. The grain size was 3.05 μm at the end of deformation, while during annealing under the same conditions it was only 1.8 μm. Hence, in this case, too, grain growth is induced by SP deformation.

A more systematic investigation of the influence of the grain size on SP parameters was undertaken by *Wakai* [10.10, 11] on a classical SP ceramic compound ZrO_2 stabilized with 3 mol.% Y_2O_3 (Y-TZP-ceramics). This ceramic material manifests remarkable ductility only at a grain size of less than 3 μm (Fig. 10.1), the ductility being strongly dependent on the grain size. If at a grain

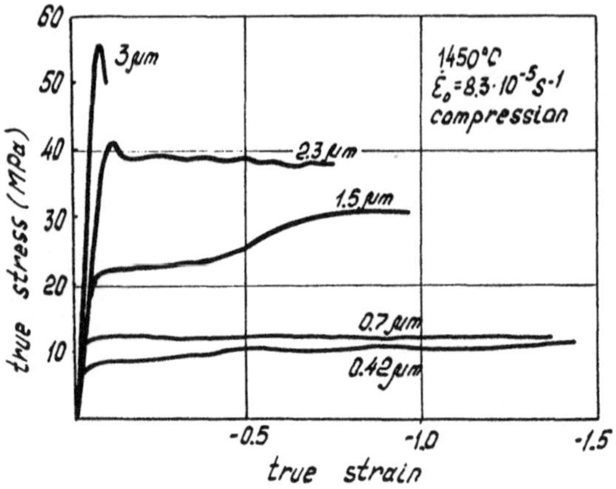

Fig. 10.1. Strain dependence of flow stress in Y-TZP ceramics with different grain sizes [10.10]

size of 3.0 μm the true deformation is only about 10%, at a grain size of 0.42 μm its value reaches 160%. When the grain size is less than 1 μm, the deformation proceeds without any noticeable strengthening and the flow stress is practically independent of the strain level. SP deformation takes place at a very small flow stress, less than 20 MPa, while in the monocrystals of the same ceramic material under the same conditions the flow stress is larger than 150 MPa [10.12].

In some cases, coarser grains can also provide for SP flow in ceramics. This is illustrated by the deformation of a spinel MgO-$2Al_2O_3$ with different initial grain sizes [10.13]. At 1680°C and a strain rate of $7 \times 10^{-5}\,s^{-1}$, a spinel with the originally microcrystalline microstructure is deformed up to 60% without failure. Contrary to the submicrocrystalline Y-TZP ceramics, the flow stress of the spinel decreases when the grains grow (Fig. 10.2). This proves that the strain conditions do not match the optimal modes for SP deformation. The ductility also decreases in this case. When the grain size is 34 μm, the material fails after 6% strain. When fine-grained samples are deformed ($d = 4.6$; 7.0; and 8.6 μm), the flow stress becomes independent of the initial temperature after 25% strain. This can be explained by the dynamic recrystallization which occurs in ceramics and by the resulting formaiton of a microstructure with grains of similar sizes.

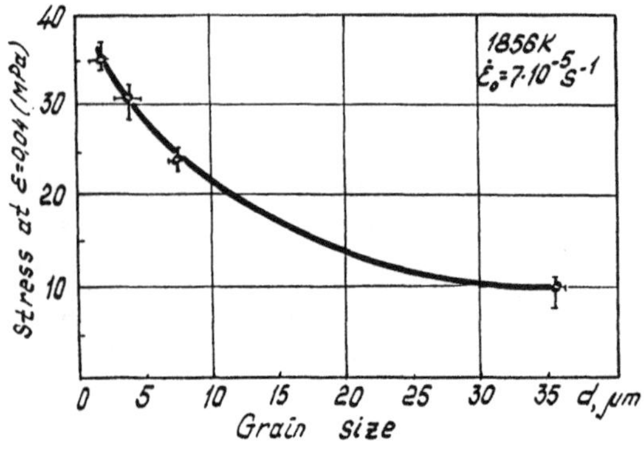

Fig. 10.2. Dependence of flow stress on grain size in spinel $MgO \cdot 2Al_2O_3$ [10.13]

The plastic deformation of coarse-grained ceramics is of practical importance because the technology of preparing starting materials is substantially simplified. Commercially produced ceramic powders are usually comprised of particles of 5–7 μm. At the final stage of manufacturing their mean grain size reaches 10 μm. Such ceramic materials can also display enhanced deformability under certain temperature-rate conditions.

We considered the influence of temperature and strain rate on the ductility of coarse-grained bismuth oxide and superplastic $YBa_2Cu_3O_{7-x}$ ceramic compound. Samples of bismuth oxide were produced by hot pressing of powders at 750°C under a pressure of 200 MPa. After this treatment, the grain size was 10 μm. At 550°C, Bi_2O_3 displays slight ductility which increases with the rising temperature and already at 750°C the deformability becomes practically unlimited. It is unusual that while at temperatures below 700°C the steady flow stage is preceded by an abrupt softening, at 750% this stage is attained after a small deformation without any softening. The microstructure of the strained samples substantially changes in this case. A similar behaviour is typical of the hot strained $YBa_2Cu_3O_{7-x}$ ceramic compound with an originally coarse-grained lamellar microstructure (Fig. 10.3a) [10.14]. Below 650°C, this material fails in the elastic region, while above 650°C, it displays remarkable ductility which also grows with a rising temperature of deformation. Figure 10.4 shows typical $\sigma - \varepsilon$ curves obtained at different test temperatures. At 700 and 800°C, even at the

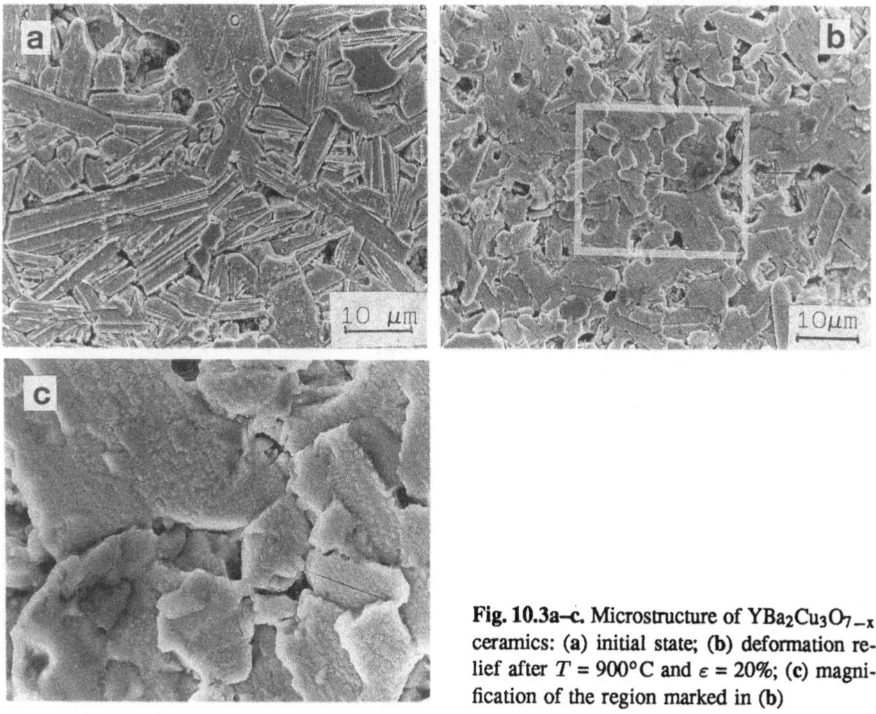

Fig. 10.3a–c. Microstructure of $YBa_2Cu_3O_{7-x}$ ceramics: (**a**) initial state; (**b**) deformation relief after T = 900°C and ε = 20%; (**c**) magnification of the region marked in (**b**)

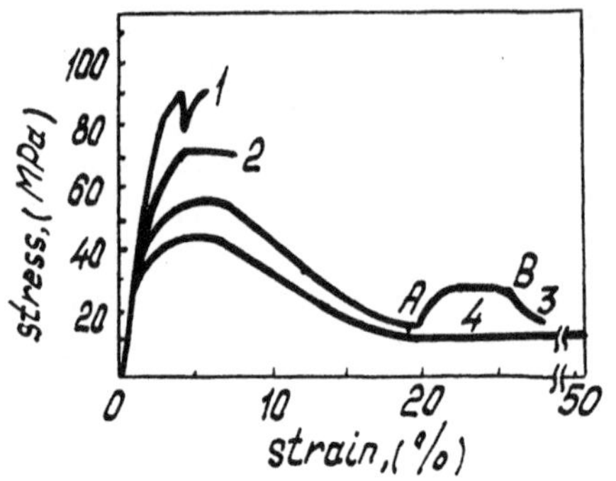

Fig. 10.4. Stress-strain curves of the ceramic compound $YBa_2Cu_3O_{7-x}$ at $\dot{\varepsilon} = 10^{-4}\,s^{-1}$. Curve (*1*) 700°C; (2) 800°C; (*3*) 900°C; and (*4*) 930°C. In the interval AB: $\dot{\varepsilon} = 2 \times 10^{-4}\,s^{-1}$

initial stage, it is possible to observe discrete drops of flow stress associated with cracking on the lateral surface of the samples (curves 1 and 2). Starting with 900°C and higher, the $\sigma - \varepsilon$ curves become monotonic. At these temperatures the steady flow stage is preceded by a flow stress peak (curves 3, 4). During deformation at 900, 930, and 950°C, cracks appear on the lateral surfaces, oriented at an angle of 45°C to the compression axis, but they appear at relatively high strain levels ($\varepsilon = 25$–30%). Under these conditions, the samples still retain their ability for plastic deformation. When the strain rate is changed from 10^{-4} to $2 \times 10^{-4}\,s^{-1}$, a high strain rate sensitivity of flow stress is displayed ($m > 0.3$).

These results indicate that even in coarse-grained ceramics SP features can be observed under certain temperature-rate conditions, i.e., large strains before failure, low flow stress, and enhanced strain rate sensitivity.

The above examples of the high-temperature deformation of ceramics indicate that grain size is an important parameter ensuring SP flow and that this is the factor that basically determines the temperature modes of deformation. In Fig. 10.5, one notes that the finer the grain size in ceramics, the lower the temperatures of transitions to SP flow, this dependence being the most explicit in the region of submicron grains.

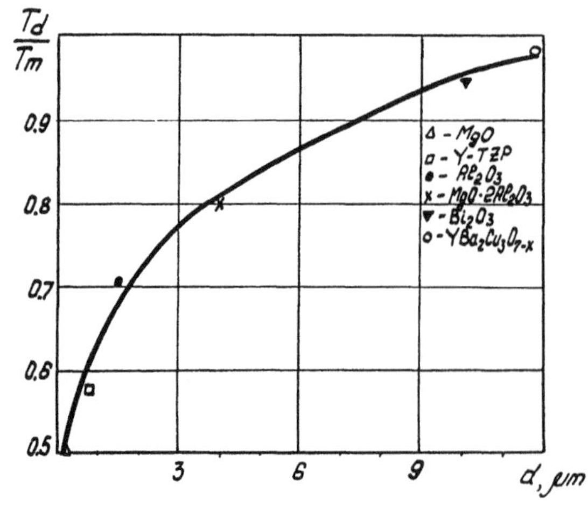

Fig. 10.5. Dependence of the homological temperature of deformation on the grain size of ceramics

Thus, the experimental data indicate that grain size strongly affects the plasticity and SP of ceramics. At the same time, a much greater refinement of the microstructure is required in ceramics in order to attain parameters of SP flow typical of metals.

Phase Composition and Phase Morphology. In many cases structural ceramics include various additives, such as strengthening particles and components for enhancing sintering, viscosity, or other specific physico-chemical and mechanical properties, and is virtually a composite ceramic material. In this case, the structure formed is usually multiphase. Contrary to metals, phases in ceramics can substantially differ from one another in their melting points, deformability and physico-chemical properties. For example, in metals the ratio of the melting points of phases is usually 1–1.2; in ceramics it is 2–3. In addition, in ceramic composites, phases may be arranged differently with reference to one another. The phase morphology can be similar to that of metals (matrix structure with dispersely distributed particles, microduplex structures), but there may be microstructures typical only of ceramics, characterized by an amorphous glass-phase or metallic layer in the grain boundaries. Such a variety of microstructures results in certain peculiarities in the way SP is displayed.

We now present some experimental data illustrating the influence of phase composition and morphology on the SP parameters of ceramic material. In Chap. 1 it was shown that in metallic alloys grains of the second phase usually raise the grain growth during deformation at high temperatures, thus enhancing ductility. A similar analysis is required in order to define the resource of ductility in ceramics as well.

In the ZrO_2(Y-TZP)-Al_2O_3 composite ceramics, the ZrO_2 phase is an SP material (the structure is stabilized with Y_2O_3), but Al_2O_3 in its pure form displays no SP properties because of a rapid grain growth. Both phases are stable below 1880°C and no other phases arise during their high temperature deformation. The ratio of the melting points of the ZrO_2 and Al_2O_3 phases is ~ 1.3, which testifies tot he similarity of the homological temperatures of deformation. This system can be used in obtaining matrix structures based on either the ZrO_2 or Al_2O_3 phases.

If these materials have a two-phase structure, they can display SP properties. For example, when the composition is 80% ZrO_2–20% Al_2O_3, the temperature is 1550°C, and strain rate is $2.8 \times 10^{-4}\,s^{-1}$, the elongation to rupture will be as large as 120% and the coefficient m will be about 0.5 [10.15, 16]. Later this material displayed an elongation of 500% at 1650°C and $\dot{\varepsilon} = 8.3 \times 10^{-4}\,s^{-1}$ [10.17].

The high ductility of this ceramic is due to the stability of its microstructure. In the process of deformation at 1550°C for 18 hours, the grain size increases from 0.5 μm to only 0.9 μm, which is much less than the grain growth in Y-TZP ceramics [10.17].

The stabilization of the Al_2O_3 microstructure with ZrO_2 also improves ductility. When the composition is Al_2O_3-14-22%ZrO_2 (Y-TZP) and the grain size is less than 1 μm, the relative elongation is more than 100%. Even if the content of the ZrO_2 phase is lower (7–14 wt.%), the ductility reaches 50–80% [10.18]. Thus, when there is a second dispersed phase, the ceramic microstructure is stabilized and, as in metals, the parameters of SP flow are improved.

A unique feature of ceramics is the distribution of the second phase in the grain boundaries in the shape of ductile or amorphous layers. This type of microstructure is never found in metals and intermetallics. A study of high-temperature behaviour of such materials, using a TiC-Ti system was undertaken.

A structure with a ductile phase along grain boundaries can be easily created during the self-propagating high-temperature synthesis (SHS) of interstitial phases with a homogeneous area [10.19]. In addition, by controlling the chemical composition of the phases, different amounts of intergranular phase can be obtained in the structure.

In the samples of nonstoichiometric titanium carbide prepared by the SHS method with subsequent force compaction, the C/Ti ratio was 0.47, which corresponds to the two-phase region of the phase equilibrium diagram. The average size of carbide grains was 10–12 μm. The samples were compression strained and some of them were tensile strained.

The microstructure of the original samples is shown in Fig. 10.6. The samples have a two-phase structure, the carbide grains are surrounded with a small number of excess titanium precipitates. Annealing, corresponding to the temperature and duration of deformation, causes hardly any substantial changes in the structure and the phase and chemical compositions are practically invariable.

We now consider the high-temperature behaviour of the material in the lower homogeneity area. In this area the phases are considerably different in their parameters. The melting point of titanium carbide is ~ 1.8 times higher than the melting point of titanium. In spite of a relatively high ductility of the titanium phase, during compression straining at temperatures below 700°C the carbide ceramic behaves like a typical ceramic material with inherent brittleness. At 700°C and at a strain rate of $10^{-4}\,s^{-1}$, the ductile-brittle transition takes place.

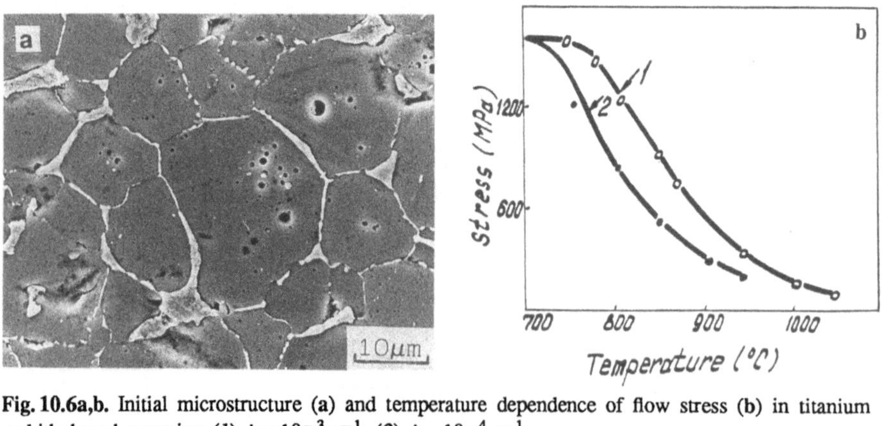

Fig. 10.6a,b. Initial microstructure (a) and temperature dependence of flow stress (b) in titanium carbide-based ceramics; (*1*) $\dot{\varepsilon} = 10^{-3}\,s^{-1}$, (2) $\dot{\varepsilon} = 10^{-4}\,s^{-1}$

As the test temperature increases, ductility grows, with the flow stress monotonically decreasing (Fig. 10.6). Hence, for the given composition of the material at $\dot{\varepsilon} = 10^{-4}\,s^{-1}$, the temperature of ductile-brittle transition is 700°C. When the strain rate increases to $10^{-3}\,s^{-1}$, the ductile-brittle transition temperature rises to 750°C. At 900–1000°C deformation proceeds at a constant value of flow stress, and the samples are deformed up to 80–90% without failure.

During tensile straining at $\dot{\varepsilon} = 10^{-3}\,s^{-1}$, the decrease of flow stress begins at 750–800°C; however, the ductility is pronounced only at temperatures above 800°C (Fig. 10.7). The maximum elongation to rupture depends only slightly on temperature, and in a range from 850 to 1000°C it is only 7–12%. When the strain rate decreases to $10^{-4}\,s^{-1}$, the flow stress rapidly decreases with rising temperature and ductility increases strongly. Thus, if at 900°C, the relative elongation to rupture is 12%, at 1000°C it increases to 40%.

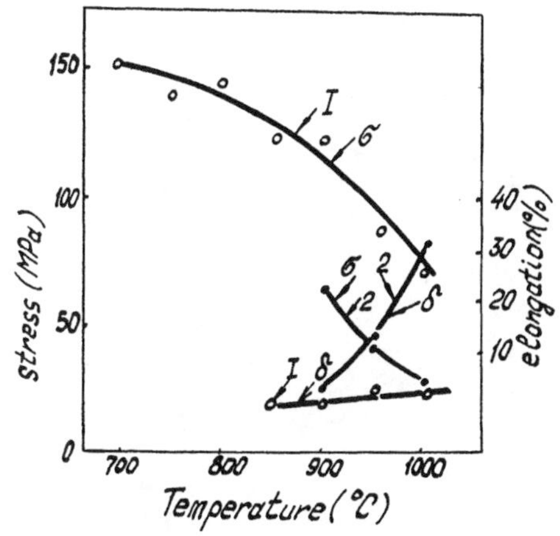

Fig. 10.7. Temperature dependence of flow stress in titanium carbide-based ceramics during tensile straining; (*1*) $\dot{\varepsilon} = 10^{-3}\,s^{-1}$, (2) $\dot{\varepsilon} = 10^{-4}\,s^{-1}$

The strain rate dependence of the flow stress of the $TiC_{0.47}$ samples under 0.2, 25, and 50% strains and at 950°C is shown in Fig. 10.8. In the strain rate interval from 10^{-3} to $10^{-1}\,s^{-1}$, a linear strain rate dependence of flow stress is observed and the strain rate sensitivity coefficient m is constant under these conditions and is equal to 0.2–0.25. As the strain level increases, the value of m slightly decreases in this interval. As the strain rate decreases to $10^{-4}\,s^{-1}$, m increases to 0.32, and at higher strain levels its value rises to 0.4.

Hence, the mechanical tests indicate that under certain temperature-rate conditions of tensile and compression straining, the $TiC_{0.47}$ based ceramic compound has a comparatively high ductility, although much lower than that of metallic materials.

To find the reasons for such behaviour, we now consider microstructural changes occurring during deformation. Substantial changes in the microstructure are observed during compression straining at 950°C in the enhanced strain rate sensitivity area at low strain levels (Fig. 10.9a). Individual grains are deformed

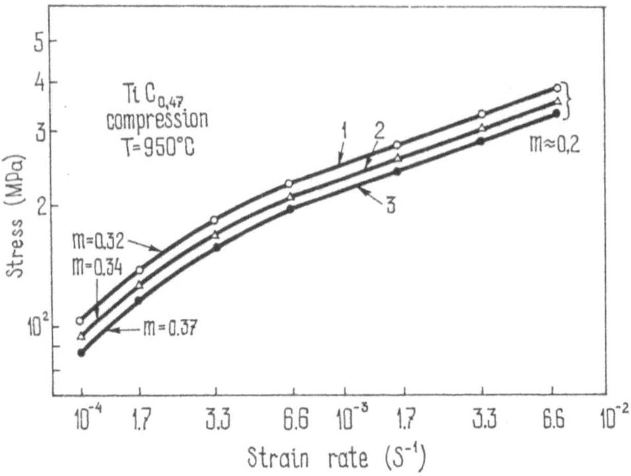

Fig. 10.8. Strain rate dependence of flow stress on temperature in titanium carbide-based ceramic; (*1*) $\varepsilon = 0,2\%$, (*2*) $\varepsilon = 25\%$, (*3*) $\varepsilon = 50\%$

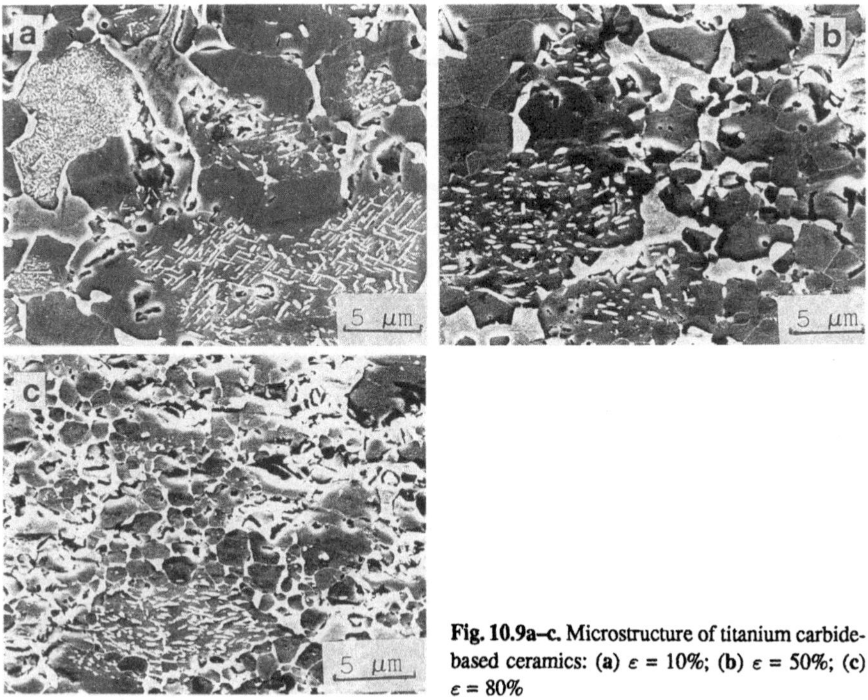

Fig. 10.9a–c. Microstructure of titanium carbide-based ceramics: (**a**) ε = 10%; (**b**) ε = 50%; (**c**) ε = 80%

yielding precipitates of titanium phase with distinct crystallographic orientations within the grain. The fact that these precipitates are not found during annealing, but appear during deformation, testifies to the deformational origin of these precipitates and to their crystallographic nature.

From Fig. 10.9b one can see that the higher the strain level, the larger the volume of fine grains separated with the layers of the titanium phase. Individual fine carbide grains are "introduced" in the titanium phase in triple junctions available in the original structure, thus creating grains of titanium carbide uniformly distributed in the microstructure. In coarse grains most of the precipitates of the titanium phase have the same orientation.

At high strain levels ($\varepsilon = 80\%$) the homogeneity of the microstructure increases, retaining however, individual coarse grains and parts of grains with the metallic phase precipitates (Fig. 10.9c).

Examination of deformation relief during high-temperature deformation at strain rates ranging from 10^{-2} to $10^{-4}\,s^{-1}$ indicated that the character of deformation is different at different strain rates. At high strain rates ($10^{-2}\,s^{-1}$) failure develops in the metallic phase located in grain boundaries. The deformation of the individual carbide phase grains is practically not observed. In decreasing the strain rate up to $10^{-3}\,s^{-1}$, the slip lines with different orientations in different grains appear in carbide grains. The grain structure is manifested distinctly, which shows the mutual displacement of grains due to grain boundary sliding. As the strain rate continues to decrease, deformation is accomplished basically due to grain boundary sliding. An intragranular titanium phase is "squeezed out" in the grain boundaries and there are individual slip lines springing predominantly from the triple junctions.

Similar changes in the microstructure are also observed during tensile straining, but low strain levels ($\varepsilon < 40\%$) make it possible to follow all the stages of the microstructure transformation in detail.

The results of mechanical tests and analysis of the as-strained sample microstructure lead to the following conclusions.

1) Below the temperature of ductile-brittle transition, the intragranular metallic phase induces no noticeable ductility.
2) The temperature of ductile-brittle transition correlates with the start of deformation in the grains of the high melting point carbide phase, the deformation of grains stimulating the intragranular structural changes.
3) The plastic deformation of the intergranular phase occurs only during the simultaneous deformation of both phases. When the strain rate sensitivity becomes too high due to the refining of carbide grains, the deformation is accomplished predominantly via interphase boundaries. Hence, the presence of the intergranular phase provides for the development of grain boundary sliding in the course of SP flow.

Grain Boundary Structure. Since the grain size in ceramic materials is smaller than in metals, the grain boundary area in ceramics is larger. In this case, the grain boundary properties must play an important role in deformation. But during

SP deformation this role is difficult to define because of the shortage of experimental data on the dependence of parameters of ceramic materials on the type of grain boundaries. According to *Raj*'s classification [10.20], grain boundaries in ceramics can be divided into three types. The first type includes high-angle grain boundaries where the exchange of atoms between the grains is the easiest, and deformation is accompanied by a growth of grains. The second type is associated with the presence of a liquid phase in the grain boundaries during deformation, which substantially enhances the diffusion properties of the boundary and affects the SP parameters. The third type includes subgrain boundaries in ceramics.

An example of the SP deformation of materials with high-angle grain boundaries is the submicrocrystalline Al_2O_3 with small additions of MgO [9]. The authors believe that the strain rate of such materials is determined by the Coble creep, i.e., the diffusion mass transfer is the dominating process accomplished in grain boundaries. However, later it was shown that, during deformation, an important role is played by grain boundary dislocations that provide for the sliding of grains [10.21]. The development of grain boundary sliding during the deformation of ceramics with high-angle grain boundaries is confirmed in a number of other papers [10.6, 22]. In general, in the presence of random type boundaries, the behaviour of ceramics under SP conditions is identical to that of metallic materials under SP, with no significant differences in the phenomenological and structural parameters observed.

The presence of the liquid phase in the grain boundaries assigns certain distinguishing features to the deformation and structural changes occurring in ceramics. In this case, at high temperatures, the amorphous phase softens and actually becomes a quasi-liquid intergranular phase.

The superplastic deformation of such ceramics was most completely studied by *Wang* and *Raj* [10.23] on the β-spodumene glass with a grain size of about 1 μm. When the content of SiO_2 in the material was 65 wt.%, the strain level of this ceramic compound at 1025°C was 400% and the content of the glass phase was 70 wt.%. A temperature of 1150°C resulted in a drop of ductility to 135%. Contrary to the materials discussed above, no grain growth is observed in glass ceramics, and the strain rate sensitivity coefficient has a very high value ($m = 1$), which corresponds to Newton's viscous flow. The same value of m was obtained in [10.24] in the Y-TZP ceramics when the amorphous phase was present in triple junctions because of impurities contained in the original powders. The growth of the deformability in the presence of the liquid phase in grain boundaries was also observed in the ceramic compounds, $MoSi_2Mo_2B_5$, ZrO_2-$3Al_2O_3 \cdot 2SiO_2$ [10.25, 26]. The authors of these papers, proceeding from the phenomenological parameters, relate the deformation of ceramics with the liquid intergrain phase either to dissolution-deposition or to diffusion creep. However, the development of "pure" grain boundary sliding seems to be the most probable in this case (Chap. 2).

Superplasticity can also be displayed in ceramics with a subgrain structure. According to *Raj* [10.20], the deformation is accomplished via the movement of dislocations from one subboundary to another through the crystalline lattice.

At the steady stage of deformation the rate of the emission of dislocations must be equal to the rate of their sinking. However, this model requires structural confirmation, although it correlates with experimental data obtained on ZrO_2 and on spinel [10.13, 16].

Grain boudaries are important elements of the ceramic structure. Therefore, one of the main tasks is further systematic investigation into the structure and properties of grain boundaries and in defining their role in the SP deformation of ceramics.

The Role of Load Scheme. Cermaics are highly sensitive to the load scheme. They can endure compression stresses that are much higher than tensile or shear stresses. This is due to the fact that, under the effect of the compressing elastic stress fields, the arising cracks and voids tend to close up, while the tensile and shear stresses cause a critical growth of cracks due to the concentration of stresses near them. In the early 1980s, *Evans* et al. [10.27] proved theoretically that if conditions for slowing down the generation and growth of voids in grain boundaries are created, ceramics can display SP properties. These conditions are easily furnished during compression tests with the "softest" straining condition without tensile stresses. This is why the first studies of the high-temperature ductility of ceramics dealt with the compression of samples [10.1–5, 8, 9]. However, compression tests fail to give a full impression of the SP flow in ceramics, for without tensile stresses the intergranular brittleness is insufficiently considered. The most complete information can be obtained during tensile tests. Experiments first performed on Y-TZP [10.28] and then on other ceramic materials [10.17, 29] indicated that, contrary to metallic materials, the SP deformation of ceramics is accomplished without necking (Fig. 10.10). The failure is caused by rapid void formation in the zone of deformation [10.16, 17]. Even in ceramics with a substantial fraction of liquid phase, the formation of voids is observed mainly in triple points [10.23].

The temperature-rate conditions of SP tensile and compression straining are identical and the value of m is practically invariable. This indicates that the mechanisms of the SP deformation of ceramics are independent of the load

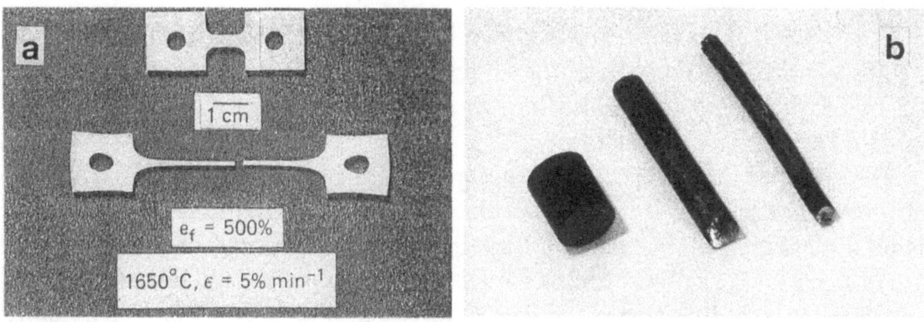

Fig. 10.10a,b. The shape of samples after superplastic deformation: (a) Y-TZP (tensile test), (b) $YBa_2Cu_3O_{7-x}$ (extrusion test)

scheme. Under different straining conditions the value of relative elongation probably depends on the stability of structure and its variations in the course of deformation and on the voids during SP deformation.

10.2 Mechanisms of Superplastic Deformation of Ceramics

To date, no detailed studies of the mechanisms of the SP deformation of ceramics have been carried out. However, there are several hypotheses which try to explain this phenomenon. It is curious that the situation was quite the same in studying the SP of metallic materials in the early 1970s, and the existing hypotheses for ceramics are similar to those proposed then. And again, in the majority of papers an attempt is made to explain the SP of ceramics proceeding from an analysis of mechanical properties and theoretical suppositions. As shown in Chap. 2, this problem cannot be solved without detailed structural investigations. We now consider some of the proposed hypotheses.

To explain the SP behaviour of the β-spodumene glass, *Raj* et al. proposed a special mechanism for deformation based on the mass transfer via a liquid phase [10.23]. At the same time, they assumed that grain boundaries had a patched structure. Such an interface ensures solid-to-solid contact and maintains the integrity of the system and also contains areas of liquid phase providing for rapid diffusion. During the isothermal annealing of this glass ceramic, the dissolution of the glass phase was observed in an area between the liquid and solid and, when cooled below the solid, the precipitation of a crystalline phase was observed. The rates of the process of dissolution-deposition and of creep deformation were similar. The authors concluded that strain rate is controlled by the rate of diffusion via the liquid phase and by the dissolution-deposition process. This special mechanism for deformation ensures the low strain rate and explains the absence of grain growth during SP flow.

However, as early as 1945, *A.A. Bochvar* [10.30] put forward a hypothesis that in SP metallic materials deformation is accomplished via the diffusional dissolution and deposition of phases. Later, numerous experimental results denied the operation of such a mechanism.

In [10.6, 9, 22] an important and dominating role in deformation is assigned to diffusion creep. This is based on an agreement of the experimentally defined rate of SP flow with the values of this parameter, calculated by the Coble or Nabarro-Herring creep equation.

The effect of diffusion creep during the deformation of ceramics is unknown and there are doubts if it actually works. Diffusion creep is to be based on the material mass transfer via the contrary diffusion of atoms and vacancies. In the case of metals this can be accomplished at high temperatures and low strain rates. For ceramics, the diffusion of individual atoms of a compound must inevitably result in a considerable change of chemical composition. For example, in Al_2O_3, due to the difference in the diffusional mobility of atoms, regions, enriched in aluminium or oxygen must appear, which has not been observed experimentally. And the cooperative diffusion of the atoms of the crystalline lattice as an unit is difficult to imagine for energy reasons.

To account for the SP of the Y-TZP ceramic compound [10.10], the Ashby-Verral model was used. In this model the main mechanism of the SP deformation is grain boundary sliding with diffusion accommodation [10.31]. The authors obtained a fairly good agreement between the theoretically calculated values of $\sigma = f(\dot{\varepsilon})$ and experimental data. As shown in Chap. 2, the agreement of the calculated values with experimental data is typical of many models, but without a detailed study of microstructure and grain boundary structure, this cannot be accepted as a proof of the existence of any mechanism of deformation.

The most probable mechanism for SP deformation of fine crystalline ceramics is grain boundary sliding with dislocation slip.

Reference [10.21] presents a comprehensive study of a dislocation structure of the fine-grained Al_2O_3 ($d \sim 0.5$–$1\,\mu m$) during SP deformation. The authors demonstrated that during SP deformation lattice dislocations are trapped by grain boundaries and those extrinsic grain boundary dislocations (EGBD) form dislocation walls. EGBD then induce grain boundary sliding.

The SP of ceramics is often associated with the development of grain boundary sliding [10.6, 21, 22]. According to *Wakai*, the contribution of grain boundary sliding to the total deformation in Y-TZP – 20%Al_2O_3 is 60–80% [10.16]. In Y-TZP the grain boundary sliding induced deformation is up to 83.5% of the total deformation [10.32]. The rapid development of grain boundary sliding, especially after about 50% of the fine grain volume has been formed, is also confirmed by the deformation relief patterns obtained on the titanium carbide and SP ceramics. Dislocation slip is observed in the grains, too.

Analysis indicates that the actual mechanisms of deformation of ceramics are basically the same as the mechanisms operating in metals and intermetallics. However, in some cases, certain unique features are observed. The development of a detailed physical model requires reliable experimental data on the mechanisms of SP deformation.

10.3 Methods for Creating Microcrystalline Structures in Ceramics

One of the important conditions which enables SP is the existence of a microcrystalline microstructure in the starting material. In metals, this structure is created by controlling the crystallization conditions and by various heat and thermomechanical treatments (Chap. 3). In ceramics it is usually produced by means of powder metallurgy, which sets special requirements for the starting powders and conditions of compaction. At the same time, the enhanced heat resistance and low oxidizing ability of the ceramic powders (especially oxide and high melting point powders) make synthesis, refining, and compaction possible without any special atmospheres, which makes processing easier and cheaper.

Methods of Powder Metallurgy. At present, there is no difficulty in producing fine disperse ceramic powders with a grain size as small as several tenths of a micron. Different means of production are described in the literature and in Chap. 3. We present only some of the methods for preparing SP ceramics.

The classical SP Y-TZP ceramic compound is usually produced by refining coarse-grained powders (0.2–0.5 mm) in ball or vibrating mills to a size of less than 1 μm [10.33]. Afterwards they are sintered in air at 1500°C [10.11]. Under this treatment the grain size does not change much and the density is equal to 96–98%.

In some cases better results can be obtained by means of hot pressing. Thus, an Al_2O_3-based microcrystalline ceramic was produced by pressing fine disperse powders at 1500°C [10.26]. An SiC ceramic compound was also produced by pressing powders with a grain size of 0.1–0.2 μm at 1200°C [10.29].

A more complicated process is used in preparing the Al_2O_3-ZrO_2 composite ceramics [10.16]. The cold isostatic pressing of the material is followed by sintering at 1400°C and then is subjected to hot isostatic pressing at a higher temperature. The same technique is used to produce the $MoSi_2$-Mo_2B_5 composite ceramics with a grain size of 0.6–1 μm [10.25].

The preparation of ceramics with a nanocrystalline structure is of special interest [10.34]. A method of obtaining these structures is described in Chap. 3 and is used to create an essentially new state of matter, i.e., a nanocrystalline material which is comprised of crystallites and grain boundaries in approximately identical fractions. *Karch* et al. [10.34] succeeded in obtaining high plasticity at 80–180°C in the nanocrystalline ceramics TiO_2 and CaF_2 with a mean grain size of about 10 nm. However, at present, the preparation of nanocrystalline materials is rather labour intensive and yields amounts sufficient only for laboratory work.

In this connection, the increase of ductility of ceramic materials during hot deformation is of interest.

Dynamic Recrystallization in Ceramics. The common phenomenology of the high temperature deformation of ceramics and metallic materials leads to an assumption that structural changes occurring in ceramics are basically the same as those occurring in metals. When the original microstructure is coarse-grained, the SP deformation in metals subjected to high temperature straining is preceded by the preliminary refining of grains, i.e., dynamic recrystallization (Chap. 3). In ceramic materials, too, as demonstrated on titanium carbide (Sect. 10.1), the initial stage of deformation is characterized by the deformation of the interior of grains and their refining. With a reduction of the grain size, the role of grain boundary sliding increases, which results in the growth of the strain rate sensitivity of the flow stress. Hence, the interaction of dynamic recrystallization and grain boundary sliding ensures high hot deformability of material. From this point of view, dynamic recrystallization is a universal method for microstructure refining providing for the subsequent SP flow.

For a more detailed study of dynamic recrystallization, we consider the microstructural changes occurring during the deformation of superconductive ce-

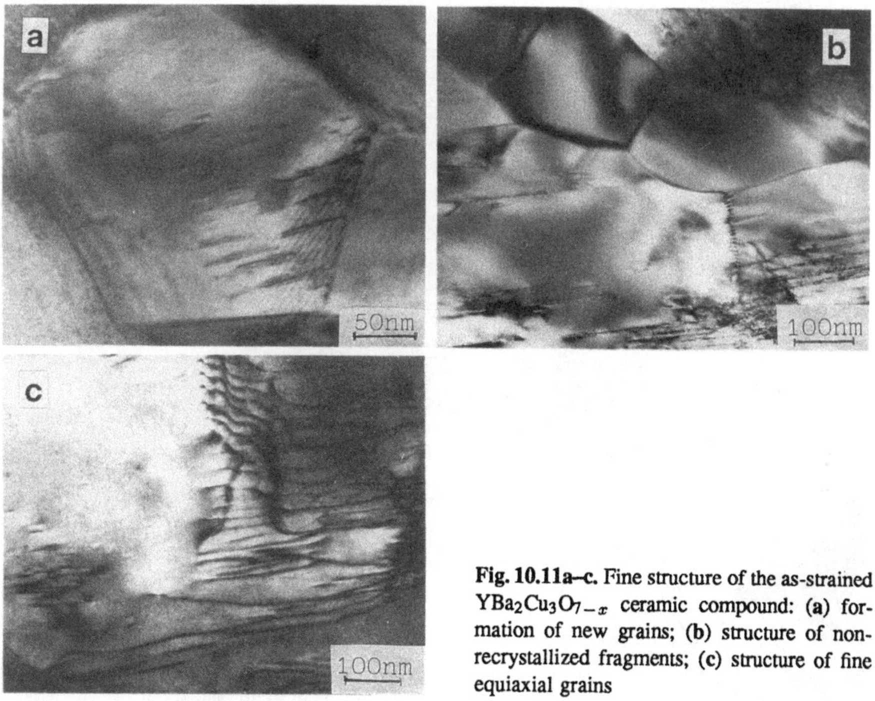

Fig. 10.11a–c. Fine structure of the as-strained $YBa_2Cu_3O_{7-x}$ ceramic compound: (**a**) formation of new grains; (**b**) structure of nonrecrystallized fragments; (**c**) structure of fine equiaxial grains

ramics $YBa_2Cu_3O_{7-x}$ whose original lamellar microstructure is unfavourable for the development of grain boundary sliding (Fig. 10.3a).

In the original state the longitudinal size of the lamellas is 15–25 μm and the transverse size is 5–10 μm. The matrix phase, comprised of $YBa_2Cu_3O_{7-x}$, coexists with voids detected metallographically and with a small amount of unreacted components (4–5 vol.%). Analysis of the fine structure indicates that the grain volume is usually twinned in two mutually normal planes. The grains have straight-line dislocations of the basal slip system. However, throughout the sample, the density of lattice dislocation ϱ is moderate and does not exceed $10^8\,cm^{-2}$.

In the course of deformation the structure undergoes considerable changes. As shown in Fig. 10.11a, the microstructure is refined due to the formation of new grains as small as several tenths of a micron. The fine grains have a globular shape, although there is a small number of lamella-shaped grains. New grains are located in colonies around coarse fragments of 20 μm or less (Fig. 10.11b). The very broad range of the fragment sizes is due to the incomplete recrystallization of the microstructure. Coarse fragments contain a high density of dislocation ($\varrho > 10^{11}\,cm^{-2}$) and are probably the nonrecrystallized portion of the original grains (Fig. 10.11b). The recrystallized volume is about 60%. In fine equiaxial grains the density of lattice dislocations is much lower than in coarse nonrecrystallized fragments and is $10^{10}\,cm^{-2}$, on the average. Fine lamellar grains have a density of dislocations lower than that in the globular grains ($\varrho = 10^8 - 10^9\,cm^{-2}$)

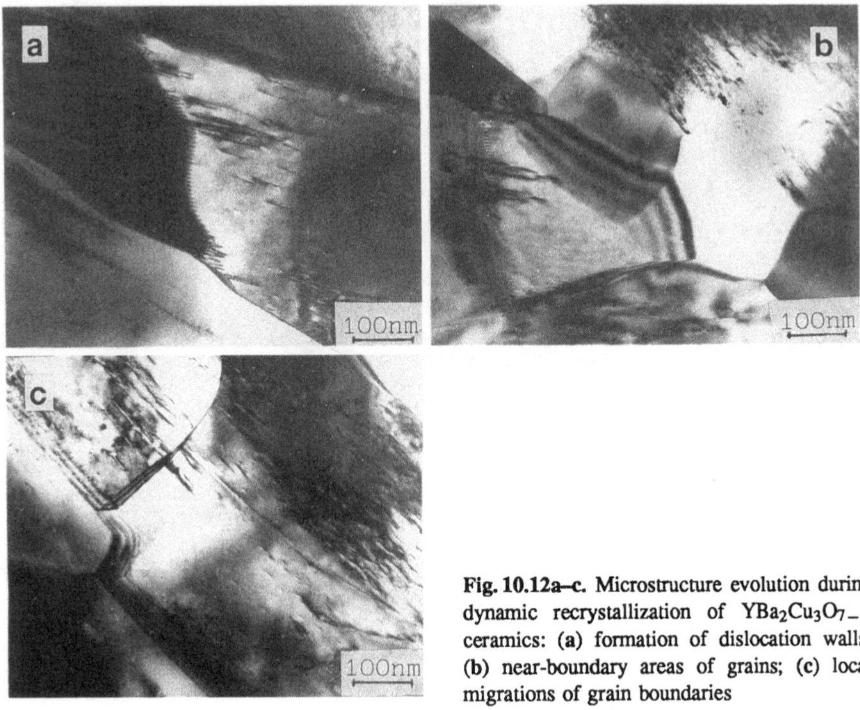

Fig. 10.12a–c. Microstructure evolution during dynamic recrystallization of $YBa_2Cu_3O_{7-x}$ ceramics: (**a**) formation of dislocation walls; (**b**) near-boundary areas of grains; (**c**) local migrations of grain boundaries

(Fig. 10.11a). Since there are well defined distinctions between the new grains, one can imply that they have different origin. The globular and lamellar grains are generated due to the dynamic and post-dynamic recrystallization, respectively.

Since the process of plastic deformation in the sample volume is nonuniform, it is possible to follow the stages of dynamic recrystallization which results in the transformation of the lamellar microstructure into the globular. At the initial stage of deformation, dislocation walls are formed within the grains (Fig. 10.12a). With an increasing strain level, the dislocation walls are localized across the lamellas and in the near boundary regions (Fig. 10.12b). The formation of the dislocation substructure in the grain interior os accompanied by a local migration of the original grain boundaries. The increasing misorientation of the dislocation walls and the amount of local migration of the original boundaries eventually result in the splitting of the lamellas into finer fragments as well as in the formation of chains of new grains on their periphery (Fig. 10.12c).

Further deformation is accompanied by a rapid grain boundary sliding (Fig. 10.3b,c). The accommodation of sliding is probably facilitated by the presence of voids in the starting material. High strain rate sensitivity observed upon changing of the strain rate and the development of grain boundary sliding indicate that SP deformation is in progress in this ceramic.

Hence, the superconductive ceramics gives an example of the transformation of the lamellar structure into globular due to the development of dynamic recrystallization. The mechanism of dynamic recrystallization is of a general character, because similar changes are also observed in the microstructures of metals and intermetallics (Chap. 3–9).

10.4 Influence of Superplastic Deformation on the Properties of Ceramics

Unfortunately, no systematic data on the influence of SP deformation on the properties of ceramics are available in the literature to date. The main task regarding ceramics, i.e., improving their plasticity at low temperatures, has not been fulfilled.

However, the available data on the SP of ceramics lead us to conclude that SP deformation stimulates diffusion processes, provides a high homogeneity in both the microstructure and chemical composition, and does not entail the accumulation of structural defects. All these features, together with a submicrocrystalline structure, point towards an improvement of low-temperature ductility and impact strength of ceramics. In the meantime, SP deformation makes it possible to apply some specifc techniques which can solve the problems of formation without worsening the desired properties.

Thus, to produce parts out of hard wear-resistant ceramics based on titanium carbide, a nonstoichiometric carbide phase should be used. As shown in Sect. 10.1, such a ceramic compound is quite easily converted to an SP state at 900–1000°C, which is acceptable for die-stamps made of superalloys. After deformation in the SP mode this ceramic can be subjected to additional treatment, i.e., carbon saturation. In this case the chemical composition of the carbide phase of the surface layers of the products may become stoichiometric, thus ensuring the required hardness (Fig. 10.13). Adjusting the temperature and duration of the process of saturation, a required distribution of hardness within the sample thickness can be attained.

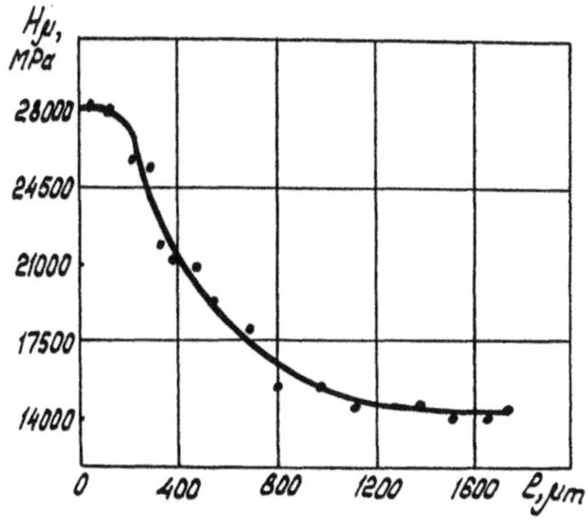

Fig. 10.13. Hardness distribution within the sample thickness in nonstoichiometric titanium carbide after saturation with carbon $YBa_2Cu_3O_{7-x}$ ceramics: (a) formation of dislocation walls; (b) near-boundary areas of grains; (c) local migrations of grain boundaries

Another specific example is provided by the SP treatment of superconductive ceramics. As the average grain size decreases, the mechanical strength of ceramics and the critical current J_c increase [10.35–37]. This is attributed to the improvement of the cohesive strength of grain boundaries. According to [10.35], in $YBa_2Cu_3O_{7-x}$, the critical grain size under which practically no cracks appear in the grain boundaries during heat treatment is 0.5–1 μm.

As shown above, the hot deformation of this ceramic yields microcrystalline structures caused by the development of dynamic recrystallization and its conversion to an SP state. However, in order to restore the superconductive properties of the ceramics, it must be subjected to a special kind of additional heat treatment.

Table 10.2 presents the methods of treatment and basic electrophysical parameters of superconductive ceramics calculated on the basis of magnetic measurements. The data indicate that, after annealing, superconductivity is restored. However, to attain high values of J_c, the modes of additional heat treatment must be optimized.

The presented data on the effect of SP deformation on the properties of ceramics are, of course, incomplete and show only some of the methods of controlling the properties. Thus, the properties of ceramics after SP deformation require further investigation.

Table 10.2. Electrophysical properties of superconductive ceramics $YBa_2Cu_3O_{7-x}$. T_i and T_f are the initial and final temperatures of transition to superconductivity, respectively; $\Delta T_f = Ti - T_f$; and J_c is the density of critical current

Sample	T_i [K]	T_f [K]	ΔT_f [K]	J_c [A/cm^2]
$YBa_2Cu_3O_{6.9\mp0.1}$ starting sample	90.3	87.8	2.5	730
$YBa_2Cu_3O_{6.9\mp0.1}$ SP deformation + air annealing: 930°C, in air, 5 h and 930°C, in oxygen, 2 h and, cooling rate 2 K/min	92.2	93.4	1.2	400

References

Abbreviations:

AIME = American Inst. of Mining, Metallurgical and Petroleum Engineers
ASM = American Society for Metals
MRS = Materials Research Society

Chapter 1

1.1 O.A. Kaibyshev: *Sverkhplastichnost' Promyshlennykh Splavov* (Metallurgia, Moscow 1984)
1.2 M.V. Grabski: *Strukturnaya Sverkhplastichnost' Metallov* (Metallurgia, Moscow 1975)
1.3 O.A. Kaibyshev: *Plastichnost' i Sverkhplastichnost' Metallov* (Metallurgia, Moscow 1975)
1.4 J.W. Edington, K.N. Melta, C.P. Culter: Progr. Met. Mater. Sci. **21**, 61–158 (1976)
1.5 I.I. Novikov, V.K. Portnoi: *Sverkhplastichnost' Splavov s Ul'tramelkim Zernom* (Metallurgia, Moscow 1981)
1.6 O.D. Sherby, J. Wadworth: "Superplasticity – Recent Advances and Future Directions", in *Progress in Material Science*, ed. by F. Neely, Vol. 33 (Plenum, New York 1989) 169–221
1.7 J. Pilling, N. Ridley: *Superplasticity in Crystalline Solids* (The Inst. of Metals, London, England 1989) p. 215
1.8 K.A. Padmanabhan, G.J. Davies: *Superplasticity* (Springer, Heidelberg, Berlin, New York 1989) p. 45
1.9 A.K. Mukherjee: "Superplasticity in Metals, Ceramic and Intermetallic", in *Plastic Deformation and Fracture of Materials*, ed. by H. Mughrabi, Materials Science and Technology, Vol. 6 (Publ. VCR Verlagsgesellschaft mbH, Germany, in press)
1.10 M. Suery, B. Baudelet: Rev. Phys. Appl. **13**(N2), 53–56 (1986)
1.11 A. Arieli, A.K.S. Yu, A.K. Mukherjee: Met. Trans. A **11A**(N1), p. 181–191 (1980)
1.12 D. Grivas, J.W. Marris, T.G. Langdon: Scripta Met. **15**(N2), 229–236 (1981)
1.13 O.A. Kaibyshev, S.Ya. Salikhov: Zavodskaya Laboratoriya N5, 612–615 (1977)
1.14 R.Z. Valiev, O.A. Kaibyshev, R.I. Kuznetsov, R.S. Musalimov, N.K. Tsenev: Dokl. Akad. Nauk SSSR **301**, 864 (1988)
1.15 O.A. Kaibyshev: "Current Problems of the Material Science of Superplasticity", in *Superplasticity and Superplastic Forming*, ed. by C.H. Hamilton, N.E. Paton (Mater., Metals and Mater. Soc. (Warrendale, Pennsylvania 1988) pp. 3–16
1.16 T.R. Bieler, T.G. Nieh, J. Wadsworth, A.K. Mukherjee: "High Rate Superplastic Behavior of Mechanically Alloyed Aluminum IN90211", in *Superplasticity and Superplastic Forming*, ed. by C.H. Hamilton and N.E. Paton (Minerals, Metals and Mater. Soc. (Warrendale, Pennsylvania 1988) pp. 161–165
1.17 T.R. Bieler, T.G. Nieh, J. Wadsworth, J.J. Stephens: Ser. Met. **22**, 81 (1988)
1.18 P. Griffiths, C. Hammond: Acta Met. **20**(N7), 935–945 (1972)
1.19 R.I. Kuznetsova, N.N. Zhykov, O.A. Kaibyshev, R.Z. Valiev: Phys. Stat. Sol. (a) **70**, 371–378 (1982)
1.20 V.K. Portnoi, I.I. Novikov, B.N. Efremov, V.A. Koganov: Izv. VUZov. TSvet. Metall. N5, 103–107 (1979)
1.21 A.K. Cash, R. Pay: Acta Met. **29**, 607–616 (1981)

1.22 O.A. Kaibyshev, I.V. Kazachkov, B.V. Rodionov: Fiz. Met. Metalloved. **39**, 308–346 (1975)
1.23 O.A. Kaibyshev, B.V. Rodionov, R.Z. Valiev: Acta Met. **26**(N12), 1877–1886 (1978)
1.24 M. Suery, B. Baudelet: Phil. Mag. **41**(N1), 41–64 (1980)
1.25 H. Naziri, P. Pearce: Scripta Met. **3**(N11), 807–810 (1969)
1.26 H. Naziri, P. Pearce: J. Inst. Met. **98**(N3), 71–77 (1970)
1.27 O.A. Kaibyshev, I.V. Kazachkov, S.Ya. Salikhov: Acta Met. **26**(N12), 1887–1894 (1978)
1.28 G.L. Dunlop, J.D. Reid, D.M.R. Taplin: Met. Trans. **2**(N8), 2308–2310 (1971)
1.29 O.A. Kaibyshev, I.V. Kazachkov: Dokl. Akad. Nauk SSSR **252**, 351–355 (1980)
1.30 O.A. Kaibyshev, I.V. Kazachkov, R.M. Galeev: J. Mater. Sci. **16**(N9), 2501–2506 (1981)
1.31 R.A. Benedek, R.D. Doherty: Scripta Met. **8**(N7), 675–679 (1974)
1.32 R.Z. Valiev, M.G. Zelin, V.I. Sergeev: Fiz. Met. Metalloved. **60**(N5), 980–985 (1985)
1.33 R.Z. Valiev, O.A. Kaibyshev, G.F. Korznikova, N.K. Tsenev: Fiz. Met. Metalloved. **62**, 180–186 (1986)
1.34 E.S. Karpov, I.I. Papirov, G.F. Tikhinski: Fiz. Khim. Obrab. Metall. N4, 96–100 (1980)
1.35 D. Lee: Met. Trans. **1**, 309–311 (1970)
1.36 R.Z. Valiev, O.A. Kaibyshev: Fiz. Met. Metalloved. **41**(N2), 382–387 (1976)
1.37 R.Z. Valiev, O.A. Kaibyshev: Phys. Stat. Sol. (a) **44**(N1), 65–76 (1977)
1.38 M.A. Clark, T.H. Alden: Acta Met. **21**(N9), 1195–1206 (1973)
1.39 R.Z. Valiev, O.A. Kaibyshev, Sh.Kh. Khannanov: Phys. Stat. Sol. (a) **52**(N2), 447–453 (1979)
1.40 R.Z. Valiev, O.A. Kaibyshev, N.K. Tsenev: Izv. VUZov. Tsvet. Metall. N3, 66–69 (1981)
1.41 R.Z. Valiev, O.A. Kaibyshev: Phys. Stat. Sol. (a) **44**(N2), 447–484 (1977)
1.42 R.Z. Valiev, O.A. Kaibyshev: Fiz. Met. Metalloved. **42**(N5), 1102–1105 (1976)
1.43 D.G. Dingley: "Direct Observation of Deformation Process in Superplastic Pb-Sn Eutectic Alloy, in *Scanning Electr. Micr.*, Proc. of the Third Annual S.E.M. Symp. (III Res. Inst., Chicago 1970) pp. 329–336
1.44 H. Naziri, R. Pearce, M.R. Brown, K.F. Hale: Acta Met. **23**(N4), 489–496 (1975)
1.45 Y. Kabayashi, Y. Ishida, M. Kato: Scripta Met. **11**, 51 (1977)
1.46 R.Z. Valiev, O.A. Kaibyshev, M.M. Myshlyaev, D.R. Chalaev: Scripta Met. **14**, 673–676 (1980)
1.47 S.W. Zehr, W. Backofen: Trans. ASM **61**, 300–313 (1968)
1.48 V.E. Ivanov, I.I. Papirov, G.F. Tikhinski, L.A. Kornienko: Dokl. Akad. Nauk SSSR **256**, 73–77 (1981)
1.49 I.I. Novikov, V.K. Portnoy, T.E. Terentieva: Acta Met. **25**(N10), 1139–1149 (1977)
1.50 D.M.R. Taplin, G.L. Dunlop, T.G. Langdon: Ann. Rev. Mater. Sci. **9**, 151–159 (1979)
1.51 Yan Ma, T.G. Langdon: "An Investigation of Cavitation and Cracking in Superplasticity", in *Proc. of the MRS International Meeting on Advanced Materials, Superplasticity*, ed by M. Kabayashi and F. Wakai, Vol. 7 (Materials Research Society, Pittsburgh, PA 1989) pp. 19–26
1.52 H. Ishikava, D.G. Bhat, F.A. Mohamed, T.G. Langdon: Met. Trans. A **8A**, 523–525 (1977)
1.53 C.W. Mumphais, N. Ridley: J. Mater. Sci. **9**, 1429–1435 (1974)
1.54 R.Z. Valiev, O.A. Kaibyshev, V.I. Sergeev: Izv. VUZov. Tsvet. Metall. N3, 81–85 (1982)
1.55 D.W. Livesey, N. Ridley: J. Mater. Sci. **14**, 825–832 (1978)
1.56 R.G. Fleck, D.M.R. Taplin, C.J. Beevers: Acta Met. **3**(N4), 415–424 (1975)
1.57 C.I. Smith, N. Ridley: Met. Technol. N4, 191–192 (1974)
1.58 W. Beere, M.V. Speight: Met. Sci. **12**, 172–176 (1978)
1.59 T.J. Chuang, R.I. Hagawa, J.R. Rice, L.B. Sills: Acta Met. **27**, 265–284 (1979)
1.60 R.I. Kuznetsova, N.N. Zhukov: Phys. Stat. Sol. (a) **50**, 385–391 (1978)
1.61 B.N. Efremov, I.I. Novikov, V.K. Protnoy, E.K. Zavarzina: Izv. VUZov. Tsvet. Metall. N5, 96 (1976)
1.62 R.A. Tait, D.M.R. Taplin: Scripta Met. N12, 77–82 (1979)
1.63 R.Z. Valiev, V.I. Sergeev: Izv. VUZov. Tsvet. Metall. N3, 81–84 (1983)

Chapter 2

2.1 R.C. Gifkins: J. Mater. Sci. **13**(N9), 1926–1936 (1978)
2.2 G. Gleiter, B. Chalmers: *Bol'sheuglovyye Granitsy Zeren* (Mir, Moscow 1975)
2.3 A.N. Orlov, V.N. Perevezentsev, V.V. Rybin: *Granitsy Zeren v Metallakh* (Metallurgia, Moscow 1970)
2.4 R.C. Gifkins: Met. Sci. **7**, 15–19 (1973)
2.5 O.A. Kaibyshev, R.Z. Valiev, V.V. Astanin: Phys. Stat. Sol. (a) **35**, 403–413 (1976)
2.6 K. Matsuki, H. Morita, M. Yamada, Y. Murakami: Met. Sci. **5**, 156–163 (1977)
2.7 R.C. Pond, D.A. Smith, P.W.J. Southerden: "Role of Grain Boundaries During High Temperature Deformation", in *Proc. IV Int. Conf. on the Strength of Metals and Alloys*, ed. by Laboratoire de Physique du Solide (Publ. ENSMIM-INPL, Nancy, France 1976) pp. 378–381
2.8 R.B. Vastava, T.G. Langdon: Acta Met. **27**, 251–257 (1979)
2.9 P. Shariat, R.B. Vastava, T.G. Langdon: Acta Met. **30**(N1), 285–296 (1982)
2.10 K.J. Kurzydlowski, K.J. Taggart, K. Tangry: Phil. Mag. **A 61**, 61–83 (1990)
2.11 R.Z. Valiev, S.N. Dudina, M.G. Zelin: "Povoroty i peremeshenia Zeren pri Deformatsii Metallov v Usloviyakh Sverkhplastichnosti", in *Teoreticheskie i Eksperimentalnyye Issledovania Disklinatsii*, ed. by V.I. Vladimirov, A.E. Romanov (Fiziko-mekhanicheskii Institut AN SSSR, Leningrad 1986) pp. 89–91
2.12 W. Beere: Phil. Trans. R. Soc., London **288A**, 177–196 (1978)
2.13 M.F. Ashby, R.A. Verall: Acta Met. **21**, 149 (1973)
2.14 P.M. Hazzledine, D.E. Newbury: "Role of Grain Boundaries During Superplastic Deformation", in *Grain Boundaries Structure and Properties*, ed. by G.A. Chadwick and D.A. Smith, N4 (Academic, New York 1976) pp. 235–264
2.15 R.Z. Valiev, O.A. Kaibyshev, M.H. Rabinovich, Yu.B. Timoshenko: Fiz. Met. Metalloved. **51**, 615–624 (1981)
2.16 D. Lee: Acta Met. **17**, 1057–1069 (1969)
2.17 R.Z. Valiev, O.A. Kaibyshev: Acta Met. **31**, 2121–2128 (1983)
2.18 R.S. Gates, R.N. Stevens: Met. Trans. **5**, 505–510 (1974)
2.19 T.G. Langdon: Met. Trans. **3**, 794–801 (1972)
2.20 K. Matsuki, Venoy, M. Yamada: J. Japan. Inst. Met. **38**, 219–226 (1974)
2.21 D.L. Holt: Trans. Met. Soc. AIME **242**, 25 (1968)
2.22 V.M. Rozenberg: *Polzuchest' Metallov* (Metallurgia, Moscow 1967)
2.23 W.R. Cannon: Phil. Mag. **25**, 1489–1497 (1972)
2.24 O.A. Kaibyshev, I.V. Kazachkov: Fiz. Met. Metalloved. **32**, 205–208 (1971)
Fiz. Met. Metalloved. **32**, 396–401 (1972)
Fiz. Met. Metalloved. **36**, 1235 (1973)
Dokl. Akad. Nauk SSSR **205**, 564–566 (1972)
2.25 O.A. Kaibyshev, I.V. Kazachkov, I.V. Aleksandrov: Acta Met. **32**, 585–589 (1984)
2.26 R.M. Bricknell, J.W. Edington: Acta Met. **25**, 447–458 (1977)
2.27 V.K. Portnoy, I.I. Novikov, B.N. Efremov: Dokl. Akad. Nauk SSSR **230**, 1142–1144 (1976)
2.28 V.K. Portnoy, I.I. Novikov, I.I. Nikiforova: Dokl. Akad. Nauk SSSR **248**, 854–856 (1979)
2.29 O.A. Kaibyshev, B.V. Rodionov, R.Z. Valiev: Dokl. Akad. Nauk **235**, 1293–1296 (1977)
2.30 J.K.L. Falk, G.L. Dunlop, T.G. Langdon: "An Observation of Dislocations During Superplasticity", in *Proc. V Int. Conf. Elect. Microscopy*, ed. by G. Tomas, Vol. 1 (Academic, Amsterdam 1980)
2.31 P.R. Nowell, J.O. Nilsson, G.L. Dunlop: Phil. Mag. **38A**, 39–47 (1978)
2.32 K.D. Fike, H.J. Rack: Trans. ASM **62**, 537–548 (1969)
2.33 Y. Ishida, K. Suzuki, M. Ichihara: Scripta Met. **12**, 999-1002 (1978)
2.34 I.I. Papirov, E.S. Karpov, M.I. Palatnik, M.B. Mileshkin: Dokl. Akad. Nauk SSSR **256**, 392–396 (1981)
2.35 R.Z. Valiev, S.N. Dudina, I.S. Obraztsova: Fiz. Met. Metalloved. **60**, 1217–1222 (1985)
2.36 O.A. Kaibyshev: Czech. J. Physics **B31**, 223–227 (1981)
2.37 S.V. Zemshy, N.E. Felin, G.K. Maltseva, V.A. Karpeliev: Zavodskaya Laboratoria N11, 1357–1360 (1977)

2.38 J.E. Harris: Met. Sci. **7**, 1–6 (1973)
2.39 A. Karim, D.L. Holt, W.A. Backofen: Trans. Met. Soc. AIME **245**, 1131–1132 (1969)
2.40 J.J. Burke, V. Weiss (eds.): *Ultrafine-Grain Metals* (Syracuse University Press, Syracuse, New York 1972)
2.41 D. Lee: Scripta Met. **3**, 893–894 (1969)
2.42 R.Z. Valiev, O.A. Kaibyshev, V.I. Sergeev: Fiz. Met. Metalloved. **49**, 1291–1298 (1980)
2.43 R.Z. Valiev, V.I. Sergeev: Izv. VUZov. Tsvet. Metall. N3, 81–84 (1983)
2.44 A. Karim, W.A. Backofen: Met. Trans. **3**, 709–712 (1972)
2.45 O.A. Kaibyshev, V.V. Astanin, R.Z. Valiev: Dokl. Akad. Nauk SSSR **245**, 1356–1368 (1979)
2.46 O.A. Kaibyshev, V.V. Astanin, R.Z. Valiev, V.G. Haimellin: Fiz. Met. Metalloved. **51**, 193–200 (1981)
2.47 R.Z. Valiev, O.A. Kaibyshev, V.I. Sergeev, M.G. Zelin: Res. Mechanica Lett. **12**, 219–228 (1984)
2.48 R.C. Gifkins, T.G. Langdon: Mater. Sci. Eng. **36**(N1), 27–33 (1978)
2.49 J.W. Edington: Metal Forum **1**(N1–2), 63–68 (1981)
2.50 R.C. Gifkins: Met. Trans. **7A**(N8), 1225–1232 (1976)
2.51 H.W. Hayden, S. Floreen, P.D. Goodell: Met. Trans. **3**(N4), 833–842 (1972)
2.52 A. Arieli, A.K. Mukherjee: Mater. Sci. Eng. **45**(N1), 61–70 (1980)
2.53 K.A. Padmanabhan: "Role of Interfaces in Superplastic Flow", in *Proc. of the Indo-USA Workshop on Interfaces: Structure and Properties*, ed. by S. Ranganathan, D.A. Smith, C.S. Pande, and B.B. Rath, Trans. Techn. Publishers, Switzerland, in press)
2.54 V.N. Perevezentsev: "K Teorii Sverkhplasticheskoi Deformatsii", in *Voprosy Teorii Defektov v Kristallakh*, ed. by V.I. Vladimirov (Nauka, Moscow 1987) pp. 85–100
2.55 R.Z. Valiev, V.Yu. Gertsman, O.A. Kaibyshev: Phys. Stat. Sol. (a) **97**(N11), 11–56 (1986)
2.56 M.W. Grabski: J. Physique **46**, Suppl. (N4), 4 (1985)
2.57 H. Gleiter: Progr. Mater. Sci. **25**(N1), 126–183 (1986)
2.58 O.A. Kaibyshev, R.Z. Valiev: *Granitsy Zeren i Svoistva Metallov* (Metallurgia, Moscow 1987)
2.59 R.C. Pond, D.A. Smith, V. Vitek: Scripta Met. **12**, 669–702 (1978)
2.60 M.F. Ashby, F. Spaenen, S. Williams: Acta Met. **26**, 1647–1663 (1978)
2.61 Ch.V. Kopetski, G.S. Shvindlerman: "Migratsia Individualnoi Granitsy v Plotnoupakovannykh Metallakh", in *Metally Vysokoi Chistoty* (Nauka, Moscow 1976) pp. 73–104
2.62 R.A. Varin, J. Wyrzykowski: Phys. Stat. Sol. (a) **46**, k79–k81 (1978)
2.63 R.Z. Valiev, V.Yu. Gertsman, O.A. Kaibyshev, V.I. Sergeev: Metallofizika **5**, 94–100 (1983)
2.64 K. Kurzydlowski, W. Lojkowski, M.W. Grabski: Res. Mechanica Lett. **1**, 275–282 (1981)
2.65 P.M. Pumphry, H. Gleiter: Phil. Mag. **30**, 563–602 (1974)
2.66 W. Lojkowski, M.W. Grabski: Scripta. Met. **13**, 511–602 (1974)
2.67 R.Z. Valiev, V.Yu. Gertsman, O.A. Kaibyshev: Phys. Stat. Sol. (a) **61**, k95–k99 (1980)
2.68 R.Z. Valiev, O.A. Kaibyshev: Dokl. Akad. Nauk SSSR **258**, 91–94 (1981)
2.69 M.W. Grabski, R.Z. Valiev, J. Wyrzykowski, W. Lojkowski: Res. Mechanica Lett. **1**, 489–497 (1981)
2.70 R.Z. Valiev, V.Yu. Gertsman, O.A. Kaibyshev, Sh.Kh. Khannanov: Phys. Stat. Sol. (a) **76**, 101–105 (1983)
2.71 R.Z. Valiev, V.G. Khairullin: J. Colloque de Physique **51**, C1–685–689 (1990)
2.72 G.R. Kegg, C.A. Horton, J.M. Silcock: Phil. Mag. **27**(N5), 1041–1055 (1973)
2.73 T. Mori, K. Tangri: Met. Trans. A **10A**, 733–740 (1979)
2.74 R.Z. Valiev, V.G. Khairullin: Fiz. Met. Metalloved. **75**, 186–191 (1990)
2.75 O.A. Kaibyshev, R.Z. Valiev, A.K. Emaletdinov: Phys. Stat. Sol. (a) **90**(N1), 197–206 (1985)
2.76 J.P. Hirth, I. Lotte: *Theory of Dislocations* (McGraw-Hill, New York 1968)
2.77 R.Z. Valiev, O.A. Kaibyshev, N.K. Tsenev, A.K. Emaletdinov: Metallofizika **6**(N5), 84–90 (1984)
2.78 G.H. Schneibel: Acta Met. **28**, 1527–1535 (1980)
2.79 O.A. Kaibyshev, I.V. Kazachkov, N.G. Zaripov: Dokl. Akad. Nauk SSSR **282**, 594–598 (1985)
2.80 O.A. Kaibyshev, R.Z. Valiev: Res. Mechanica Lett. **3**(N1), 23–33 (1981)
2.81 P. Malek, P. Lukac: Czech. J. Physics **B37**, 975–992 (1987)

2.82 R.Z. Valiev, M.G. Zelin, V.I. Sergeev: Fiz. Met. Metalloved. **66**(N5), 980–985 (1985)
2.83 P. Lukac: "The Role of Matrix Dislocations in the Superplastic Deformation", in *Proc. of Int. Conf. on Superplasticity and Superplastic Forming*, ed. by E.H. Hamilton, N.E. Paton (Minerals, Metals and Mater. Soc. (Warrendale, Pennsylvania 1988) pp. 109–115

Chapter 3

3.1 R. Birringer, U. Herr, H. Gleiter: Trans. Japan Inst. Met. **27**, 43–52 (1986)
3.2 R.Z. Valiev, R.R. Mulyukov, V.V. Ovchinnikov: Phil. Mag. Lett. **62**, 253–256 (1990)
3.3 A. Lasalmoney, T.L. Strudel: Mat. Science **21**, 1837–1852 (1986)
3.4 R.Z. Valiev, N.K. Tsenev: "Structure and Superplasticity of Al-based Submicron Grained Alloys", in *Hot Deformation of Aluminium Alloys*, ed. by T.G. Langdon, H.D. Marchant, J.G. Morris, M.A. Zaidi (Minerals, Metals, Mater. Soc. Publ., Warrendale, Pennsylvania 1991) pp. 319–329
3.5 G.A. Chapman, D.V. Wilson: J. Inst. Metals, XCI **39** (1962)
3.6 E.P. Abrahamson: "The Influence of Grain Refinement on Some Mechanical Properties", in *Surfaces and Interfaces II*, ed. by J.J. Burke, N.L. Reed, V. Weiss (Syracuse University Press, Syracuse, New York 1968) pp. 262–269
3.7 H.G.F. Wilsdorf, O.T. Inal, L.E. Murr: Z. Metallkunde **69**, 701 (1978)
3.8 H. Karch, R. Birringer, H. Gleiter: Nature (London) **330**, 556–558 (1987)
3.9 R.W.K. Honeycombe: *The Plastic Deformation of Metals* (Edward Arnold, London 1968) pp. 1–408
3.10 A.W. Thompson: Met. Trans. **8A**, 833–842 (1977)
3.11 A.M. Chohshi, A. Rosen, H. Karch, H. Gleiter: Scripta Met. **23**, 1679–1683 (1989)
3.12 J.S.C. Jang, C.C. Koch: Scripta Met. **24**, 1599–1604 (1990)
3.13 O.A. Kaibyshev, R.Z. Valiev: *Granitsy Zeren i Svoistva Metallov* (Metallurgia, Moscow 1987)
3.14 R.Z. Valiev, R.R. Mulyukov, Kh.Ja. Mulyukov, V.I. Novikov, L.I. Trusov: JETP Lett. **15**, 78–81 (1989)
3.15 R. Birringer: Mater. Sci. Engng. **117A**, 33–44 (1989)
3.16 A.W. Thompson, W.A. Backofen: Acta Met. **19**, 597–606 (1971)
3.17 G.S. Koul, G.F. Bolling: "Polucheniye Sverkhmelkozernistoi Struktury Metallov i Splavov pri Kristallizatsii", in *Sverkhmelkoye Zerno v Metallakh*, ed. by L.K. Gordiyenko (Metallurgia, Moscow 1973) pp. 41–81
[English transl.: *Ultrafine-Grain Metals*, ed. by J.J. Burke and V. Weiss (Syracuse University Press, Syracuse, New York 1970)]
3.18 O.A. Kaibyshev: *Sverkhplastichnost' Promyshlennykh Splavov* (Metallurgia, Moscow 1984)
3.19 V.S. Shipilova, S.P. Molodchinina, L.M. Kofman: Tsvetnyye Metally **11**, 40–41 (1981)
3.20 P.B. Crossley, L.F. Mondolfo: Modern Casting **49**, 89 (1966)
3.21 M.L. Khazin, B.N. Smirnov, G.A. Kitaev: Tsvetnyye Metally **1**, 84–86 (1985)
3.22 M.D. Merz, S.D. Dahlgren: J. Appl. Phys. **46**, 3235–3237 (1975)
3.23 A.F. Belov, N.F. Anoshkin, O.H. Fatkullin. *Struktura i Svoistva Granulirovannykh Nikelevykh Splavov* (Metallurgia, Moscow 1984)
3.24 B. Gunther, A. Boalmann, H. Weiss: Mat. Res. Soc. Symp. **190** (1991)
3.25 J.S. Benjamin, M.J. Bromford: Met. Trans. **8A**, 1301 (1977)
3.26 T.C. Nich, P.S. Gilman, J. Wadsworth: Scripta Met. **19**, 1375 (1985)
3.27 P.W. Bridgman: *Studies in Large Plastic Flow and Fracture* (McGraw-Hill, New York and Maidenhead, 1952)
3.28 S. Reusswig, R. Gleichmann, P.G. Zielinski, D.G. Ast, R. Ray: Acta Met. **32**, 1553–1560 (1984)
3.29 I.I. Novikov, V.K. Portnoy, N.S. Zhuravleova, E.G. Shmurygina: Dokl. Akad. Nauk USSR **230**, 642–644 (1976)
3.30 R.A. Grange, A. Szirmal: Trans. Met. Soc. AIME **245**, 1063 (1963)
3.31 I.I. Novikov, V.K. Portnoy: "Optimization of Heterogeneity as General Principle of Controlling

Alloys Structures for Superplastic Forming, in *Superplasticity and Superplastic Forming*, ed. by C.H. Hamilton and N.E. Paton (Minerals, Metals and Mater. Soc. Publ. (Warrendale, Pennsylvania 1988) pp. 401–406
3.32 O.A. Kaibyshev, R.Ya. Lutfullin, G.A. Salishchev: Fiz. Met. Metalloved. **66**, 966–972 (1988); Fiz. Met. Metalloved. **66**, 1163–1171 (1988)
3.33 M.I. Mazursky, G.A. Salishchev: Fiz. Met. Metalloved. **68**, 1104–1107 (1989)
3.34 V.K. Fedyukin, M.E. Smagorinsky: *Termotsiklicheskaya Obrabotka Metallov i Detaley Mashin* (Mashinostroenie, Leningrad 1989) pp. 1–255
3.35 M. Otsuka, J. Miura, R. Horiuchi: Scripta Met. **8**, 1405–1408 (1974)
3.36 H.W. Schadler: Trans. Met. Soc. AIME **242**, 1281 (1968)
3.37 L.F. Porter, D.S. Dabkovsky: "Regulirovanie Razmera Zerna Putem Termotsiklirovania", in *Sverkhmelkoe Zerno v Metallakh*, ed. by L.K. Gordienko (Metallurgia, Moscow 1973) pp. 135–164. [English transl.: *Ultrafine-Grain Metals*, ed. by G.G. Burke and V. Weiss (Syracuse University Press, Syracuse, New York 1970)]
3.38 A.A. Baranov: *Fazovyye Prevrashchenia i Termotsiklirovanie Metallov* (Naukova Dumka, Kiev 1974)
3.39 S.S. Gorelik: *Rekristallizatsia Metallov i Splavov* (Metallurgia, Moscow 1978)
3.40 R.D. Doherty: "Zarozhdenie Novykh Zeren", in *Rekristallizatsia Metallicheskikh Materialov*, ed. by Ch.V. Kupetsky (Metallurgia, Moscow 1982) pp. 33–70
[English transl.: *Recrystallization of Metallic Materials*, ed. by F. Haessner (Riederer, Stuttgart 1978)]
3.41 V.I. Trefilov, Yu.V. Milman, S.A. Feerstov: *Fizicheskie Osnovy Prochnosti Tugoplavkikh Metallov* (Naukova Dumka, Kiev 1975)
3.42 E.P. Abrahamson: "Poluchenie Sverkhmelkozernistykh Splavov Metodom Kontroliruemogo Legirovania", in *Sverkhmelkoe Zerno v Metallakh*, ed. by L.K. Gordienko (Metallurgia, Moscow 1973) pp. 82–89
[English transl.: *Ultrafine-Grain Metals*, ed. by J.J. Burk and V. Weiss (Syracuse University Press, Syracuse, New York 1970)]
3.43 E. Hornbogen, U. Kester: "Rekristallizatsia Dvukhfaznykh Splavov", in *Rekristallizatsia Metallicheskikh Materialov*, ed. by Ch.V. Kopetsky (Metallurgia, Moscow 1982) 165–201
[English transl.: *Recrystallization of Metallic Materials*, ed. by F. Haessner (Riederer, Stuttgart 1978)]
3.44 F.G. Humphreys: Met. Sci. **13**, 136–145 (1979)
3.45 I.L. Dillamor: "Regulirovanie Protsessov Rekristallizatsii", in *Rekristallizatsia Metallicheskikh Materialov*, ed. by Ch.V. Kopetsky (Metallurgia, Moscow 1982) pp. 230–251
[English transl.: *Recrystallization of Metallic Materials*, ed. by F. Haessner (Riederer, Stuttgart 1978)]
3.46 N. Hansen, B. Bay: Acta Met. **29**, 65–77 (1981)
3.47 J.A. Wert, N.E. Paton, C.H. Hamilton, M.W. Machoniu: Met. Trans. **12A**, 1267–1276 (1981)
3.48 J.W. Martin: *Micromechanisms in Particle-Hardened Alloys* (Cambridge University Press, Cambridge, London 1980)
3.49 C. Brock, E. Hornbogen, P. Stratmann, H. Thommeck: Arch. Eisenhütt. **47**, 513 (1976)
3.50 R.C. Gibson, H.W. Hayden, G.H. Brophy: Trans. Qaurt. ASM **61**, 85–88 (1968)
3.51 G.A. Salishchev, R.M. Galeev: "Dynamic Recrystallization and Superplasticity of Metals and Alloys", in *Recrystallization'90*, Proc. Int. Conf., ed. by T. Chandra (Minerals, Metals and Mater. Soc. Publication (Warrendale, Pennsylvania 1990) pp. 603–608
3.52 N.A. Smirnova, V.I. Levit, V.P. Pilyugin, R.I. Kuznetsov, M.V. Degtyarev: Fiz. Met. Metalloved **62**, 566-570 (1986)
3.53 K. Gardner, R. Grimes: Met. Sci. **13**, 216–222 (1979)
3.54 R.J. Wray: Met. Trans. **15A**, 2009–2019 (1984)
3.55 H.J. McQueen, J.J. Jonas: J. Appl. Met. **3**, 233–241 (1984)
3.56 J.J. Jonas, J. Weiss: Met. Sci. **18**, 238–245 (1979)
3.57 J. Crawford: Met. Sci. **18**, 395–402 (1984)

3.58 S.Z. Bokshtein, S.S. Ginzburg, S.T. Kishkin, I.M. Razumovsky: Poverkhnost. Fizika, Khimia, Mekhanika **1**, 5–19 (1984)
3.59 G.A. Salishchev, M.I. Mazursky, I.E. Levin: Fiz. Met. Metalloved. **72**, 149–155 (1990)
3.60 O.R. Valiahmetov, R.M. Galeev, G.A. Salishchev: Fiz. Met. Metalloved. **72**, 204–206 (1990)

Chapter 4

4.1 S.L. Kuzmin, V.A. Likhachev: Scripta Met. **12**, 735 (1978)
4.2 B.A. Kolachev, V.A. Livanov: *Metallovedenie i Termicheskaya Obrabotka Zvetnykh Metallov* (Metallurgia, Moscow 1972)
4.3 T.V. Reinor: *Metallovedenie Magniya i Ego Splavov* (Metallurgia, Moscow 1964)
4.4 M.B. Altman, A.F. Belov, V.I. Dobatkin, M.E. Drits, F.I. Kodakov: *Magnitevyye Splavy*, Vol. 1 (Metallurgia, Moscow 1978)
4.5 O.A. Kaibyshev: *Sverkhplastichnost' Promyshlennyh Splavov* (Metallurgia, Moscow 1984)
4.6 P.E. Hauser, P.P. Zandon, I.E. Dorn: Trans. ASM **50**, 856 (1958)
4.7 A. Karim, D.L. Holt, W.A. Backofen: Trans. Met. Soc. AIME **245**, 2421 (1969)
4.8 W. Backofen, G.S. Murty, S.W. Zehr: Trans. Met. Soc. AIME **242**, 329 (1968)
4.9 S.H. Reichman, J.W. Smythe: Powder Metals **6**, 67 (1970)
4.10 O.A. Kaibyshev, J.V. Gusev, R.R. Salikhov: Metalloved. Term. Obrab. **45**(N1) (1975)
4.11 O.A. Kaibyshev: *Plastichnost' i Sverkhplastichnost' Metallov* (Metallurgia, Moscow 1975)
4.12 S.E. Ion, F.J. Humphreys, S.H. White: Acta Met. **30**, 1909 (1982)
4.13 H.G. Aaronson, H.B. Aaron, K.R. Kinsman: Metallography **4**, 1 (1971)
4.14 F. Haessner (ed.): *Recrystallisation of Metallic Materials* (Riederer, Stuttgart 1978) Chapter 2
4.15 H.J. McQueen, B. Baudelet: "Comparison and Contrast of Mechanisms, Microstructures, Ductilites in Superplasticity and Dynamic Recovery Recrystallization", in *Proc. of the 5th Int. Conf. on Strength of Metals and Alloys*, Vol. 1, ed. by P. Haasen, V. Gerold, and G. Kostorz (Pergamon Press, Toronto, 1979) p. 329
4.16 S.S. Gorelik: *Rekristallizatsiya Metallov i Splavov* (Metallurgia, Moscow 1978)
4.17 H.J. McQueen, J.J. Jonas: Appl. Met. **3**, 233 (1984)
4.18 N.G. Zaripov, A.R. Vagapov, R.O. Kaibyshev: Fiz. Met. Metalloved. **63**, 774 (1987)
4.19 N.G. Zaripov, R.O. Kaibyshev: "Dynamic Recrystallization and Superplasticity of a Magnesium Alloy", in *Superplasticity and Superplastic Forming*, ed. by C.H. Hamilton and N.E. Paton (Minerals, Metals and Mater. Soc. Publ., Warrendale, Pennsylvania 1988) p. 91–95
4.20 I.Sh. Valeev, N.G. Zaripov, R.O. Kaibyshev, G.A. Salishchev: Tsvet. Met. **7**, 91 (1988)
4.21 O.A. Kaibyshev, R.R. Salikhov: Metalloved. Term. Obrab. **3**, 33 (1981)
4.22 M.E. Drits, Z.A. Sviderskaya, F.M. Elkin, V.F. Trokova: *Sverkhlegkiye Konstruktsionnyye Splavy* (Nauka, Moscow 1972) p. 34

Chapter 5

5.1 O.A. Kaibyshev: *Sverkhplastichnost' Promyshlennykh Splavov* (Metallurgia, Moscow 1984)
5.2 R. Grimes, C. Backer, M.J. Stowell, B.M. Watts: Aluminium **51**, 720 (1975)
5.3 R. Grimes, M.G. Stowell, B.M. Watts: Met. Tech. **3**, 154 (1976)
5.4 K. Matsuki, M. Yamada: J. Japan Inst. Met. **37**, 448 (1973)
5.5 A. Juhasz, N.Q. Chinh, P. Tasnadi, J. Kovacs: J. Mat. Sci. **22**, 137 (1987)
5.6 R.Z. Valiev, O.A. Kaibyshev, N.K. Tsenev: Metallofizika N6, 117 (1980)
5.7 A.A. Alalykin, B.G. Pavlov, A.M. Diskin: Tsvetn. Met. N2, 85 (1990)
5.8 O.S. Bochvar, Yu.P. Pshenichnov: Izv. VUZov. Tsvetn. Met. N4, 115 (1978)
5.9 Yu.P. Pshenichnov, V.D. Makarov, N.K. Gribova, Yu.V. Gusev: Metalloved. Term. Obrab. N3, 31 (1981)
5.10 B.M. Watts, M.J. Stowell, B.L. Baikie, D.G.E. Owen: Met. Sci. **10**, 198 (1976)
5.11 B.A. Wilcox, A.H. Clauer: "Superplastic Tendencies of Two Melt Spun Aluminium Alloys of

Commercial Composition", in *The Microstructure and Design of Alloys*, Proc. of the 3rd Int. Conf. on Strength of Metals and Alloys, Institute of Metals, London (1973) pp. 227–231
5.12 J.J. Blandin, G.J. Lacroix, M. Suery: "Superplasticity and Cavitation of the 2091 Al-Cu-Li-Mg Alloy", in *Proc. Int. Conf. Superplasticity and Superplastic Forming*, ed. by C.H. Hamilton and N.E. Paton (Minerals, Metals and Mater. Soc. (Warrendale, Pennsylvania 1988) p. 221
5.13 A.O. Nikiforov, O.B. Makova, N.J. Kolobnev, G.B. Chohlatova: "Sverkhplastichnost' Litiisoderzhashchikh Aluminievykh Splavov", in *Tez. Dokl. IV Konf. Sverkhplastichnost Metallov* (Inst. Probl. Sverkhplast. Met., Ufa 1989) p. 107
5.14 J.W. Wert, N.E. Paton, C.H. Hamilton, M.W. Mahoney: Met. Trans. A **12A**, 1267 (1981)
5.15 V.K. Portnoy, A.O. Nikiforov, J.P. Konstantinov, A.V. Kumanin: Tekh. Leg. Spl. N1, 5 (1984)
5.16 C.H. Hamilton, C.C. Bampton, N.E. Paton: "Superplasticity of High Strength Aluminium Alloys", in *Superplastic Forming of Structural Alloys*, ed. by N.E. Paton, C.H. Hamilton (Met. Soc. of AIME, California 1982) p. 165
5.17 O.V. Ievleva, V.K. Portnoy, V.S. Levchenko, A.O. Nikiforov: "Povyshenie Temperatury Solidusa Vysokoprochnogo Aluminievogo Splava 1973 v Protsesse Sverkhplasticheskoi Deformatisii", in *Tez. Dokl. IV Konf. Sverkhplastichnost Metallov* (Inst. Probl. Sverkhplast. Met., Ufa 1989) p. 120
5.18 T.R. Bieler, T.G. Nieh: Scripta Met. **22**, 81 (1988)
5.19 K. Matsuki, M. Tokizawa, G. Staniek: "Superplasticity of Rapidly Solidified 7475-0.7 wt%Zr Alloys", in *Proc. Int. Conf. Superplasticity and Superplastic Forming*, ed. by C.H. Hamilton and N.E. Paton (Minerals, Metals and Mater. Soc. (Warrendale, Pennsylvania 1988) p. 395
5.20 R.Z. Valiev, O.A. Kaibyshev: Dokl. Akad. Nauk SSSR **301**, 846 (1988)
5.21 E. Nes: Met. Sci. **13**, 211 (1979)
5.22 S.S. Gorelyk: *Rekristallizatsia Metallov i Splavov* (Metallurgia, Moscow 1978)
5.23 M.Kh. Rabinovich: *Termomekhanicheskaya Obrabotka Aluminievykh Splavov* (Mashinostroenie, Moscow 1972)
5.24 V.K. Portnoy: Tsvetn. Met. **5**, 79 (1987)
5.25 M.Kh. Rabinovich: *Sposoby Poluchenia Ultramelkozernistoi Struktury v Aluminievykh Splavakh* (VINITI Report N7944, Moscow 1987)
5.26 V.I. Dobatkin, V.I. Elagin: *Granuliruemyye Aluminievyye Splavy* (Metallurgia, Moscow 1981)
5.27 O.S. Bochvar, Yu.P. Pshenichnov, S.A. Belov: Tekh. Leg. Spl. N10, 6 (1978)
5.28 G.S. Murty, M.J. Kocjak: Scripta Met. **20**, 1721 (1986)
5.29 M.C. Pandey, J. Wadsworth, A.K. Mukherjee: Mater. Sci. Eng. **78**, 115 (1986)
5.30 Z.N. Archakova, G.A. Balakhontsev, I.G. Basova, V.N. Bobovnikov, I.V. Butusova, Yu.M. Vainblat, E.N. Vasilyeva, N.M. Vinogradova et al.: in *Struktura i Svoistva Polufabrikatov iz Aluminievykh Splavov*, ed. by V.I. Elagin and V.A. Livanov (Metallurgia, Moscow 1984)
5.31 V.I. Dobatkin: "Svoistva Konstruktsionnykh Aluminievykh Splavov so Sverkhmelkim Zernom", in *Metallovedenie Legkikh Splavov* (Vsesoyuz. Inst. Legk. Splavov, Moscow 1985)
5.32 M.Kh. Rabinovich, V.G. Trifonov, M.V. Markushev: Izv. VUZov. Tsvet. Metall. N4, 75 (1987)
5.33 Yu.M. Vainblat, S.S. Khayurov, L.V. Ber: Tekh. Leg. Spl. N3, 18 (1976)
5.34 O.G. Senatorova, N.A. Ryazanova, V.I. Kopnov, G.M. Golovina, I.P. Shegina, N.I. Novosiltseva: "Zerennaya Struktura i Svoistva Listov iz Splava V95", in *Metallovedenie Legkikh Splavov* (Vsesoyuz. Inst. Legk. Splavov, Moscow 1985) p. 92
5.35 H. Kosuge, K. Takeuti: Bull. J. Inst. Met. **21**, 104 (1982)
5.36 V.S. Ivanova, V.F. Terentyev: *Priroda Ustalosti Metallov* (Metallurgia, Moscow 1975)
5.37 A. Lasalmonie, J.L. Strudel: J. Mater. sci. **21**, 301 (1986)

Chapter 6

6.1 U. Zwicker: *Titan und Titan-legierungen* (Springer, Berlin, Heidelberg, New York 1974)
6.2 E.A. Borisova, H.A. Bochvar, M.J. Brun, S.G. Glazunov, B.A. Kolakchev, O.S. Korobov, A.V. Malkov, V.N. Moiseev et al.: *Titanovyye Splavy. Metallographia Titanovykh Splavov* (Metallurgia, Moscow 1980) p. 464

6.3 D. Lee, W.A. Backofen: Trans. Met. Soc. AIME **239**, 1034–1040 (1967)
6.4 P. Griffiths, C. Hammond: Acta Met. **20**, 935–945 (1972)
6.5 N.J. Nadirashvili, M.Kh. Shorshorov, V.K. Antipov, A.S. Tikhonov: Fiz. Khim. Obrabotki Mater. N5, 134–137 (1971)
6.6 L.A. Elagina, B.F. Brailovskaya, B.A. Kapitonov: Tsvetnye Metally N2, 63–66 (1979)
6.7 F.H. Froes, C.F. Yolton, J.C. Cheshutt, C.H. Hamilton: "The Preparation of the Desired Microstructure in Titanium Alloys for the Superplastic Deformation", in *Forging and Properties of Aerospace Materials* (The Metals Society, London 1978) pp. 371–398
6.8 O.A. Kaibyshev, R.M. Galeev, G.A. Salishchev: Metalloved. Term. Obrabot. Metal. N1, 37–40 (1984)
6.9 O.A. Kaibyshev, R.M. Galeev, G.A. Salishchev: Fiz. Met. Metalloved. **57**, 788–794 (1984)
6.10 H. Javin, A. Rosen: Scripta Met. **20**, 833–837 (1986)
6.11 N. Furushiro, S. Hori: "Strain-rate Sensitivity of Flow Stress and Effective Stress in Superplastic Deformation of Ti-8Mn Alloy", in *Titanium 80, Science and Technology*, ed. by H. Kimura and O. Izumi, Vol. 2 (The Metallurgical Soc. of AIME (Warrendale, Pennsylvania 1980) p. 1067
6.12 G.C. Morgan, C. Hammond: "The Role of Dislocation Creep Mechanisms in the Superplastic Deformation Properties of Metastable β-Titanium Alloys", in *Titanium Science and Technology* (Deut. Gesell. Metall. Proc. 2, München, FRG 1984) pp. 717–724
6.13 O.A. Kaibyshev: *Sverkhplastichnost' Promyshlennykh Splavov* (Metallurgia, Moscow 1984) p. 184
6.14 N.E. Paton, C.H. Hamilton: "Superplasticity in Titanium Alloys", in *Titanium, Science and Technology*, ed. by G. Lutjering, U. Zwicker, and W. Bunk (Deut. Gesell. Metall. Proc. 2, München, FRG 1984) pp. 649–672
6.15 K. Khemmond: "Sverkhplastichnost' Splavov na Osnove Titana", in *Sverkhplasticheskaya Formovka Konstruktsionnykh Splavov* (Metallurgia, Moscow 1985) pp. 129–142 [English transl.: *Superplastic Forming of Structural Alloys*, ed. by N.E. Paton and C.H. Hamilton (The Metallurgical Soc. of AIME 1982)]
6.16 A.S. Tikhonov: *Effekt Sverkhplastichnosti Metallov i Splavov* (Nauka, Moscow 1978) p. 142
6.17 M.A. Gimin, R. Kent, C. Hammond: J. Mat. Sci. **21**, 475–487 (1986)
6.18 M. Majdic, G. Ziegler: Z. Metallkunde **65**(N8), 173–178 (1974)
6.19 P.K. Sengupta, W.T. Roberts, D.V. Wilson: Met. Technol. **5**, 171–179 (1981)
6.20 I.I. Novikov, V.K. Portnoy: *Sverkhplastichnost' Splavov s Ultramelkim Zernom* (Metallurgia, Moscow 1981)
6.21 O.R. Valiakhmetov, R.M. Galeev, G.A. Salishchev: Fiz. Met. Metalloved **72**, 204–206 (1990)
6.22 N.E. Paton, C.H. Hamilton: Met. Trans. A **10A**, 241–250 (1979)
6.23 A.K. Ghosh, R. Raj: Acta Met. **29**, 607 (1981)
6.24 M.T. Cope, D.R. Evetts, N. Ridley: J. Mater. Sci. **21**, 4003–4008 (1986)
6.25 S.P. Agrawal, R.R. Boyer, E.D. Weisert: "Effect of Small Amounts of Yttrium on the Superplastic Behavior of Ti-6Al-4V", in *Titanium 80, Science and Technology*, ed. by H. Kimura and O. Izumi, Vol. 2 (The Metallurgical Soc. of AIME (Warrendale, Pennsylvania 1980) pp. 1057–1066
6.26 J.A. Wert, N.E. Paton: Met. Trans. A **14A**, 2535–2544 (1982)
6.27 P.J. Lederich, S.M.L. Sastry, J.E. O'Neal: "Influence of Hydrogen Addition on High-Temperature Superplasticity of Titanium Alloys", in *Advanced Processing Methods for Titanium*, ed. by D.C. Hasson and C.H. Hamilton (The Metallurgical Soc. of AIME, Louisville 1981) pp. 115–128
6.28 M. Yamazaki: "Alloy Design of Superplastic Nickel-base and Titanium-base Alloy", in *Superplasticity and Superplastic Forming*, ed. by C.H. Hamilton and N.E. Paton (Minerals, Metals and Mater. Soc. Publ. (Warrendale, Pennsylvania 1988) pp. 353–363
6.29 J.B.I. Moore, J. Tequesta, R.L. Athey: *Fabrication Method for the High Temperature Alloys*. U.S. Patent 3,519,503 (July 1970)
6.30 O.A. Kaibyshev, G.A. Salishchev, R.Ya. Lutfullin: Metalloved. Termich. Obrab. Metal. N3, 27–31 (1981)

6.31 K.M. Kulkarni, M.M. Parikh, T. Watmough: J. Inst. Metals **100**(N5), 146–151 (1972)
6.32 G.V. Greenwood, W.E. Seeds: "Isothermal Forging of Titanium Alloys for the Airspace Technics", in *Titanium and Titanium Alloys. Scientific and Technological Aspects*, Vol. 1, ed. by J.C. Williams and A.F. Belov (Plenum Press, New York 1982) pp.327–337
6.33 D.S. McDarmaid, P.G. Partridge: "Mechanical Properties of Titanium and Aluminium Alloys after Superplastic Deformation", in *Superplasticity and Superplastic Forming*, ed. by C.H. Hamilton and N.E. Paton (Minerals, Metals and Mater. Soc. Publ. Warrendale, Pennsylvania 1988) pp.529–540

Chapter 7

7.1 M. Yoshiyuki, J. Nobihiro, H. Musaki: Netzu shori. J. Jap. Soc. Heat. Treat. N2, 84 (1985)
7.2 A.P. Gulyaev: *Sverkhplastichnost' Staley* (Metallurgia, Moscow 1982) p.56
7.3 O.A. Kaibyshev, I.N. Nugmanov: Izv. VUZov. Chem. Met. N3, 107 (1986)
7.4 N.I. Nadirashvili, M.H. Shorshorov, V.K. Antipov: Fiz. Khim. Obrab. Metall. N5, 134 (1971)
7.5 M.J. Stewart: Met. Trans. **7A**, 399 (1976)
7.6 E. Pohl, M.L. Bernshtein, L.M. Kaputkina: Izv. Akad. Nauk SSSR, Metally, N4, 159 (1980)
7.7 J. Wadsworth, O.D. Sherby: J. Mater. Sci. **13**, 2645 (1979)
7.8 J. Wadsworth, L.E. Eiselstein, O.D. Sherby: Mater. Eng. Appl. N1, 143 (1979)
7.9 A.A. Bochvar, V.A. Davydov, L.K. Druginin: Dokl. Akad. Nauk SSSR **230**(N2), 318 (1976)
7.10 W.J. Kim, G. Frommeyer, O.A. Ruano, J.B. Wolfelstein, O.D. Sherby: Scripta. Metal. **23**, 1515 (1989)
7.11 A.P. Surovtsev, V.E. Sukhanov: Metalloved. Term. Obrab. Metal. N10, 15 (1984)
7.12 O.A. Kaibyshev: *Sverkhplastichnost' Promyshlennykh Splavov* (Metallurgia, Moscow 1984) p.264
7.13 C.I. Smith, N. Ridley: Metals Technol. **1**, 191 (1974)
7.14 H.W. Shadler: Trans. Met. Soc. AIME **242**, 1281 (1974)
7.15 O.D. Sherby, B. Walser, C.M. Yong, E.M. Cady: Scripta. Metal. **9**, 569 (1975)
7.16 K. Watanabe, J. Torisaka, H. Suzuki: J. Iron Steel Inst. Jap. **5**, 1193 (1989)
7.17 O.A. Bannykh, T.K. Pedan: Izv. Akad. Nauk SSSR, Metally N6, 138 (1975)
7.18 M. Hulderbrand, T. Kreybing: Neue Hütte **2**, 89 (1974)
7.19 D. Treheux, D. Nguyen, C. Renault, P. Guiraldeng: Mem. Sci. Rev. Met. **12**, 765 (1977)
7.20 J. Yashuhiro: Curr. Adv. Mater. Process **1**(N5), 1309 (1988)
7.21 C.R. Yoder, V. Weiss: Met. Trans. **3**, 675 (1972)
7.22 V.M. Umnova, V.D. Sadovski: Fiz. Met. Metalloved. **45**(N5), 1028 (1978)
7.23 V.K. Fedukin, M.E. Smagorinsky: *Termotsiklicheskaya Obrabotka Metallov i Detaley Mashin* (Mashinostroenie, Leningrad 1989) p.255
7.24 V. Fedukin, Ju. Ganeva: Techn. mysl' **12**(N1), 8 (1979)
7.25 M.N. Bodjako, S.A. Astapchik: *Elektrotermoobrabotka Splavov s Osobymi Svoistvami* (Nauka i tekhnika, Minsk 1977) p.256
7.26 D. Spinelly: Metallurgica ABM **31**, 449 (1975)
7.27 O.A. Bannykh, U.K. Kovneristy, T.N. Pedan: Izv. Akad. Nauk SSSR, Metally N6, 123 (1972)
7.28 G. Carfi, C. Pedrix, D. Bouleau, C. Donadille: "Microstructural Evolution in AISI 304L Grade Stainless Steel During Isothermal Straining by Hot Torsion", in *Proc. 7th Int. Conf. on Strength of Metals and Alloys*, Vol.2, ed. by H.J. McQueen, J.-P. Bailon, J.I. Dickson, J.J. Jonas, and M.G. Akben (Pergamon Press,Oxford, 1986) p.929–934
7.29 T. Maki, K. Akasaka, K. Okuno, I. Tamura: Trans. Iron. Steel Inst. Jap. **82**, 253 (1982)
7.30 T. Maki, S. Okagishi, I. Tamura: "Dynamic Recrystallization in Ferritic Stainless Steel", in *Proc. 6th Int. Conf. on Strength of Metalls and Alloys*, Vol.2, ed. by R.C. Gifkins (Pergamon Press, Oxford 1983) p.529
7.31 Gao Fei: Acta Met. Sin. **24**, 195 (1988)
7.32 H. Yoshihira, M. Ishii: J. Iron Steel Inst. Jap. **69**, 1440 (1983)
7.33 M. Mantel, B. Baroux, D. Gex, P. Pedarre: Met. Etud. Sci. Rev. Met. **86**, 381 (1989)

7.34 O.A. Bannykh, M.I. Zakharova, A.I. Vasilyeva: Metalloved. Term. Obrab. Metal. N1, 38 (1984)
7.35 J. Mikhel, M. Burshak: Fiz. Metal. Metalloved. **64**, 815 (1981)
7.36 N. Matzumura, M. Tokizane: Trans. Iron Steel Inst. Jap. **26**, 315 (1986)
7.37 M. Okado, M. Tokisane, O.D. Sherby: Trans. Iron Steel Inst. Jap. **22**, 143 (1982)
7.38 J. Wadsworth, J. Lin, O.D. Sherby: Metals Technology **8**, 190 (1981)
7.39 T. Oyame, J. Wadsworth, M. Korchinsky, O.D. Sherby: "Superplasticity UHC Steels", in *Proc. 5th Int. Conf. on Strength of Metalls and Alloys*, Vol. 2 (Pergamon Press, Toronto 1979) p. 381–386
7.40 Y. Maehara, Y. Ohmori: Met. Trans. **18A**, 663 (1987)

Chapter 8

8.1 F.F. Khimushin: *Zharoprochnyye stali i splavy* (Metallurgia, Moscow 1969)
8.2 A.S. Kleshchev: Tekhnologia Legkikh Splavov **11**, 44–48 (1978)
8.3 R.F. Decker, Ch.T. Sims: in *The Superalloy*, ed. by Ch.T. Sims, W.C. Hagel (Wiley-Interscience, New York 1972)
8.4 V.M. Rozenberg: *Osnovy Zharoprochnosti* (Metallurgia, Moscow 1973)
8.5 O.A. Kaibyshev: *Sverkhplastichnost' Promyshlennykh Splavov* (Metallurgia, Moscow 1984) pp. 228–252
8.6 A.A. Bochvar: Izv. Akad. Nauk SSSR, Otdel. Tekhn. Nauk **5**, 743–752 (1946)
8.7 O.M. Smirnov: *Obrabotka Metallov Davleniem v Sostoyanii Sverkhplastichnosti* (Mashinostroenie, Moscow 1978)
8.8 O.A. Kaibyshev, O.Kh. Fatkullin, R.Z. Valiev, F.Sh. Sharifianov: Tekhnologia Legkikh Splavov **11**, 51–55 (1978)
8.9 G.N. Alyoshin, R.Z. Valiev, O.D. Molodtsova, E.T. Razuvaev, V.N. Trofimenko: Tekhnologia Legkikh Splavov **3**, 57–63 (1982)
8.10 V.F. Kalugin, E.I. Razuvaev, P.R. Medvedovskaya: Metalloved. Termich. Obrab. Metal. **10**, 32–35 (1975)
8.11 O.A. Kaibyshev, F.Z. Utyashev, R.Z. Valiev: Izv. VUZov. Chern. Metall. **7**, 120–125 (1982)
8.12 O.A. Kaibyshev, F.Z. Utyashev, V.A. Valitov: Metalloved. Termich. Obrab. Metall. **7**, 40–44 (1989)
8.13 O.A. Kaibyshev, F.Sh. Sharifianov, R.Z. Valiev, A.B. Notkin: Izv. VUZov. Chern. Metall. **9**, 64–68 (1982)
8.14 G.N. Alyoshin: "Povedenie Granits Zeren i Faz pri Rekristallizatsii i Sverkhplasticheskoi Deformatsii Nikelevogo Splava JS6KP", Ph. D. Dissertation (Inst. Probl. Sverkhplast. Met., Ufa 1987)
8.15 G.N. Alyoshin, R.Z. Valiev, V.U. Kazykhanov: Izv. VUZov. Fizika **6**, 40–45 (1986)
8.16 B.I. Beresnev, K.I. Ezerski, E.V. Trushin, B.I. Kamenetski: *Vysokie Davleniya v Sovremennykh Tekhnologiakh Obrabotki Materialov* (Nauka, Moscow 1988) Chap. 1
8.17 G.B. Moore, K.L. Athey: Des. News **5**, 17 (1970)
8.18 A.I. Kolpashnikov, V.A. Vyalov, A.A. Fedorov, A.G. Petrov: *Goryachee Pressovanie Metallurgicheskikh Materialov* (Mashinostroenie, Moscow 1977) 271 p.
8.19 V.A. Livanov, N.S. Rakhmanov, A.V. Kashcheeva, N.M. Berezko: Tekhnologia Legkikh Splavov **2**, 48–52 (1975)
8.20 G.N. Alyoshin, R.Z. Valiev, V.Yu. Gertsman, O.A. Kaibyshev: Dokl. Akad. Nauk SSSR **289**, 1116–1119 (1986)
8.21 J.B. Moore, J. Tequesta, R.L. Athey: *Fabrication Method for High Temperature Alloys*, U.S. Patent 3,519,503 (July, 1970)
8.22 G.F. Merrick: in *Superplastic Forming of Structural Alloys*, ed. by N.E. Paton, C.H. Hamilton (The Metallurgical Society of AIME, San Diego, California June 21–24, 1982)
8.23 F.Sh. Sharifianov: "Sverkhplastichnost' i Struktura Nikelevogo Splava Tipa JS6K, Poluchaemogo iz Granul",in *Sbornik Tez. Dokl. I Vsesoyuz. Konf. Sverkhplastichnost' Metallov* (Ufa 1978) (Conference proceedings)

8.24 N.E. Zorov: Tekhnologia Legkikh Splavov **7**, 61–73 (1981)
8.25 C.E. Shamblen, R.E. Allen, F.E. Valcer: Met. Trans. A **6A**, 2073–2083 (1975)
8.26 I.A. Maevski: *Obrabotka Zaroprochnykh Splavov* (Mashinostroenie, Moscow 1964)
8.27 N.S. Rakhmanov, V.A. Livanov: Electronnaya Tekhnika N2, 72 (1976)
8.28 S.S. Gorelik: *Rekristallizatsiya Metallov i Splavov* (Metallurgia, Moscow 1978)

Chapter 9

9.1 N.S. Stoloff, R.G. Devis: *Mekhanicheskie Svoistva Uporyadochivayuchikhsya Splavov* (Metallurgia, Moscow 1969)
9.2 L.E. Popov, E.V. Kozlov: *Mekhanicheskie Svoistva Uporyadochennykh Tverdykh Rastvorov* (Metallurgia, Moscow 1970)
9.3 L.E. Popov, N.A. Koneva, I.V. Tereshko: *Deformatsionnoe Uprochnenie Uporyadochennykh Splavov* (Metallurgia, Moscow 1979)
9.4 S.M. Barinov, Z.A. Samoilenko: Izv. Akad. Nauk SSSR, Metally **3**, 164–168 (1984)
9.5 O. Izumi: Materials Trans., Japan Inst. Metals **30**, 639-647 (1989)
9.6 M.D. Martsinkovskii: "Teoria i Pryamoye Nablyudenie Antifaznykh Granits i Dislokatsii v Sverkhstrukturakh", in *Elektronnaya Mikroskopiya i Prochnost' Kristallov* (Metallurgia, Moscow 1968) pp. 215–320
9.7 B.A. Grinberg, B.I. Siutkina: *Novyye Metody Uprochneniya Uporyadochennykh Splavov* (Metallurgia, Moscow 1985)
9.8 K. Aoki, O. Izumi: Nippon Kinzoku Gakkaishi Jap. Inst. Metals **43**, 1190 (1979)
9.9 T.G. Nieh, W.C. Oliver: Scripta Metal. **23**, 851 (1989)
9.10 A.K. Mukherjee, T.R. Bieler, A.H. Chokshi: "Superplasticity in Metals and Ceramics", in *Materials Architecture*, Proc. 10th Risø Int. Symposium, ed. by J.B. Bilde-Sorensen et al. (Risø National Laboratory, Roskilde, Denmark 1989) pp. 207–231
9.11 M.S. Kim, S. Hanada, S. Watanabe, O. Izumi: Materials Trans., Japan Inst. Metals **30**, 77–85 (1989)
9.12 R.Z. Valiev, O.A. Kaibyshev, A.V. Korznikov, V.V. Stolyarov: Fiz. Met. Metalloved. **63**, 165–171 (1987)
9.13 R.M. Imayev, O.A. Kaibyshev, G.A. Salishchev. Fiz. Met. Metalloved. **9**, 180–188 (1990)
9.14 R.M. Imayev, G.A. Salishchev: Col. Phys. Sup. **51**, 519–524 (1990)
9.15 O.A. Kaibyshev, S.G. Glazunov, G.A. Salishchev, R.M. Imayev, V.I. Ivanov: Fiz. Met. Metalloved. **64**, 1005–1010 (1987)
9.16 G.A. Salishchev, R.M. Imayev, A.B. Notkin, D.V. Elagin: Tsvetnye Metally **7**, 95–98 (1988)
9.17 O.A. Kaibyshev, G.A. Salishchev, R.M. Imayev, A.B. Notkin: Fiz. Met. Metalloved. **65**, 998–1006 (1988)
9.18 R.M. Imayev, O.A. Kaibyshev, G.A. Salishchev: Fiz. Met. Metalloved. **9**, 180–188 (1990)
9.19 O.A. Kaibyshev, A.Kh. Nurislamov: Izv. VUZov. Chem. Metal. **5**, 108–111 (1989)
9.20 R.Z. Valiev, O.A. Kaibyshev, V.U. Kazykhanov, A.V. Korznikov, V.V. Stolyarov: Izv. VUZov. Chem. Metal. **9**, 84–89 (1987)
9.21 H.A. Lipsitt, D. Shechtman, R.E. Schafrik: Metal. Trans. **6A**, 1991–1996 (1975)
9.22 J.B. McAndrew, H.D. Kesseler: J. Metals **8**, 1348 (1956)
9.23 D. Shechtman, M.J. Blackburn, H.A. Lipsitt: Metal. Trans. **5**, 1373–1381 (1974)
9.24 B.A. Grinberg: JETP Lett. **I4**, 1655-1658 (1988)
9.25 F.H. Froes, C. Suryanarayna: J. Metals **6**, 12–18 (1989)

Chapter 10

10.1 C.A. Johnson, R.C. Bradt, J.H. Hoke: J. Am. Ceram. Soc. **58**, 37 (1975)
10.2 J. Crampon, B. Escaig: J. Mat. Sci. **13**, 2619 (1978)
10.3 L.A. Winger, R.C. Bradt, J.H. Hoke: J. Am. Ceram. Soc. **63**, 291 (1980)
10.4 T.E. Chung, T.J. Davies: Acta Met. **27**, 627 (1979)

10.5 C. Carry, A. Mocellin: J. Am. Ceram. Soc. **69**, C215 (1986)
10.6 Y. Maehara, T.G. Langdon. J. Mat. Sci. **25**, 2275 (1990)
10.7 O.D. Sherby, J. Wadsworth: Prog. Mater. Sci. **33**, 2275 (1989)
10.8 J. Crampon, B. Escaig: J. Am. Ceram. Soc. **63**, 680 (1980)
10.9 K.R. Venkatachari, R. Raj: J. Am. Ceram. Soc. **69**, 135 (1986)
10.10 F. Wakai: "Superplasticity of ZrO_2 Toughened Ceramics", in *Superplasticity and Superplastic Forming*, ed. by C.H. Hamilton and N.E. Paton (Minerals, Metals and Mater. Soc. Publ. Warrendale, PA 1988) pp. 619–623
10.11 F. Wakai, N. Murayama, S. Sakaguchi: "Deformation of Superplastic Tetragonal ZrO_2 Polycrystals", in *Advances in Ceramics*, ed. by S. Somiya, N. Yamamoto, and H. Yanagida, Vol. 24 (American Ceramic Society, Westerville, OH 1988)
10.12 A. Dominguez-Rodriguez, V. Lanteri, A.H. Heuer: J. Am. Ceram. Soc. **69**, 281 (1986)
10.13 P.C. Panda, R. Raj, P.E.D. Morgan: J. Am. Ceram. Soc. **68**, 522 (1985)
10.14 O.A. Kaibyshev, R.M. Imaev, M.F. Imaev: Dokl. Akad. Nauk SSSR **305**, 1120 (1989)
10.15 F. Wakai, H. Kato, S. Sakaguchi, N. Murayama: J. Ceram. Soc. Jap. (Yogyo-Kyokai-Shi) **94**, 1017 (1986)
10.16 F. Wakai, H. Kato: Adv. Ceram. Mat. **3**, 71 (1988)
10.17 T.G. Nieh, C.M. McNally: Scripta Met. **23**, 457 (1989)
10.18 T. Kuroishi, K. Uno, F. Wakai: "Characterization of Superplastic ZrO_2 Toughened Al_2O_3 Prepared by Slip Casting", in *Proc. MRS Int. Meeting on Advanced Materials*, ed. by M. Kobayashi and F. Wakai, Vol. 7, 267 (Materials Research Society, Pittsburgh, PA 1989)
10.19 A.G. Merzhanov, I.P. Borovinskaya, V.I. Yukhvid, V.I. Ratnikov: "Novyye Metody Polucheniya Vysokotemperaturnykh Materialov, Osnovannyye na Gorenii", in *Nauchnyye Osnovy Materialovedenia* (Nauka, Moscow 1981) p. 193
10.20 R. Raj: "Mechanisms of Superplastic Deformation in Ceramics", in *Superplasticity and Superplastic Forming*, ed. by C.H. Hamilton and N.E. Paton (Minerals, Metals and Mater. Soc. Publ. Warrendale, PA 1988) pp. 583–594
10.21 S. Lartigue, L. Priester: J. Am. Ceram. Soc. **71**, 430 (1988)
10.22 A.K. Mukherjee, T.R. Bieler, A.H. Chokshi: "Superplasticity in Metals and Ceramics", in *Materials Architecture*, Proc. 10th Risø Int. Symposium, ed. by J.B. Bilde-Sorensen, N. Hansen, D. Juul Jensen, T. Leffers, H. Lilhot, and O.B. Pedersen (Risø National Laboratory, Roskilde, Denmark 1989) pp. 207–231
10.23 J.G. Wang, R. Raj: J. Am. Ceram. Soc. **67**, 399 (1984)
10.24 F. Wakai: "Non-Newtonian Flow and Superplasticity in Ceramics", in *Proc. MRS Int. Meeting on Advanced Materials*, ed. by M. Kobayashi and F. Wakai, Vol. 7 (Materials Research Society, Pittsburg, PA 1989) p. 225
10.25 T. Watanabe, K. Shobu: "The Development of Plastic $MoSi_2$-Boride Ceramics", in *Proc. MRS Int. Meeting on Advanced Materials*, ed. by M. Kobayashi and F. Wakai, Vol. 7 (Materials Research Society, Pittsburg, PA 1989) p. 303
10.26 F. Wakai: J. Iron Steel Inst. Jap. **75**, 389 (1989)
10.27 A.G. Evans, J.R. Rice, J.P. Hirth: J. Am. Ceram. Soc. **63**, 368 (1980)
10.28 F. Wakai, S. Sakaguchi, N. Murayama: Adv. Ceram. Mat. **1**, 259 (1986)
10.29 T. Nose, T. Fujii, H. Kubo: "High Temperature Plastic Deformation of Hot-Pressed SiC with B and C Sintering Additives", in *Proc. MRS Int. Meeting on Advanced Materials*, ed. by M. Kobayashi and F. Wakai, Vol. 7 (Materials Research Society, Pittsburgh, PA 1989)
10.30 O.A. Kaibyshev: Sverkhplastichnost' Promyshlennykh Splavov (Metallurgia, Moscow 1984)
10.31 M.F. Ashby, R.A. Verrall: Acta Met. **21**, 149 (1973)
10.32 T. Hermansson, K.P.D. Lagerfof, G.L. Dunlop: "Superplasticity of Y-TZP Zirconia", in *Superplasticity and Superplastic Forming*, ed. by C.H. Hamilton and N.E. Paton (Minerals, Metals and Mater. Soc. Publ., Warrendale, PA 1988) pp. 631–635
10.33 A.G. Evans, T.G. Langdon: *Konstruktsionnaya Keramika* (Metallurgia, Moscow 1980)
10.34 J. Karch, R. Birringer, H. Gleiter: Nature **330**, 556 (1987)
10.35 T.M. Shaw, S.L. Shirde, D. Dimos, R.F. Cook, P.R. Duncombe, C. Kroll: J. Mater. Res. **1**, 248 (1989)

10.36 M. Kuwabara, H. Shimooka: Appl. Phys. Lett. **55**, 2781 (1989)
10.37 D.S. Smith, S. Suasmoro, C. Gaubt, F. Cailland, A. Smith: Rev. Phys. Appl. **25**, 61 (1990)
10.38 O.A. Kaibyshev, A.G. Merzhanov, N.G. Zaripov, L.V. Petrova, O.Yù. Efimov, A.N. Pityulin: J. Mater. Shaping Technol. **9**, 77 (1991)

Subject Index

Author Index

SPRINGER NATURE

GPSR Compliance

The European Union's (EU) General Product Safety Regulation (GPSR) is a set of rules that requires consumer products to be safe and our obligations to ensure this.

If you have any concerns about our products, you can contact us on ProductSafety@springernature.com

In case Publisher is established outside the EU, the EU authorized representative is:

Springer Nature Customer Service Center GmbH
Europaplatz 3
69115 Heidelberg, Germany

MIX
Papier | Fördert gute Waldnutzung
FSC® C083411

Kolibri 360 GmbH
Ferdinand-Jühlke-Straße 7
99095 Erfurt, Deutschland
produktsicherheit@kolibri360.de